T0295427

Control, Communication, Monitoring and Protection of Smart Grids

Other related titles:

You may also like

- PBPO132 | Casella, A. Anpalagan | Power Line Communication Systems for Smart Grids | 2018
- PBPO094 | S.K. Salman | Introduction to the Smart Grid: Concepts, technologies and evolution | 2017
- PBTR016 | Kishor *et al.* | ICT for Electric Vehicle Integration with the Smart Grid | 2019
- PBPO095 | S.M. Muyeen and Saifur Rahman | Communication, Control and Security Challenges for the Smart Grid | 2017
- PBPO097 | Dusmanta Kumar Mohanta and M. Jaya Bharata Reddy | Synchronized Phasor Measurements for Smart Grids | 2017

We also publish a wide range of books on the following topics:
Computing and Networks
Control, Robotics and Sensors
Electrical Regulations
Electromagnetics and Radar
Energy Engineering
Healthcare Technologies
History and Management of Technology
IET Codes and Guidance
Materials, Circuits and Devices
Model Forms
Nanomaterials and Nanotechnologies
Optics, Photonics and Lasers
Production, Design and Manufacturing
Security
Telecommunications
Transportation

All books are available in print via https://shop.theiet.org or as eBooks via our Digital Library https://digital-library.theiet.org.

IET ENERGY ENGINEERING SERIES 244

Control, Communication, Monitoring and Protection of Smart Grids

Edited by
Bidyadhar Subudhi, Pravat Kumar Ray and
Vinod Khadkikar

The Institution of Engineering and Technology

About the IET

This book is published by the Institution of Engineering and Technology (The IET).

We inspire, inform and influence the global engineering community to engineer a better world. As a diverse home across engineering and technology, we share knowledge that helps make better sense of the world, to accelerate innovation and solve the global challenges that matter.

The IET is a not-for-profit organisation. The surplus we make from our books is used to support activities and products for the engineering community and promote the positive role of science, engineering and technology in the world. This includes education resources and outreach, scholarships and awards, events and courses, publications, professional development and mentoring, and advocacy to governments.

To discover more about the IET please visit https://www.theiet.org/

About IET books

The IET publishes books across many engineering and technology disciplines. Our authors and editors offer fresh perspectives from universities and industry. Within our subject areas, we have several book series steered by editorial boards made up of leading subject experts.

We peer review each book at the proposal stage to ensure the quality and relevance of our publications.

Get involved

If you are interested in becoming an author, editor, series advisor, or peer reviewer please visit https://www.theiet.org/publishing/publishing-with-iet-books/ or contact author_support@theiet.org.

Discovering our electronic content

All of our books are available online via the IET's Digital Library. Our Digital Library is the home of technical documents, eBooks, conference publications, real-life case studies and journal articles. To find out more, please visit https://digital-library.theiet.org.

In collaboration with the United Nations and the International Publishers Association, the IET is a Signatory member of the SDG Publishers Compact. The Compact aims to accelerate progress to achieve the Sustainable Development Goals (SDGs) by 2030. Signatories aspire to develop sustainable practices and act as champions of the SDGs during the Decade of Action (2020-2030), publishing books and journals that will help inform, develop, and inspire action in that direction.

In line with our sustainable goals, our UK printing partner has FSC accreditation, which is reducing our environmental impact to the planet. We use a print-on-demand model to further reduce our carbon footprint.

British Library Cataloguing in Publication Data

A catalogue record for this product is available from the British Library

ISBN 978-1-83953-804-9 (hardback)
ISBN 978-1-83953-805-6 (PDF)

Typeset in India by MPS Limited

Cover image credit: Hiroshi Watanabe/Photodisc via Getty Images

Contents

Preface

With the growing technological revolution in the past two decades, Smart grid (SG) is unfolding a plethora of perspectives towards the evolution of the power sector. Fueled by digitalization, SG technology has emerged as a game changer by generating feasible solutions for the challenges faced in the power sector. It integrates innovative tools and technologies from generation, transmission, and distribution to household appliances and consumer equipment. However, this integration poses new hurdles, especially concerning energy operation and management, particularly within the distribution segment of the grid. To effectively address these challenges, it is essential to implement appropriate power electronic controls, communication technologies, and robust monitoring and protection systems. Subsequently, enhancing the reliability and economic efficiency of smart grids is very important, emphasizing optimal power quality whilst facilitating the integration of renewables and electric vehicles (EVs), advanced fault direction identification strategy, robust control schemes with cyber resiliency and condition monitoring and health prognosis applications.

The contents and structure of this book are organized considering the recent trends in SG technology, for the transformation of the power grid, bringing promising benefits to both producers and customers. The book is structured into 13 chapters. Each chapter provides an overview of the corresponding topic with simulation/experimental validation for basic ideas on the practical feasibility of smart grid operation, control and protection. Case studies on different issues are included for the broader idea of the relevant context. Primarily, emphasis is given to experimental or test bed validations of the proposed algorithms or methods for the clear outlook of the readers.

Chapter 1 introduces microgrids and smart grids, exploring their key features, benefits, and challenges. It briefly covers different types of microgrids and their varied control aspects. It discusses various evolutionary initiatives like distributed generation, renewable energy integration, energy storage, and demand-side management for smart grid implementation. Brief discussions on cybersecurity measures and sustainable energy management are also included.

Chapter 2 presents different aspects of renewable energy integration in smart grids, thus, enhancing the reliability and quality of today's power demand. This chapter is organized into two subsections, focusing on a distinct approach to integrating PV with efficient power management and power quality improvement techniques. One is the integration of PV through power converters supported with hybrid energy storage system (HESS). Another one is the integration of PV through custom power devices (CPD). It also includes simulation as well as experimental analysis under various dynamic scenarios of varying PV power, uncertain grid and load disturbances.

Integration of microgrids with the existing grid imposes numerous difficulties such as low inertia, sporadic nature of RERs, supply-demand mismatch, uncertain converter switching, communication-related vulnerabilities, etc. Chapter 3 explores the challenges faced in microgrid integration and its possible solution. It includes a discussion on modern adaptive control approaches to address the aforementioned challenges with respect to microgrid topologies. A robust-adaptive distributed secondary control strategy for PV-based islanded ac microgrid is presented with control objectives to restore voltage and frequency along with active and reactive power sharing among distributed generations (DGs) amidst possible uncertainties.

Chapter 4 investigates the stability of renewable energy sources integrated power systems. Wide area damping controllers (WADCs) based on wide area measurement systems (WAMS) can boost power system stability and provide adequate damping, but their performance significantly deteriorates as a result of actuator saturation. This study develops a model-free control algorithm by taking into account actuator saturation and latency in communication channels and improvisation of power system stability by implementing reinforcement learning techniques.

Chapter 5 delves into the examination of secondary control methods like consensus-based sliding mode control (SMC) to address voltage fluctuations from converter current distribution. It establishes fault-tolerant control through optimal computation aligned with event-triggering regulations. Islanded DC microgrids are evolving into complex cyber-physical systems (CPS) with intelligent controllers and communication networks making them vulnerable to cyber-attacks. Thus, identifying and mitigating threats within microgrid CPS is critical. This chapter examines an AI-driven approach to detect false data injection (FDI) instances in microgrid CPS.

Chapter 6 introduces an advanced control strategy for virtual synchronous generators in hybrid AC–DC microgrids. It ensures a decentralized mode of precise power-sharing among microgrids to counter the destabilization effects of the transient state and also facilitates plug-and-play capability. The chapter offers valuable insights into the effectiveness and adaptability of the proposed control strategy in network-connected hybrid microgrids.

Chapter 7 addresses the different challenges for maintaining power quality within microgrids, it also explores various strategies to maintain power quality, including droop control, centralized and decentralized load sharing, negative virtual harmonic impedance (NVH-Z) technique for harmonic current sharing, and dynamic voltage restorer (DVR) which are aimed at preserving power quality. Furthermore, it incorporates different case studies to illustrate the efficacy of these strategies in improving the performance and power quality of microgrids across multiple scenarios, including linear, nonlinear, and unbalanced load conditions.

The sporadic nature of renewable energy sources and the increased use of distributed energy resources (DERs) significantly alter the level and direction of power flow and fault current in the distribution systems. Traditional protection devices such as fuse, circuit breakers, etc., designed for unidirectional fault current detection, struggle to provide proper service towards the reliability and safety of microgrids (MGs). It is crucial to ascertain the direction of fault currents precisely. Chapter 8 discusses key challenges in microgrid protection and proposes an advanced fault direction estimation scheme for ac microgrids based on the dynamic value of superimposed positive sequence impedance angle (SPSIA) to enhance fault detection efficiency.

Chapter 9 delves into various aspects of the application of communication in power systems protection, highlighting key considerations for effectiveness and security. It explores performance factors like availability, redundancy, temporal assessment and reliability, crucial for system integrity. Additionally, the chapter examines communication systems in power applications like EMS, SCADA, pilot protection, security, substation, distribution automation, and wide-area protection. It compares communication mediums such as fiber-optic, Ethernet, and wireless systems. Various protocols from physical-based like RS232 and RS485 to layer-based protocols like DNP-3.0, ModBus, and IEC 61850 are also explored.

The surges in distributed energy resources (DERs) and IoT devices in power systems have increased cyber components. An automated control architecture is needed to address uncertainty amid DER penetration in distribution systems for better situational awareness and decision-making. Chapter 10 introduces holonic control architecture, offering flexible adaptation to evolving grid conditions with centralized, decentralized, distributed, and local modes. It discusses cyber needs and vulnerabilities for each mode, with a seamless transition between different modes. Two cases of holonic control—the volt–watt control and service restoration are presented.

Chapter 11 highlights the importance of condition monitoring and health prognosis in smart grids. The chapter emphasizes real-time data-driven prognostics for predictive maintenance, reliability improvement, and resilience enhancement in smart grids. It underscores the transformative potential of these techniques towards a sustainable and intelligent future of smart grid. However, the critical challenge lies in managing vast data volumes, ensuring data quality, and achieving real-time processing. The incorporation of explainable AI, cybersecurity measures, and edge computing is the key to unlocking the full potential of condition monitoring and prognosis to attain sustainability.

Vehicle-to-grid (V2G) technology empowers electric vehicles (EVs) to act as distributed energy resources (DERs). For V2G applications, EV batteries and the AC power grid must exchange energy bi-directionally. Hence, Chapter 12 reviews widely used dc-dc converter and bidirectional power factor correction (PFC) topologies in V2G and G2V applications along with a discussion on challenges and future directions in bidirectional converter topology development. It evaluates the performance of PFC topologies using total harmonic distortion (THD) of grid current and, compares the efficiency and THD characteristics of each topology.

Chapter 13 evaluates smart grid development in the US, Australia, India, China, the EU, and other countries, emphasizing the key drivers of smart grid implementation like renewable energy integration and innovations in grid infra-structure. Concrete energy policies and collaborative research efforts are pivotal, focusing on interoperability and technology standards. Recent advancements in smart meters, demand-side management, self-healing technologies, and big data analytics show promising progress in Smart Grid technology.

It is expected the book will immensely benefit undergraduates, postgraduates, research scholars, faculty members, engineers, and scientists in electrical, electronics and computer engineering from different industries and organizations. In particular, the target audience is a broad domain of engineers in the power industry and computer scientists focusing on cybersecurity design and machine learning.

The book will help readers enrich their technical knowledge on several aspects including power quality, adaptive and robust control schemes, communication and computations, condition monitoring and health prognosis applications in smart grids with V2G and G2V operations in EV charging schemes as well as the idea of smart grid scenarios of different countries.

Bidyadhar Subudhi
Pravat Kumar Ray
Vinod Khadkikar

About the editors

Bidyadhar Subudhi is the Director of the National Institute of Technology Warangal and is currently on deputation from the Indian Institute of Technology Goa, where he is a Professor in the School of Electrical Sciences. He was a recipient of the Outstanding Leadership Award by the IEEE Computer Society, Bio-inspired Computing, in 2022. He is a Fellow of the Indian National Academy of Engineering, the Asia-Pacific Artificial Intelligence Association, the IET, and the Industry Academy of the International Artificial Intelligence Industry Alliance. His research interests include systems and control, applications of power electronics and control to power systems, microgrids, and electric vehicles. He has supervised 44 PhD theses.

Pravat Kumar Ray is a professor at the Department of Electrical Engineering, National Institute of Technology Rourkela, India. From January 2016 to June 2017, he was a postdoctoral fellow at Nanyang Technological University, Singapore. He has more than 20 years of experience in research on estimation and filtering in power systems, power quality, hybrid AC/DC grids, solar irradiance forecasting and renewable integration. He received the excellent research paper award of the CSEE *Journal of Power and Energy System* in 2021 and the best research paper award for the *Journal of Modern Power System and Clean Energy* in 2022.

Vinod Khadkikar is a professor in the Electrical Engineering Department at Khalifa University, Abu Dhabi, UAE. From December 2008 to March 2010, he was a postdoctoral fellow at the University of Western Ontario, Canada, and in 2010, a visiting professor at MIT, Cambridge, USA. He is an IEEE Fellow and a prominent lecturer for the IEEE Industry Applications Society. He also serves as Co-Editor-in-Chief of *IEEE Transactions on Industrial Electronics*. His research interests include the application of power electronics in distribution systems and renewable energy resources, grid interconnection, power quality, active power filters, and electric vehicles.

Chapter 1

Introduction to smart grids and microgrids

*Pravat Kumar Ray[1], Shobhit Nandkeolyar[1],
Bidyadhar Subudhi[2] and Chinmaya Jagdev Jena[1]*

1.1 Background

With the increased awareness of climate change, the use of non-conventional energy sources has gained popularity. Also, the scarcity of adequate energy sources, the need to reduce greenhouse gas emissions, and the deregulation of the electricity market opened the path for the usage of Distributed Generation (DG). The popularity of DG gave rise to the idea of smart grids and microgrids. Understanding the concept of microgrids (MG) and smart grids is essential for modern power systems as these are crucial for enhancing grid reliability, efficiency, and resiliency. Both are interconnected concepts. While they share some similarities, they also have distinct characteristics and purposes. A significant difference between these two is the scale of operations. The microgrid handles energy generation and distribution on a small scale. Whereas the smart grid is engineered to handle power distribution on a larger scale and is equipped with communication channels between the consumer and utility grid.

The microgrid is often described in the literature as an arrangement of groups of loads and distributed energy sources (DERs) that operate within specified electrical boundaries, and function as a single unit [1]. Microgrids can function in both grid-connected and island modes. Energy sources and loads present in the microgrid can be disconnected and reconnected at any time with minimal disruption to the overall system. Some of the important characteristics of the microgrid lie in its ability to balance local energy supply and demand, integration of non-conventional energy sources, and reliable power supply even during grid disruption. These are very useful in providing power to isolated regions, communities, or essential establishments such as hospitals, Military bases, etc. Microgrids provide financial gain, sustainability, and can also optimize power quality and reliability.

The microgrids are flexible towards their total capacity i.e., more DGs can be integrated as load demand increases. After fulfilling the local load demand, the MG

[1]Department of Electrical Engineering, National Institute of Technology Rourkela, India
[2]School of Electrical Sciences, Indian Institute of Technology Goa, India

operator can sell surplus power generation or buy in case of deficit power generation from the utility grid. Thus, the MG operator can maximize the economic benefits of the DERs. Figure 1.1 shows a typical structure of the microgrid. This includes different generation units: renewable sources (such as PV, wind, fuel cell), nonrenewable sources (such as combined heat and power, diesel generator), storage devices (battery, supercapacitor, flywheel), and various types of loads (linear load, nonlinear load, variable load). These are connected to the utility grid at the point of common coupling (PCC) via a breaker. The breaker helps to operate the MG in both grid-connected mode and islanded mode.

Compared to the concept of microgrid, a smart grid is a digital technology-equipped grid system that incorporates communication networks, control and automation technology. The smart grid concept was introduced to address some of the inadequacies of the traditional grid system. It is a significant upgrade of the traditional power grid. With real-time communication and data interchange between different grid components, the smart grid makes it possible for the utilities to efficiently control the flow of electricity enabling better power demand and supply management.

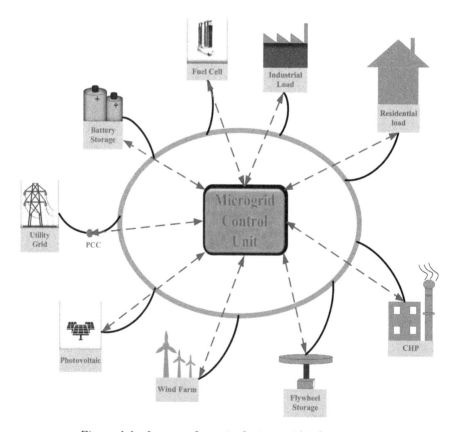

Figure 1.1   Layout of a typical microgrid infrastructure

1.2 Features of microgrid

Microgrids are becoming more popular and practical due to constraints in using conventional fossil fuels and their adverse effects on the environment. Being a small-scale energy network equipped with an advanced control unit and incorporating various DERs, and storage devices, the Microgrid exhibits various key features that ensure efficient operation. Some of these features of the Microgrid are as follows:

- ❖ **Better energy management and control:** Microgrids are well-planned and equipped with advanced control technologies incorporating energy management at a local level. Furthermore, the use of storage units in Microgrids enhances energy management.
- ❖ **Flexibility:** The flexible nature of Microgrid ensures us to meet the ever-increasing load demand. With the increase in load demand, more generating units can be added to the microgrid and can also be scaled down to lower capacity if needed. Making the microgrid more adaptable to various scenarios.
- ❖ **Decentralized:** As the microgrids are decentralized systems, they bring the control of power flows nearer to the consumption, unlike the traditional grid system. This gives more freedom to manage the generation and consumption.
- ❖ **Off-grid and on-grid capability:** Microgrid can be operated in both grid-connected mode and Islanded mode. In the event of a grid failure or any power quality problem, the microgrid is operated independently of the grid. This is a prominent feature of the microgrid. Local loads are benefited from the continuous power supply.
- ❖ **Resilience and reliability:** The usage of localized power generation and distribution, microgrid is able to enhance the grid resilience by minimizing the impact of extensive outages. With the features of operating independently, the microgrid can supply power without interruption. This makes it more reliable than the traditional grid.
- ❖ **Loss minimization:** Microgrids reduce the transmission and distribution losses significantly that are present in the case of traditional grids due to long-range transport of power. As the generation and consumption of electricity happen in proximity, the efficiency of the microgrid increases and the utilization of resources is optimized.

1.3 Components of microgrids

The microgrid is comprised of different components from generations, consumption, storage, and power electronic converters all within an enclosed network. These various components are briefly discussed below.

1.3.1 DERs/DG

The DERs are smaller power generation units and are the core component of a microgrid. DERs can be classified into two types: Dispatchable and non-dispatchable. The operator does not have full control over the output of non-dispatchable DERs since they are of an intermittent nature [2]. Wind energy and solar energy are examples of these types of DERs. DERs can also classified into AC sources and DC sources.

1.3.2   *Energy storage units (ESUs)*

Energy storage systems are useful for the enhancement of stability and reliability of the microgrid. ESUs are used for accumulating the excess generated power and supplying the deficit power to the microgrid. Different examples of ESUs are batteries, flywheels, supper capacitors, compressed air, fuel cells, and so on. One or a combination of different ESUs (hybrid storage units) can be implemented depending on the requirement.

1.3.3   *Converters*

Converters are implemented in microgrids for interfacing between AC systems and DC systems. These converters enable bidirectional power flow between the utility grid and MG and also help step up or step down the voltage levels according to the load requirement.

1.3.4   *Loads and load management system*

Loads can be classified into different categories, such as: AC and DC loads, Linear and non-linear loads, and critical and non-critical loads. Some MG operators have the freedom to actively manage the non-critical loads. The operator reduces the power demand by switching off the non-critical loads in case of deficit power generation

1.3.5   *MG control unit*

The control unit is the brain of the MG. It controls every component of the MG and handles the power flow implementing efficient energy and load management and switches between grid-connected mode and islanded mode of operation. A brief discussion about the MG control is presented in Section 1.6.

1.4   Classification of microgrid

Microgrids can categorized according to various criteria, such as operating modes, power capacity, distribution systems, applications, etc. Figure 1.2 shows the classification of microgrids under these categories. Some of the categories are briefly discussed in the following sub-sections.

1.4.1   *Classification based on modes of operation*

Based on operation with integration to the utility grid, the MG can be classified as grid-connected and stand-alone. Grid-connected MG can operate in both on-grid and islanded modes in case of grid outages. In the case of the Islanded operation of the MG, it operates independently of the grid and must maintain the desired level of voltage and frequency. In the grid-connected mode of operation, the MG can supply or absorb power from the utility grid thereby enhancing stability and reliability. By selling excess power generated to the utility grid the MG can make some financial benefits. The stand-alone MG operates entirely independently of the grid. It is also called isolated MG. These MG are majorly used for remote areas where the utility grid has not reached yet.

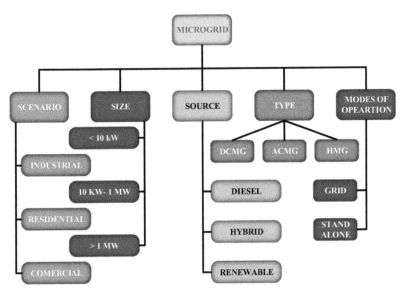

Figure 1.2   Classification of microgrids based on different characteristics

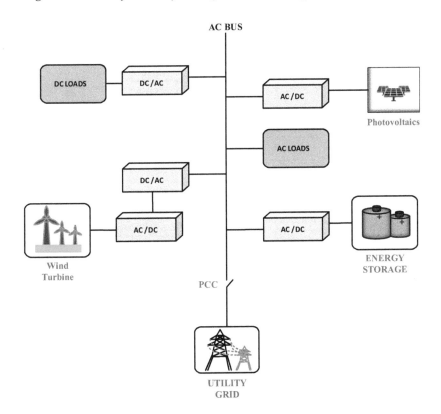

Figure 1.3   Schematic representation of an AC microgrid system

1.4.2   Classification based on distribution system

On the basis of power, the MG can be classified into three categories: AC micro-grid (ACMG), DC microgrid (DCMG), and Hybrid microgrid (HMG) [3]. Figures 1.3–1.5 illustrate these types of MG respectively. Each of these groups of MG has its own advantages and disadvantages.

1.4.2.1   AC microgrid

In the case of an AC microgrid, all the generating sources and loads are integrated into a common AC bus. Figure 1.3 shows a typical structure of the ACMG. Integration to the utility grid is easy for this type of MG. However, there is an increase in complexity in the control of ACMG as we have to maintain the phase and frequency at the desired level. Along with active power, reactive power control is also needed for ACMG. For integrating the DC components to the MG, DC/AC converters are used. As a result, the overall efficiency of the system decreases [4]. ACMG can be further categorized into: single-phase common bus ACMG, and three-phase common bus ACMG (with a neutral line or without a neutral line).

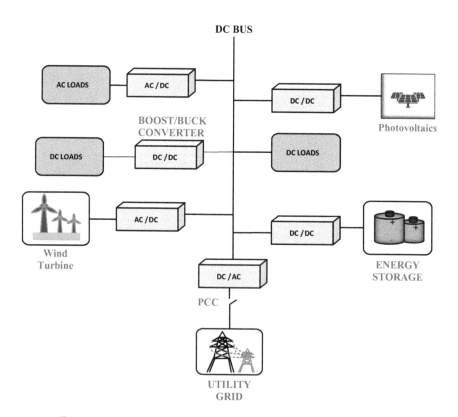

Figure 1.4   Schematic representation of a DC microgrid system

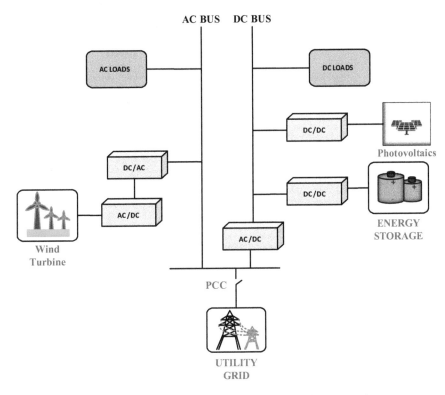

Figure 1.5   Schematic representation of a hybrid microgrid system

1.4.2.2   DC microgrid

In the case of DCMG a common DC bus is present for interfacing all the components of the MG. A DC/AC converter is used for the integration of MG to the utility grid. DC loads with the same voltage rating as the DC bus can be directly connected. Otherwise, a DC/DC converter is used for the connection. In comparison to the ACMG, the DCMG requires fewer conversion stages, which results in higher efficiency and lower conversion losses [4]. Reactive power, phase, and frequency control are not needed for DCMG. Figure 1.4 shows a typical structure of the DCMG. A brief comparison between ACMG and DCMG is presented in Table 1.1.

1.4.2.3   Hybrid microgrid

A hybrid microgrid can be visualized by integrating both DCMG and ACMG into one distribution system. HMG (hybrid microgrid) is interfaced with the utility grid via a static transfer switch (STS) [5]. A typical structure of HMG is illustrated in Figure 1.5. The HMG carries the advantages of both the AC and DC MG. Both DC and AC devices can be directly connected to the MG as both AC and DC buses are present here. Hence, fewer conversion stages are needed, resulting in lower power loss and higher reliability.

Table 1.1   Tabular comparison between ACMG and DCMG on different aspects

Aspect	ACMG	DCMG
MG control	Simpler control	Complexity in control due to frequency and phase
Number of converters required	Low	High
Power quality	High	Low
Energy management	Easy	Complex
Protection	Costly and complex	Simple and cheap
Efficiency	High	Low
Power loss	Low	High
Grid Integration	No conversion stage required	Conversion stage required

1.4.3   Classification based on other factors

There are other types of classifications as well. If we consider the types of consumers an MG is catering to then the MG can be classified into: Industrial, Residential, and Commercial MGs. If we consider the overall generation capacity of the MG, then they can be classified into: sub-10 kW, 10 kW–1 MW, and over-1 MW MG systems. If we consider the types of energy generation sources in the MG, then they can be classified as: Diesel, Hybrid, or Renewable MGs. Hybrid MG in this classification refers to an MG that has both diesel generators as well as renewable sources of energy.

1.5   Addressing technical challenges encountered by microgrid

Despite MGs having so many advantages compared to conventional utility grids, it is not very convenient to switch to the MG. One must do careful planning and overcome different issues posed before implementing the MG. With different DERs, power electronic devices interfacing the MGs to the power grid may cause adverse effects on the system such as reliability, power quality, etc. Some of these challenges and their countermeasures are discussed below.

1.5.1   Fluctuation in power

Due to the intermittent nature of some DERs, power generation is not in the MG operator's control. This leads to an imbalance of power within the microgrid. Energy storage units are to be used to address this issue. An effective energy management scheme should be implemented along with a sophisticated control strategy of the MGs to cope with this fluctuation in power.

1.5.2   Harmonics

Several power electronics converters are utilized in an MG. But they are a source of harmonics injection into the power system. These harmonics increase the safety risk for the ESUs [6]. The harmonics can be reduced by the use of active and passive filtering techniques [7].

1.5.3 Transition from grid connected to island mode

In case of any fault or grid failure, an MG should have the capability to switch from grid-connected to islanded mode of operation without affecting its stability.

1.5.4 Protection

A protection scheme should be implemented with the MG to provide a quick response to any fault. In the event of a fault at the grid side, it should detect and isolate the MG from the grid for its protection. In the event of a fault within the microgrid, the faulty parts should be cut off from the remaining microgrid.

1.5.5 Stability of MG

Stability in MGs refers to the voltage and frequency stability [8]. The voltage and frequency operating range of the MG may be violated due to disturbance corresponding to load changes, and unbalanced power sharing between DERs. These issues can be addressed by implementing supply decentralization, effective power allocation, and demand and supply management.

1.5.6 Regulatory barriers of MG

To integrate the MG into the utility grid, it must adhere to all regulations and requirements of the utility companies and regulatory authorities. Some standards related to voltage and frequency range have been set by the authorities during the operation of the MG for it to be allowed to connect to the grid. These standards may vary from region to region. In some regions, they might not be well defined also. This is to be considered at the planning stage of the MG.

1.6 Advanced control strategies for microgrid management

Microgrid control strategies are essential for maintaining a microgrid's stability, reliability, and efficiency. Efficient MG control approaches are intended to make the best use of energy resources, manage the load demand, and allow for smooth transitions between grid-connected and islanded modes while maintaining stable frequency and voltage. Microgrid control strategies can be classified based on various factors. Figure 1.6 illustrates these classifications. Based on the architecture, the control system of MGs can be classified as: centralized, distributed, and decentralized [9]. Figure 1.6 illustrates the architectures of these control systems.

- **Centralized**: In the centralized control of the MG, as the name suggests, a single unit of central control system is implemented to manage all the components of the MG. This has the advantage of achieving proper coordination among all the DERs and loads. This leads to faster control of the MG. However, the drawback of the centralized control system is if the control unit fails the whole MG shuts down. For critical loads, the implementation of the centralized control system is not advisable.

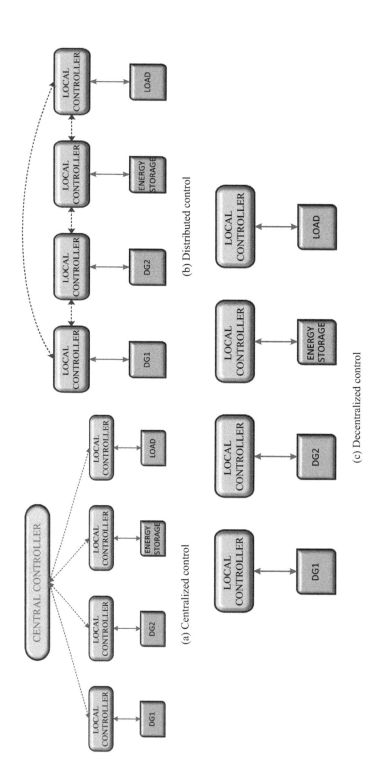

(a) Centralized control

(b) Distributed control

(c) Decentralized control

Figure 1.6 Types of microgrid control architecture

- **Distributed**: In the case of a distributed control system, several controllers are used in the MG. Each controller can share information with other control units. If any one of the controllers fails, it can be separated from the rest of the MG.
- **Decentralized**: Decentralized control architecture also uses several controller units as in the case of distributed systems. But there is no communication among these controllers.

In MG, hierarchical control refers to the organization of control levels, with each level having specific responsibilities and functions. A microgrid typically has three hierarchical control levels: Primary, secondary, and tertiary. Each layer of control can be differentiated based on its control objective and time intervals. These layers of controls are briefly explained below.

- **Primary layer:** This layer of control is concerned with local control and is the fastest among the three. Its objective is to maintain the voltage and frequency levels within the desired limit. It ensures the power demand and supply in the MG are balanced.
- **Secondary layer:** The secondary control layer deals with optimal economic dispatch and distributes power among the DERs. It also deals with voltage and frequency disturbances due to the primary layer control. The secondary layer works on a slower time scale than the primary layer. The control variable of the primary layer is the output values of the secondary layer.
- **Tertiary layer:** The tertiary control layer is concerned with long-term planning and decision-making procedures. Tertiary control contributes to demand-side management (DSM) strategies, energy exchange with the main grid, and other MGs. It provides the reference values for the secondary layer of control. Its time scale is the longest among the three layers.

The hierarchical control application of the MG allows us to manage various DERs efficiently, and quickly adapt to the changes of generation or load. All the layers of hierarchic control coordinate and communicate among themselves to provide better stability and reliable control of the MGs, as depicted in Figure 1.7.

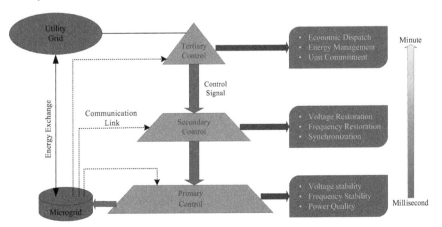

Figure 1.7   Illustration of a hierarchical control architecture

When designing the control for a MG, the operator must decide the objective of the control system i.e., active and reactive power (P-Q) sharing, voltage and frequency (V-f) control, droop control, energy management system (EMS), etc. The objective of implementing this active and reactive power-sharing control of a microgrid is to match the generation and demand. It helps in power exchange between the MG and the grid and also acts as a partial EMS [10]. Any disturbances in the MG in terms of load or generation may cause a deviation from the system voltage and frequency operating range. This may lead to system stability problems or power quality issues. V-f regulation is another aspect of MG control. In microgrids, when the active power supply decreases there will be a drop in the frequency of the system and similarly, an increase in active power leads to a rise in frequency. This is also true in the case of reactive power and the voltage of the system. Another type of control is the droop control. Droop control is a reliable and efficient approach to managing power distribution and preserving frequency or voltage stability. The implementation of EMS in a microgrid ensures smooth distribution of the power among the loads and storage units, and exchanges the surplus or deficit power with the utility grid. Considering the energy market, the EMS can make a profit for the MG owner by selling the excess power available in the MG. Table 1.2 illustrates some of the recent research work on these control aspects of the MG.

Table 1.2   Tabulated summary of various control techniques in the existing literature

Control objective	Reference	Control technique	Remarks
Active and reactive power control	[11]	Artificial neural network-based controller	ANN is used for generating the reactive power reference for improved power-sharing. It uses all three layers of control for both grid-connected mode and islanded mode.
	[12]	PI controller	The decentralized control of individual DGs is implemented for active power balancing in the microgrid. The PI controller is used for maintaining the DC bus voltage.
	[13]	Event-based distributed consensus control	Active power-sharing among a cluster of microgrids is achieved. Communication is reduced among the controllers of the DGs.
Voltage and frequency control	[14]	Model predictive control	A hierarchical control structure for controlling the frequency and voltages of an island-based microgrid. Droop control is used in primary layers for reducing voltage and frequency fluctuation.
	[15]	Event-triggered distributed control algorithm	Each DG checks its triggering condition relaying its own clock. This improves their efficiency in frequency regulation
	[16]	PI and PR controller	Decentralized control for individual DGs with two compensation levels for voltage unbalance.

(Continues)

Table 1.2 (*Continued*)

Control objective	Reference	Control technique	Remarks
Droop control	[17]	Adaptive droop control	An adaptive droop controller for power sharing is used in an LVDC microgrid based on a superimposed frequency.
	[18]	f-P/Q droop control	Droop control is developed for the synchronization of DERs and to maintain the power balance between cascaded MGs.
EMS	[19]	PI controller	Implements a power management scheme for grid-connected DCMG. The hybrid storage system is used for reducing transient stress on the battery.
	[20]	MPC	A double–layer algorithm based on MPC for optimal EMS for MG is used. Multiple generation, storage, and loads are considered for the verification of the algorithm.
	[21]	PI controller	A power management scheme is proposed for PV power smoothing, DC voltage, and frequency regulation

1.7 Reliable microgrid protection: challenges and solutions

The protection aspect is a crucial component of microgrids (MG) while designing and operating. This ensures the stability of the system, the safety of the workers, and the equipment in the MG. Different protective mechanisms are in place for grid-connected MGs and isolated MGs. The protection scheme for the isolated MGs is for protecting the equipment within the microgrid. But for a grid-connected MG, the protection is implemented for both internal and external protection.

Overcurrent protection is needed to safeguard the components of an MG from short-circuit faults. MGs should be equipped with fault detection mechanisms to identify a fault, and circuit breakers for isolating the faulty part. Islanding detection and anti-islanding protection should be implemented for unintentional islanding of a grid-connected microgrid. Anti-islanding protection safeguards utility grid workers. Frequency-based relays are provided within the microgrid for monitoring the operating frequency of the MGs. These can initiate load shading or shut down of some of the DGs if the frequency differs from the normal operating range. Similar to the frequency relays, the voltage relays monitor the voltage levels of the microgrid.

When designing the microgrid, it is required to follow the industry, and grid standards to the best of our ability. The specific protection to be utilized in a microgrid may vary depending upon the size of the MG, types of DG units, and location of the MG.

1.8   Integration of communication technologies in microgrid infrastructure

The use of a communication channel is critical in a microgrid to ensure reliable and efficient operation. Microgrids can function independently or in coordination with the utility grid. To achieve seamless operation and coordination among the microgrids as well as the utility grid, the application of a communication network is necessary. Implementing the correct communication technology, which offers low latency, low losses, and good bandwidth is necessary for reliable operation and performance of the MGs. As more and more MGs are interconnected, the planning of a proper communication strategy becomes increasingly difficult. Communication channels are of two types, namely wired and wireless communication. For a wired communication channel twisted coaxial cable, fiber optics, telephone communication networks, etc. can be used. In the case of a wireless communication topology GPRS, a wireless local area network (WLAN) or worldwide interoperability for microwave access system (WiMAX) can be used.

The communication infrastructure implemented in MGs has the responsibility of exchanging information. A poor communication topology can lead to various performance issues, including reduced energy efficiency, grid and MG synchronization problems, and decreased service quality. The communication infrastructure must have the following characteristics: low latency, reliability, security, and time synchronization. The latency of the communication infrastructure refers to the time taken for data exchange. The lower the latency, the faster the data transfer thus lowering the time delay. The control system for the microgrid is heavily dependent on the data from every node of the MG. A communication failure will affect the reliability of the MGs. Apart from failure, noise and interference in communication leads to performance degradation of control. So, it is necessary to strengthen the communication reliability. Security of the communication refers to protection from both physical damage and cyber-attack. MG communication channels may have a geographical spread in tens and hundreds of kilometers. This exposes them to a risk of physical damage. However, when wireless technology is used, it becomes more susceptible to cyber-attacks. Therefore, it is necessary to implement communication security solutions. Many pieces of equipment in MGs are required to operate in time synchronization. Time synchronization can be achieved via various ways such as: Precision Time Protocol (PTP), Global Positioning System (GPS), or Simple Network Time Protocol (STNP) [22].

1.9   Moving towards a smarter grid

The conventional power grid has been predominantly operating using a top-down approach until now. Bulk power generation occurs through centralized sources positioned at a few specific geographical points. This power is then transmitted through power lines and then distributed to small and large consumers. However, the conventional power grid offers limited flexibility for grid expansion or the integration of newer technologies. It is a well-known fact that the electricity demand is increasing every year. Consequently, there is a need to expand the generation capacity of conventional power plants. To mitigate the environmental impact, we have harnessed Renewable Energy Sources for power generation. Since this growing demand also strains the existing transmission and distribution network, the power infrastructure has to be upgraded every year. This imposes a significant financial burden on the government which has to invest huge capital towards the enhancement of the power grid.

Conventional power grid offers rigid solutions to today's growing issues. These solutions lack long-term sustainability and the grid has to be constantly upgraded in response to the aforementioned issues. This led us to transition towards a grid that is smarter and capable in all aspects. A smarter grid promises enhanced energy efficiency, improved grid reliability, and increased integration of renewable resources.

1.9.1   Typical Smart Grid infrastructure

A smart grid is a technologically advanced electricity distribution network that employs digital communication and automation. It is aimed at enabling real-time monitoring, and optimization of power generation, transmission, distribution as well as consumption. Figure 1.8 depicts the fundamental components comprising a typical infrastructure for a Smart Grid.

Broadly speaking a typical Smart Grid infrastructure incorporates conventional power plants which may be classified into peak load plants and base load plants;

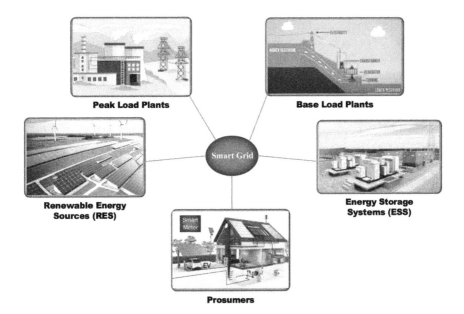

Figure 1.8   Constituents of a typical Smart Grid infrastructure

RES (Renewable Energy Sources) generation plants, Energy Storage Systems (ESS), and Prosumers. Unlike conventional power grid, each of these components has a larger responsibility to improve the grid's efficiency, stability, and reliability. There is a two-way communication between the Smart Grid and consumers. This facilitates better load forecasting. Peak load plants are designed to meet the fluctuating high-demand periods. They provide a rapid and flexible response to supply additional power when needed, ensuring grid stability during times of increased usage. Base load power plants run continuously and these plants act as a reliable baseline power producer. The strategy of power production depends on the forecasted load demand. With the exception of hydroelectric power generation, conventional power plants have a detrimental impact on the environment. This prompted us to harness RES for our power generation needs. Renewable energy sources, such as solar, wind, hydro, and geothermal, play a crucial role in the Smart Grid. They reduce the dependency on fossil fuels and produce clean and sustainable energy. However, due to their intermittent nature, they require advanced forecasting, monitoring, and integration mechanisms to ensure consistent energy supply to the grid. Energy Storage Systems, including batteries and other storage technologies, enable the Smart Grid to efficiently manage the variable nature of renewable energy generation. It can also cater to the increased load demand during the peak period to reduce stress on the grid. A few examples of ESS in a Smart Grid include a Battery Energy Storage System (BESS), Flywheel Energy Storage System (FESS), and Electric Vehicle (EV) in a Vehicle-to-Grid (V2G) mode of operation. These ESS can play a major role in improving the overall energy reliability. Not only that, but they can also enhance grid stability and support demand-response initiatives.

Finally, the term "Prosumers" is distinctly associated with the context of the Smart Grid. "Prosumers" is a term that combines the words "producer" and "consumer." When an individual consumer of electricity can also generate power using renewable energy sources or other methods, they are referred to as a prosumer. Within a Smart Grid, prosumers have the ability to feed their surplus power into the grid, resulting in financial benefits. Apart from engaging in energy trading, they can also participate in demand response programs, thus enhancing grid flexibility. Demand Response is another crucial aspect of Smart Grid which will be discussed later on in this Chapter.

1.9.2   Smart Grids vs. traditional power grids

In the previous section, we explored the Smart Grid concept and the rationale behind transitioning to a Smarter Grid. Before delving any deeper, it is apt to provide a comprehensive comparison between the conventional grid and the Smart Grid. Table 1.3 offers this comparison.

Utilities and governments need to plan for the future of energy infrastructure. Studying the differences between Conventional grids and Smart Grid helps stakeholders make informed decisions about upgrading existing grids or building new smart grid systems.

Table 1.3 Tabular comparison between conventional grid and Smart Grid

Differentiation factor	Conventional grid	Smart Grid
Structure	Conventional grid features a hierarchical structure.	Smart Grid features a network-type arrangement.
Technology	Conventional grid employs traditional infrastructure for electricity distribution. They lack real-time data monitoring and advanced technologies, which limits efficiency.	Smart grid integrates advanced digital technologies, such as sensors and communication systems, enabling real-time data monitoring.
Generation	In a traditional energy setup, power generation is restricted to a central site, preventing the seamless integration of alternative energy sources into the system.	Smart grid infrastructure allows for the distribution of power from various plants and substations, aiding load balancing, reducing strain during peak periods, and minimizing instances of power disruptions.
Distribution	Only the central power plant has the capability to distribute electrical energy within the framework of a traditional energy infrastructure. Power flow is unidirectional.	Independent Power Producers (IPP) or Prosumers with alternate sources of generation like solar panels can inject surplus energy into the grid.
RES integration	Seamless integration of Renewable Energy Sources (RES) can be challenging. It lacks the necessary infrastructure like a distributed energy storage system, market structure, and regulatory mechanism. Only large Independent Power Producers (IPP) can produce power using RES.	It supports the integration of Renewable Energy Sources as Smart Grids offer hybrid storage and other readily available alternative solutions to effectively deal with the intermittent nature of RES.
Monitoring	Manual monitoring approach is necessary for energy distribution in a traditional infrastructure due to its inherent limitations.	The smart grid has inherent self-monitoring capabilities, allowing it to efficiently detect and resolve outages, as well as autonomously manage distribution, thereby removing the need for direct human involvement.
Control	It primarily consists of electromechanical controllers, manual relay settings, SCADA (Supervisory Control and Data Acquisition) with partial automation of distribution system (DS), traditional energy meters, and a few sensors throughout the grid to manage and control operations.	On the other hand, smart grids consist of electronic or digital controllers, online relay settings, SCADA with complete automation of DS, advanced meter infrastructure, and lots of sensors with advanced communication throughout the grid to manage and control operations.

(Continues)

Table 1.3   (Continued)

Differentiation factor	Conventional grid	Smart Grid
Energy tariff	The energy tariff is fixed by the utility operator.	Real-time pricing can be implemented here. Incentives can be provided for prosumers participation in Demand Response programs.
Power quality	Conventional grid assumes low responsibility for power quality issues in the grid.	High level of responsibility for power quality with various mechanisms in place to improve power quality.
Performance	Conventional grids are less efficient, less reliable, and have high losses during power distribution. Their impact on the environment is unsatisfactory and they offer lower consumer satisfaction.	Smart Grids are energy efficient, more reliable, and have lower losses in power distribution. They offer higher consumer satisfaction. In comparison to conventional grids, Smart Grid has a positive impact on the environment.

1.10   Key constituents of a Smart Grid framework

Within the Smart Grid framework, several essential components work in tandem to create a sophisticated energy-efficient system. Each of them has a crucial role in bringing smartness to the grid. In the following sections, we will delve into these key constituents, exploring their functions, benefits, and impacts on shaping the future of the power grid.

1.10.1   Smart systems for renewable energy

Integrating renewable energy sources (RES) with the power grid is not a new concept. However, the Smart Grid brings a significant advantage by offering seamless RES integration, thanks to its extensive range of technologies that eliminate the limitations of RES. Unlike conventional grids, which tend to be more centralized in power generation, modern grids incorporate Distributed Energy Resources (DERs), encompassing solar, wind, small hydroelectric sites, and sometimes fossil fuel-based small-scale generation. Smart systems for RES refer to the infrastructure that enables effortless RES integration.

The Smart Grid not only supports DERs but also incentivizes prosumers to generate power using RES. The main issue associated with RES is its intermittent nature. Furthermore, it decreases the system's inertia since they are connected to the grid through Power Converters. As a result, when there is a sudden change in generation or demand, RES cannot provide the same inertial response as conventional plants. Both these factors lead to a problem of system stability and generation-demand mismatch might lead to excursions in voltage and frequency. Smart Grids employ sophisticated monitoring and forecasting tools that provide valuable insights into RES generation patterns, weather conditions, and energy demand.

Another prominent issue specifically associated with the integration of solar PV, which accounts for a significant share of renewable energy generation, is evident in Figure 1.9. This issue is commonly referred to as the "Duck Curve" in power systems [23]. As the installed capacity of solar PV increases each year, it affects the shape of the graph representing the load that conventional sources of generation must provide. During daytime when solar irradiance is available, the burden on the grid decreases as solar PV supplies the load. However, in the evening when PV generation stops, the burden on the grid rises rapidly. This sudden increase in demand necessitates ramping up conventional power generation, resulting in strain on thermal or nuclear power plants designed for continuous operation. Thermal or Nuclear power plants are designed to operate throughout the day. Ramping up these plants may lead to less economic operation and pose stability issues due to limited grid flexibility.

Smart Grids are capable of effectively dealing with these types of scenarios. Through advanced communication between solar PV generation and grid operators, it is possible to ensure that if solar PV generation surpasses a predetermined baseline, the DER will cut off generation from some panels. This will reduce the burden on conventional sources of generation. Another viable option is to implement Energy Storage Solutions (ESS) to store solar energy during the daytime and utilize it during other parts of the day. This approach maximizes the utilization of the total PV installed capacity. Smart grids provide access to efficient ESS, enabling energy storage even at various geographical locations for later use i.e., Distributed Energy Storage (DES).

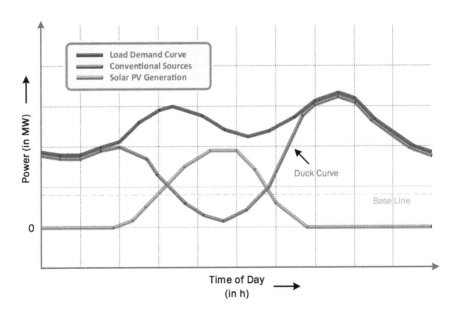

Figure 1.9   Challenges of solar PV integration: The Duck Curve

Another effective solution is Demand Response, which allows for leveling the yellow curve more evenly throughout the day, preventing sudden fluctuations in load demand. Additionally, incentivizing energy consumption during evening hours for consumers can help reduce peak load demand. These last two solutions fall under Demand Side Management, which will be further explored in Section 1.11.3.

In conclusion, a Smart Grid serves as a facilitator for seamless RES integration into the grid by providing comprehensive solutions to mitigate the limitations of RES generation.

1.10.2   Energy Storage Systems

The Energy Storage System (ESS) is not a recent introduction to power grids. Battery Energy Storage System (BESS) and Pumped Hydro Energy Storage (PHES) are a few common types of ESS [24]. They have been in use for a long time. Even in a conventional grid, constituting mostly non-renewable sources of generation, electrical energy is stored in these ESS during off-peak periods. Subsequently, this stored energy is utilized during peak load periods or in the event of an outage. By this arrangement, the burden on the peak load plants is reduced.

Based on the mechanism utilized to store electrical energy, ESS can be classified into the following types:

- **Electrochemical:** Lead Acid Batteries, Lithium-Ion Batteries, Zinc Batteries, Flow Batteries (Redox Flow, Hybrid Flow)
- **Electrical:** Super Capacitor Bank, Superconducting Magnetic Energy Storage (SMES)
- **Mechanical:** Pumped Hydro Energy Storage (PHES), Flywheel Energy Storage System (FESS), Compressed Air Energy Storage (CAES)
- **Thermal:** Sensible-Molten Salt, Latent-Ice Storage, Phase Change Materials (PCM), Thermochemical Storage, Concentrating solar power (CSP)
- **Hydrogen-based:** Hydrogen-based Fuel Cells

The Energy Storage System (ESS) is a fundamental component of the Smart Grid. It is required due to the extensive integration of RES with the grid. RES generation is inherently variable in nature, and ESS serves as an efficient means to store this produced energy for future utilization [25]. As discussed in the previous section, ESS helps to level out the power that ought to be generated from conventional power plants. When these ESS are strategically positioned across different geographic locations, they are often referred to as Distributed Energy Storage (DES). Smart Grids equipped with DES can store electrical energy generated by RES situated in remote areas. This stored energy becomes accessible to meet local demands, expanding the scope of RES generation beyond what conventional grids could achieve. This expansion is facilitated by sophisticated Energy Management Schemes in a Smart Grid that regulates the charging and discharging of ESS as well as DES.

Energy Storage Systems can efficiently store surplus energy during low-demand periods and release it during high-demand instances. This ensures an equilibrium between power generation and consumption. This process enhances load management, grid stability, and the effective integration of intermittent renewable sources. ESS also contributes to peak shaving, microgrid establishment, voltage regulation, and demand response facilitation, and acts as a reliable backup during grid disruptions. Peak shavings can be attained since ESS helps to reduce peak demand by supplying stored energy during times of high consumption. This not only prevents strain on the grid but also reduces the need for expensive peaking power plants that are often less efficient and more polluting. By fulfilling these roles, ESS collaborates harmoniously with smart grid technologies, promoting sustainable energy utilization, optimizing grid efficiency, and facilitating the transition towards an eco-friendly energy landscape.

Apart from RES integration, ESS can participate in the energy market as an ancillary service, providing a short-duration power supply to address stability concerns. There is an additional category of ESS exclusive to Smart Grids, known as Demand Side Storage (DSS). These DSSs are essentially Virtual Energy Storage Systems that can emulate the functionality of an ESS. Their distribution across vast geographical areas and substantial storage capacity contribute to their effectiveness as an energy storage solution. Further insights into DSS will be provided in Section 1.11.3.

1.10.3 Internet of Things (IoT)

The Internet of Things (IoT) functions as a network of physical devices similar to the internet's structure as a network of computers. In the IoT network, diverse devices, such as household appliances like toasters, air conditioners (a/c), room heaters, and coffee makers, are interconnected to communicate and exchange data, earning them the designation of "Things" [26].

Thanks to advancements in technology, modern electrical appliances have become not only smarter but also more efficient. Power electronics plays a significant role in this transformation, as many appliances, ranging from LED bulbs to inverter air conditioners and washing machines, now incorporate power electronic components to control the load more efficiently. For instance, inverter air conditioners can vary the speed of the compressor, a capability lacking in earlier models. This enables them to smoothly adjust their power consumption between off mode and fully operational mode (lowest temperature setting). At the consumer level, IoT has already demonstrated various practical applications. For instance, an IoT-enabled toaster can coordinate with a coffee maker to have your breakfast prepared precisely when needed. These interconnected devices can also relay information to your car, prompting it to position itself on the driveway for convenience.

On a commercial scale, IoT showcases an even broader range of applications, with many implementations having proven their worth over decades. For instance, in the context of oil grids, IoT facilitates real-time pressure monitoring, enabling early warning signals to prevent potential disasters. In the case of power-generating plants, the IoT system can provide real-time information about the overall status of the plant. This contributes to the automation of the entire plant.

The power generation plant can enhance sustainability and operational availability by analyzing vast quantities of data in real time, while simultaneously reducing maintenance costs.

The continued integration of IoT capabilities, particularly within Smart Grid systems, promises to revolutionize the way we manage and optimize energy distribution and consumption in the modern world.

1.10.4   Advanced Metering Infrastructure (AMI) and Smart Meters

The "Advanced Metering Infrastructure" (AMI) represents the culmination of Smart Meters and a two-way communication network. A Smart Meter is a digital energy measuring and monitoring device that serves multiple purposes in a Smart Grid. In contrast to conventional energy meters, which were essentially of the electromechanical induction type, a Smart Meter can gather numerous pieces of information beyond just energy consumption. The list of parameters that the Smart Meter can collect includes:

- Instantaneous real and reactive power
- Instantaneous voltage and current
- Peak voltage and current
- RMS voltage and current
- System operating frequency
- Phase angle between voltage and current
- Power factor
- Total Harmonic Distortion (THD)

Both the utility and the consumer can track all this data and use it to their benefit. A significant advantage of smart meters is their contribution to energy conservation, as consumers are aware of their real-time energy consumption.

In the past, meters required manual tracking of customers' power consumption, which was a costly and inefficient process. The introduction of Automatic Meter Reading (AMR) technology was an initial step toward smart grid implementation. However, even though AMR-enabled utility meters enabled remote monitoring, they lacked two-way communication. Modern Smart meters enable utilities to collect data from the meters while also sending commands, updates, and information back to the devices [27]. Utilities can remotely disconnect or regulate service or implement time-based pricing through these smart meters.

Smart meters have a significant role to play in the development of a sustainable community energy system, which is a subset of distributed generation. This sustainable community energy system integrates RES and high-efficiency co-generation sources to fulfill the energy needs of a local community. Energy storage systems (ESS) can also be integrated with this system to enhance reliability and mitigate various issues related to RES integration. Figure 1.10 displays how smart meters are installed at various points in the distribution network. Through these smart meters, utilities can monitor the amount of power delivered to each entity. They can also gain

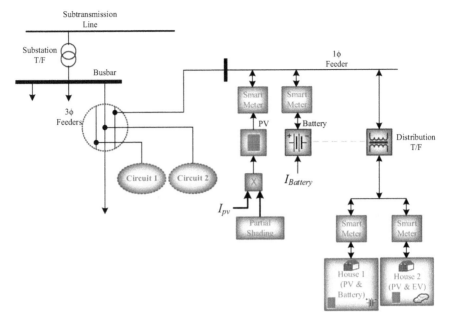

*Figure 1.10   Smart meters in a sustainable community energy system
infrastructure*

insight into the energy injected into the grid from these community energy systems, as they might have RES generation or residences with rooftop photovoltaics (PV) and Electric Vehicles (EVs) in Vehicle-to-Grid (V2G) mode.

The AMI is the output of Smart Grid IoT technology and employs both wired and wireless communication techniques [28]. Various communication protocols are employed to facilitate interactions with the opposite end of the meter. In residential settings, typical protocols utilized for establishing a Home Area Network (HAN) are ZigBee, Z-Wave, HomePlug, and many more. These protocols establish a mesh network architecture, wherein smart devices do not exclusively connect to a central hub (router). Instead, each smart device functions as a node within this network, interconnecting with other devices to create an intricate web of interconnected smart components. Every smart device integrated into the HAN possesses the ability to function as both a signal endpoint and a signal redistribution point. This signifies that each individual smart device can receive a signal and subsequently respond to it. Furthermore, it can serve as a route for relaying this signal to the nearest smart device within the network, enhancing the signal's reach and robustness across the network. For instance, the Byram Labs EnergyAxis REX2-EA smart meter utilizes both 900 MHz and 2.4 GHz ZigBee wireless communications to act as a mesh-network-connected smart meter endpoint and gateway for other smart energy devices in the home. Smart meters connected via a wireless mesh network can be linked to hubs wired via landline, fiberoptic, or other telecom infrastructure, ultimately connecting to the utility's IT servers.

In summary, the advanced metering infrastructure is more than just an array of smart meters. It constitutes a sophisticated system comprising smart meter hardware and software, communication networks linking all subnetworks and hubs, meter data management software, IT systems, and utility operational systems to enable efficient data processing from the smart grid.

1.10.5   Distribution automation and grid monitoring systems

To enhance the efficiency and reliability of the Smart Grid, Distribution Automation is implemented. This involves the integration of intelligent devices with advanced technologies. Distribution Automation facilitates effective monitoring and remote control of Smart Grid components. The key objectives of Distribution Automation include the following:

- **Remote monitoring and control:** This is done with the help of sensors and an advanced communication network. Apart from these sensors, Supervisory Control and Data Acquisition (SCADA) also assist in monitoring the status and performance of the grid components. All of this is done remotely. With this data, grid operators can make informed decisions and remotely control devices to optimize the distribution system's operations [29].
- **Conservation Voltage Reduction (CVR):** It is a method to reduce power consumption through a controlled reduction in voltage levels. The application of CVR involves regulating voltages to align with the lower threshold of the allowed voltage range specified by the Distribution System Operator. Apart from reduction in energy consumption, it helps to reduce the peak load, and losses happening in the transformers and transmission lines. This in turn helps in reducing the impact of conventional power generation on the environment.
- **Volt/VAR control:** Based on the distribution system operational status as sensed by the SCADA and AMI, voltage and VAR levels can be controlled in the network. This can be done through capacitors. They can also help to improve the power factor.
- **Fault Location, Isolation, and Service Restoration (FLISR):** It aims to identify the location of faults in the distribution grid and isolate the faulty sections to prevent further damage [30]. Thereafter, it swiftly restores service to the unaffected areas. By automating this process, power outages are minimized, and service restoration times are significantly reduced. When a fault is detected, the system analyzes the grid's topology to find alternative paths for power flow, bypassing the faulty section. This self-healing capability allows for the rapid restoration of power to affected customers. This minimizes downtime and greatly enhances the reliability of the grid.

Distribution Automation helps in enhancing the overall performance of the power grid. The integration of these Distribution Automation functions transforms traditional power distribution grids into modern and intelligent systems.

1.11 Fundamentals of smart grid operation: core concepts

The Smart Grid's infrastructure is extensive and intricate. It is a culmination of diverse components and technologies. Within this section, our focus will be directed towards exploring the operational aspect of the Smart Grid.

1.11.1 Strategies for achieving energy efficiency and grid modernization

In the context of Smart Grids, achieving energy efficiency and grid modernization involves adopting various strategies to enhance the overall performance and sustainability of the system.

The Smart Grid comprises diverse Distributed Generation sources in conjunction with ESS. This makes the energy management a challenging task. Optimal Load Dispatch (OLD) scheduling ensures efficient energy management across various components of the Smart Grid. The objectives of this OLD can be to minimize grid losses, maximize profit, or minimize emissions.

In the consumer's end, rooftop PV generation, battery storage, and V2G (EV) may be present which makes them self-reliant and energy efficient. Smart devices connected to HAN optimize their operation according to pre-set rules. This helps to shift their operation during off-peak hours.

In terms of grid modernization, Smart Grids have IoT-enabled generation for both conventional power plants and RES generation, advanced Power Electronic converters, ESS, ancillary services to provide grid support, AMI, HAN, electronics controllers, digital sensors, SCADA, communication system, and Unified Power Quality Controller (UPQC).

1.11.2 Renewable energy integration

The global energy challenge is a pressing issue that has prompted a significant upsurge in the installation of renewable-energy power plants. Projections indicate that the overall capacity expansion on a global scale can experience an almost twofold increase within the forthcoming 5-year span. This surge in capacity growth will not only surpass coal as the primary source of electricity generation but will also contribute significantly to sustaining the prospect of containing global warming within the confines of a 1.5 °C increase.

In the context of electricity generation, utility-scale solar photovoltaic (PV) and onshore wind systems emerge as economically favorable choices across a substantial majority of nations worldwide. According to the Renewable Energy Policy Network for the 21st Century (REN21), global capacity for solar PV is anticipated to nearly triple over the period from 2022 to 2027 [31]. It has the potential to surpass coal and become the predominant source of power generation across the globe. Table 1.4 provides an overview of the current RES generation installed capacity. Additionally, it outlines the targets that have to be reached for achieving net zero global carbon dioxide emissions from energy generation by 2030 and 2050. To achieve this projected target,

*Table 1.4   Overview of current renewable energy generation capacity and
           projected targets [31]*

RES	RES share for 2022 (in %)	Installed capacity as of 2021 (in GW)	Increment in installed capacity for 2022 (in %)	Target for 2030 (in Thousand GW)	Target for 2050 (in Thousand GW)
Solar PV	6.9	942	+25	4.9	14.4
Wind power	5.2	829	+9	3.1	8.3
Hydroelectric power	15.1	1198	+2	1.8	2.6
Other renewable sources	2.7	161	+4	0.5	1.4

strong measures have to be taken to expand power distribution networks, explore new sites for renewable energy generation, and address supply chain issues.

The schematic diagram of RES-energy generation in a Smart Grid is quite similar to that of the AC microgrid system shown in Figure 1.3. The major difference is the distant location where the RES generation sites are situated compared to microgrids. ESS might be present to mitigate the effects of variable generation. The control techniques of RES-energy generation in a Smart Grid will be similar to those discussed in Section 1.6. Additionally, IoT-based technologies will be implemented in the RES-energy generation plants to manage the demand-supply balance. This is crucial because RES integration reduces grid inertia. An advanced communication infrastructure connects these distributed generation units to the grid operator. They can forecast energy generation based on weather conditions and communicate it to the grid so that it can implement Demand Response measures if required.

1.11.3   Demand Side Management (DSM)

Demand Side Management (DSM) is a concept that is implemented on the demand-side of the system, just like the name suggests. It comprises a plethora of measures, which may span from the replacement of outdated electrical appliances with modern power-efficient counterparts to the integration of a dynamic load management scheme. DSM puts a halt on the continual expansion of generation installed capacity and transmission line capacity upgrades. It achieves this by introducing flexibility to the grid and improving its existing limits. In this Section, we will explore different types of DSM classified based on their operational time window and the impact they have on the grid and consumer processes. Figure 1.11 offers a concise visual representation of the DSM concept.

Loads can be categorized as either critical or non-critical. Critical loads are those loads that should not be controlled through external means, as doing so could result in significant inconvenience for consumers. On the other hand, non-critical loads are loads that can be controlled when required for DSM operation. Demand

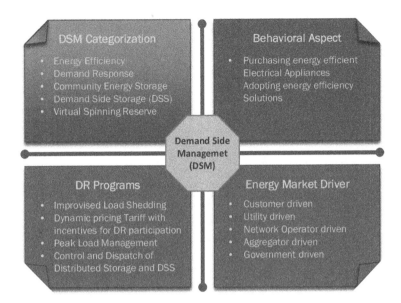

Figure 1.11   Diagrammatic overview of the Demand-Side Management (DSM) concept

Response (DR), and Demand Side Storage (DSS) are an integral part of DSM. Room heaters, a/c, and similar appliances fall under the category of non-critical loads. Using these approaches, the Smart Grid aggregator can curtail certain non-critical loads. This curtailing of loads reduces power consumption by a certain degree. DSM leverages this concept to showcase its capacity to effectively "generate" power in a virtual manner. If necessary, the aggregator can also increase the power consumption by turning on certain non-critical loads. In this way, the distribution system will act like a sink during periods of surplus power generation.

A useful utilization of DSM is to shape the daily load demand curve at the distribution level. Depending on specific requirements, it is possible to reduce consumption during daily peak load periods if the electricity distribution companies (DISCOMs) perceive the energy rates to be high during those time frames. Alternatively, the load can be adjusted to align with the generation curve. This synchronization facilitates a balanced match between energy supply and demand, potentially mitigating a range of grid-related challenges. Beyond these methods, there exist various other techniques for shaping load demand. The impact of these DSM operations on the peak load demand and energy demand is detailed in Table 1.5. Figure 1.12 depicts a visualization of these procedures on a standard daily load demand curve characterized by two peaks occurring in the late morning and evening hours.

1.11.3.1   Energy conservation and efficiency

Energy Conservation and Energy Efficiency are fundamental concepts of Demand Side Management (DSM), which yield lasting and sometimes permanent

Table 1.5   Analyzing peak load and energy demand changes due to DSM initiatives

Goal of DSM operation	Peak load demand	Energy demand
Reduce the total energy demand (reduced energy consumption)	↓	↓
Boosting load demand (strategic load growth)	Might increase	↑
Shifting load demand to off-peak period (load shifting)	↓	No change
Curtail load demand during peak hours (peak load clipping)	↓	↓
Boosting load demand during off-peak period (valley filling)	No change	Increase
Trigger load variations in response to supply (load following)	↓	Might decrease

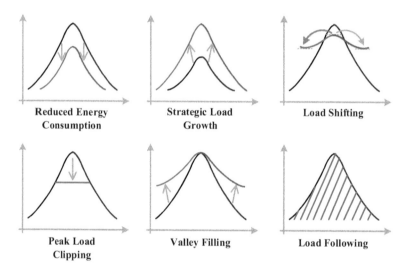

Figure 1.12   Smart meters in a sustainable community energy system infrastructure

effects. In today's era of significant advancements in electronics and electrical engineering, the market offers a range of energy-efficient devices. If all the consumers in the distribution network start making use of these energy-efficient devices, then we have already made the demand side more energy-efficient. Meanwhile, industries can optimize energy consumption by adopting energy-efficient technologies, refining production processes, and conducting regular energy audits to pinpoint areas for enhancement. Behavioral shifts encourage individuals to reduce energy consumption, curtail the use of non-essential appliances, and actively engage in environmentally friendly initiatives. This forms the basis of Energy Conservation.

Over the last decade, the government has been promoting the adoption of LED lighting over incandescent bulbs. Manufacturing industries, too, have the potential to significantly contribute to the promotion of Energy Efficiency measures. They can introduce various schemes in the market, allowing consumers to exchange outdated electrical appliances for new, energy-efficient ones. For instance, a consumer could replace their old air conditioning and heating units with an energy-efficient Heating, Ventilation, and Air Conditioning (HVAC) system. Additionally, they can take other measures such as improving house insulation to reduce the need for heaters or incorporating more windows in newly constructed homes to enhance natural ventilation.

The major aim is to transition towards a more energy-efficient environment and alleviate the strain on our current energy sources, which currently have to generate more energy than is necessary.

Figure 1.13 provides a visual representation of how various DSM measures impact system performance duration and operational timeframe.

1.11.3.2 Demand Response (DR)

In response to any grid-related issues, the Smart Grid aggregator can implement Demand Response Programs (DRP) on the distribution side of the grid. Their objective can range from simply shifting or reducing the peak load demand to improving the power quality of the distribution grid's supply. In return, consumers who participate in DRP receive monetary benefits from the distribution grid utility through adjustments in their energy tariffs [32]. Depending on the nature of these monetary benefits, DR can be categorized into the following classifications:

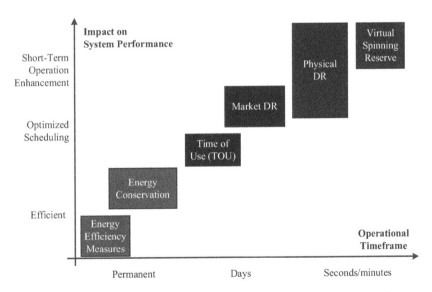

Figure 1.13   Duration of impact on system performance and operational timeframe across differing DSM measures

❖ **Energy price-based DRP:**

These DRPs provide monetary benefits to participants through a time-based pricing scheme. They are also called Time Varying Rates based programs (TVR). The DRPs that fall into this category are Critical Peak Pricing (CPP), Peak Time Rebate Pricing (PTR), Time-of-Use Pricing (TOU), and Real Time Pricing (RTP).

❖ **Incentive-based DRP:**

These DRPs provide monetary incentives that are proportionate to the level of contribution from participants. The DRPs that fall into this category are Direct Load Control (DLC), Curtailable/Interruptible Services (I/C), Emergency DR Program (EDRP), Demand Bidding (DB), Capacity Market Program, and Ancillary Services Market.

In TOU Pricing, energy prices are pre-decided hourly values which are different for peak and off-peak periods. In RTP, the price varies according to the day-ahead market rates for electricity. The rates fluctuate throughout the day. In CPP Pricing, the rates are static throughout the day except utilities assign higher rates for peak periods. In contrast to CPP, Peak Time Rebate (PTR) pricing sets a comparatively higher rate during off-peak periods and subsequently reduces the rate during peak periods.

On the other hand, the implementation of incentive-based DRPs is quite different. In the case of DLC, Smart Grid aggregators have free access to consumer processes. In the case of I/C, consumers engage in a distinct contract that involves a controlled reduction of their electricity consumption. In EDRP, consumers voluntarily respond to emergency signals by curtailing their energy usage. In Capacity Market Programs, consumers commit to providing additional power to the grid during periods of high demand. In the case of DB, consumers have the opportunity to submit bids for curtailing their energy consumption at appealing prices. Lastly, in the Ancillary Services Market, load controllers with fast responsive non-critical loads can participate and provide ancillary services to the grid.

1.11.3.3   Demand Side Storage (DSS)

If the power ratings of specific loads within the distribution network are considerably higher, we can conceptualize another aspect of DSM known as Demand Side Storage (DSS). Certain industrial loads fall into this category due to their time-shiftable nature. This implies that their operations can be rescheduled to different time periods based on certain criteria. For instance, consider a water storage plant responsible for treating and storing water in tanks, achieved through centrifugal pumps powered by induction motors. When the water levels in the storage tanks are satisfactory, controlling the motors' speed and hence power consumption will not adversely affect the process. These water storage plants can be viewed as a DSS. When the load controller controlling these motors receives instructions from the aggregator that the power supply in the distribution network is insufficient, these DSS can contribute "energy" by reducing their demand, akin to the discharging process of ESS. Conversely, during periods of surplus power availability, these DSS can absorb the surplus power provided the storage tanks are not already full

[33]. The operational timeframes of these DSSs typically span from a few minutes to a few hours.

In a Sustainable Community Energy system integrated within a Smart Grid, the presence of a Battery Energy Storage (BES) installation or a community-scale Electric Vehicle (EV) charging station can be considered as DSS entities. The BES will provide real energy in this case. While its primary role is to offer energy during outages or meet peak demand, it can also be harnessed as a DSS to supply power to the grid as needed. In scenarios where EVs are operating in Vehicle-to-Grid (V2G) mode, they also possess DSS capabilities. They can deliver actual power from their batteries during power deficits and be switched to charging mode during periods of surplus generation. Even when EVs are not operating in V2G mode, their charging process can be remotely controlled, allowing charging to be postponed and resumed later, owing to the inherently time-shiftable nature of their charging process. As a result, a community-scale EV charging station can function as a DSS, even when not connected in the V2G mode of operation.

1.11.3.4 Virtual Spinning Reserve

Spinning Reserves (SR) refer to the surplus generation capacity available within a grid, ready to provide power in the event of a power deficit or stability concerns. Similarly, within Demand Side Management (DSM), the concept of Virtual Spinning Reserves is introduced through load management strategies. These reserves exhibit swift response characteristics akin to conventional SR. Traditionally, this role has been fulfilled by regulating power plants.

Loads have the potential to act as virtual spinning reserves by adjusting their own power consumption based on grid conditions. This kind of dynamic load control can emulate the droop control method used in conventional grids. Essentially, this entails reducing the power consumption of participating loads when the grid frequency decreases, and conversely, increasing consumption when the grid frequency rises. When this load control is done autonomously, it resembles primary control; whereas, if conducted in a coordinated manner, it aligns with secondary control.

Communication capabilities might be helpful for implementing a Virtual Spinning Reserve. It will bring "fairness" into the process. In scenarios where multiple devices have the capacity to reduce their load and the contribution of a single device is sufficient, an uncommunicative approach might lead to preferential treatment to the quickest responder and it will receive the monetary incentives always. Others might never get their turn. By incorporating communication, a rotational approach can be followed, ensuring that each participant receives their fair share of opportunities. This is also good for the system stability. If all the loads react to the grid changes in a similar fashion without any communication among themselves then it might lead to stability issues in the Smart Grid [34].

Communication capabilities can be coupled with model-predictive controls to enhance performance. When a controller possesses its own load model, it can precisely ascertain the extent to which it can reduce or increase its power consumption and for how long, without significantly disrupting its processes. This awareness equips the controller to respond optimally to grid fluctuations, making it a superior and more efficient approach for implementing Virtual Spinning Reserves.

1.11.4   Energy market and business model

Transactive energy is a term associated with Smart Grids which refers to an intelligent multi-level communication system that helps in coordinating power generation, distribution, and consumption. Here, prosumers can also confidently engage in buying or selling electricity. This marketplace operates in coordination with utility grid operations, ensuring a secure and reliable power supply for consumers.

In a conventional grid, a generation utility can either self-schedule their generation in response to prices or they can participate in the energy market. Energy trading takes place in a deregulated market, where competitive generation utilities submit bids to a market pool at specified time intervals. Their goal is to maximize their profits. The pool operator makes decisions based on these bids. Smart Grid has an identical energy market for competitive power generation utilities.

A similar approach can be followed at the distribution network. The current wholesale electricity market for large-scale transactions is replicated at the distribution grid level to facilitate smaller-scale transactions. The participants in these markets are ESS, DSS, small DER generations, microgrids, prosumers, and smart buildings. This democratizes the currently closed electricity market. However, to accommodate millions of participants, compared to hundreds or thousands, this market demands far greater flexibility, robustness, and scalability.

In Smart Grids, economic benefits to the DR participants can be provided via price-based programs or incentive-based programs. Based on these two different monetary compensation models, DR programs can be classified as shown in Figure 1.14. All this is facilitated by the Smart Grid aggregator.

Transactive energy markets aim to expand DR programs through the adoption of Automated Demand Response (ADR) technologies, and an initiative similar to OpenADR which creates a base for the exchange of interoperable information to facilitate ADR.

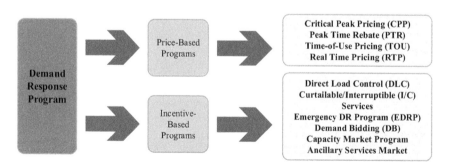

Figure 1.14   Classification of DR programs based on monetary compensation models

1.11.5   Cybersecurity and data privacy considerations

Similar to any other system utilizing a communication network, confidentiality, integrity, and availability are the key security requirements of a smart grid. Confidentiality prevents any leak of private consumers' information. Integrity refers to the accuracy of information while Availability ensures continuity of service. Cybersecurity measures are crucial to ensure the grid's security, resilience, and robustness against potential threats.

Targeted threats involve malicious activities aimed at specific areas of the smart grid, such as the infamous "Stuxnet" worm that targeted Supervisory Control and Data Acquisition (SCADA) systems, causing significant damage to Iranian nuclear centrifuges. Other malware like "Havex" and "BlackEnergy" have also been identified as threats to industrial control systems and power grids, respectively. Defense against such targeted threats and the development of effective countermeasures are critical priorities [35].

Hybrid threats combine cyber and physical attacks. They create complex challenges for defense and protocol definition as the attackers can target any of the major components in the Grid.

Key cybersecurity considerations in smart grids include:

- **Secure communication:** Appropriate encryption and authentication protocols are necessary to ensure the confidentiality and authenticity of the exchanged data.
- **Intrusion detection:** Advanced intrusion detection systems should be deployed to continuously monitor the network for potential security threats. Machine learning and AI algorithms may be utilized in real time to detect anomalous system behavior.
- **Access control:** Unauthorized access to critical grid components has to be prevented which can be ensured by granting role-based access. Remote access to grid infrastructure for monitoring and maintenance must be secure.
- **Regular firmware updates:** Vulnerabilities can be addressed through regular firmware updates.
- **Cyber hygiene:** Awareness regarding cybersecurity has to be promoted among grid personnel. They should adhere to good cyber hygiene such as strong password management and prevention of phishing attempts.

Data Privacy is another key consideration in a Smart Grid. In a Smart Grid, "Data Subject" refers to an individual consumer who can be identified, either directly or indirectly, through an identification number or specific factors related to their physical, economic, cultural, or social identity, or through their activities. "Data Controller" refers to the organization or individual responsible for collecting or creating personal information obtained from the data subject. This role entails making decisions regarding how and why data about the data subjects will be processed, adhering to legal requirements and privacy considerations. Moreover, the data controller holds the responsibility to safeguard the data securely during the time it is under their custodianship. At times, the Data Processor/Manager may assume the role of Data Controller or they may act on its behalf. Through this strategy, data privacy is ensured within a Smart Grid.

1.12   Summary

This chapter goes through the concepts of microgrids and smart grids. The micro-grid can be considered as a small-scale grid that uses distributed energy resources like solar PV systems, wind turbines, and Combined Heat and Power (CHP) with a centralized control system to implement the Energy Management Scheme. They can make use of energy storage systems for reliable power supply. Microgrids promote the use of RES for clean and cost-effective energy generation. An efficient EMS can take care of the power quality issues that arise due to power electronic converters. The chapter explores key features, benefits and challenges to overcome during its implementation. Different types of microgrids are discussed, and certain control aspects are also briefly covered.

A smart grid constitutes an electrical infrastructure that employs digital tech-nology and other cutting-edge advancements to effectively monitor and regulate the transmission of electricity from diverse power generation sources. This intricate network is designed to efficiently address the dynamic electricity demand of end consumers. It involves initiatives aimed at modernizing the infrastructure of the grid, which encompasses generation, transmission, and distribution networks. This modernization facilitates the effective integration of concepts such as Distributed Generation, Renewable Energy Source (RES) integration, Energy Storage, and Demand Side Management. Due to the extensive reliance on communication sys-tems within the smart grid, it is essential to have protocols in place to prevent any cybersecurity attacks.

Microgrids and smart grids are modern-day energy infrastructures that are primed for the future and actively support sustainable energy distribution.

References

[1]   Kroposki, B., Lasseter, R., Ise, T., Morozumi, S., Papathanassiou, S., and Hatziargyriou, N.: "Making microgrids work," *IEEE Power Energy Mag.* 6(3), 40–53 (2008).

[2]   Baker, K., Hug, G., and Li, X.: Optimal integration of intermittent energy sources using distributed multi-step optimization. *Proceedings of the 2012 IEEE Power and Energy Society General Meeting, San Diego*, CA, USA, 22–26 July 2012.

[3]   Yoldaş, Y., Önen, A., Muyeen, S.M., Vasilakos, A.V., and Alan, İ.: "Enhancing smart grid with microgrids: Challenges and opportunities," *Renew. Sustain. Energy Rev.*, 72, 205–214 (2017).

[4]   Justo, J.J., Mwasilu, F., Lee, J., and Jung, J.-W.: "AC-microgrids versus DC-microgrids with distributed energy resources: A review," *Renew. Sustain. Energy Rev.* 24, 387–405 (2013).

[5]   Sur, U., Biswas, A., Bera, J.N., and Sarkar G.: "A modified holomorphic embedding method based hybrid AC–DC microgrid load flow," *Elect. Power Syst. Res.*, 182(106267), 1–9 (2020).

[6] Abbasi, M., Abbasi, E., Li, L., Aguilera, R.P., Lu, D., and Wang, F.: "Review on the microgrid concept, structures, components, communication systems, and control methods," *Energies*. 16(1), 484 (2023).

[7] Leggate, D., and Kerkman, R.J.: Adaptive Harmonic Elimination Compensation for Voltage Distortion Elements. U.S. Patent 10,250,161, 2 April 2019.

[8] Farrokhabadi, M., Canizares, C.A., Simpson-Poroc, J.W., *et al.*: "Microgrid stability definitions, analysis, and examples," *IEEE Trans. Power Syst.* 35(1), 13–29 (2020).

[9] Vasilakis, A., Zafeiratou, I., Lagos, D.T., and Hatziargyriou, N.D.: "The Evolution of Research in Microgrids Control," *IEEE Open Access J. Power Energy*, 7, 331–343, (2020).

[10] Sen, S., and Kumar, V.: "Microgrid control: A comprehensive survey," *Annu. Rev. Control*, 45, 118–151, 2018.

[11] Baghaee, H., Mirsalim, M., and Gharehpetian, G.: "Power calculation using RBF neural networks to improve power sharing of hierarchical control scheme in multi-DER microgrids," *IEEE J. Emerging Sel. Top. Power Electron.* 4(4), 1217–1225 (2016).

[12] Divshali, P.H., Alimardani, A., Hosseinian, S.H., and Abedi, M.: "Decentralized cooperative control strategy of microsources for stabilizing autonomous VSC-based microgrids," *IEEE Trans. Power Syst.* 27(4), 1949–1959 (2012).

[13] Zhou, J., Zhang, H., Sun, Q., Ma, D., and Huang, B.: "Event-based distributed active power sharing control for interconnected AC and DC microgrids," *IEEE Trans. Smart Grid* 9(6), 6815–6828 (2018).

[14] La Bella, A., Cominesi, S.R., Sandroni, C., *et al.*: "Hierarchical predictive control of microgrids in islanded operation," *IEEE Trans. Autom. Sci. Eng.*, 14(2), 536–546 (2016).

[15] Mohammadi, K., Azizi, E., Choi, J., Hamidi-Beheshti, M.-T., Bidram, A., and Bolouki, S.: "Asynchronous periodic distributed event-triggered voltage and frequency control of microgrids," *IEEE Transactions on Power Systems*, 36(5), 4524–4538 (2021).

[16] Lizhen, W., Xiaohong, H., Zhengzhe, L., *et al.*: "Decentralized control approach for voltage unbalance compensation in islanded microgrid," in *IEEE 8th Int. Power Electronics and Motion Control Conf. (IPEMC-ECCE Asia)*, Hefei, China, 2016.

[17] Peyghami, S., Mokhtari, H., and Blaabjerg, F.: "Decentralized load sharing in a low-voltage direct current microgrid with an adaptive droop approach based on a superimposed frequency," *IEEE J. Emerging Sel. Top. Power Electron.*, 5(3), 1205–1215 (2017).

[18] Sun, Y., Shi, G., Li, X., *et al.*: "An f-P/Q droop control in cascaded-type microgrid," *IEEE Trans. Power Syst.* 33, 1136–1138 (2018).

[19] Jena, C.J., and Ray, P.K.: "Power management in three-phase grid-integrated PV system with hybrid energy storage system," *Energies*. 16(4), 1–20 (2023).

[20] Cominesi, S.R., Farina, M., Giulioni, L., Picasso, B., and Scattolini, R.: "A two-layer stochastic model predictive control scheme for microgrids," *IEEE Trans. Control Syst. Technol.* 26(1), 1–13 (2017).

[21] Bharatee, A., Ray, P.K., and Ghosh, A.: "A power management scheme for grid-connected PV integrated with hybrid energy storage system," *J. Mod. Power Systems Clean Energy*, 10(4), 954–963 (2022).

[22] Yang, H., Li, Q., and Chen, W.: "Microgrid communication system and its application in hierarchical control," in *Smart Power Distribution Systems*, New York: Academic Press, pp. 179–204 (2019).

[23] Calero, I., Canizares, C.A., Bhattacharya, K., and Baldick, R.: "Duck-Curve mitigation in power grids with high penetration of PV generation," *IEEE Trans. Smart Grid*, 13(1), 314–329 (2022).

[24] Calero, F., Canizares, C.A., Bhattacharya, K., *et al.*: "A review of modeling and applications of energy storage systems in power grids," *Proc. IEEE*, 111 (7), 806–831 (2023).

[25] Li, X., Wang, L., Yan, N., and Ma, R.: "Cooperative dispatch of distributed energy storage in distribution network with PV generation systems," *IEEE Trans. Appl. Supercond.*, 31(8), 1–4 (2021).

[26] Halba, K., Griffor, E., Lbath, A., and Dahbura, A.: "IoT capabilities composition and decomposition: A systematic review," *IEEE Access*, 11, 29959–30007 (2023).

[27] Sun, Q., Li, H., Ma, Z., *et al.*: "A comprehensive review of smart energy meters in intelligent energy networks," in *IEEE Internet Things J.*, 3(4), 464–479 (2016).

[28] Ghosal, A., and Conti, M.: "Key management systems for smart grid advanced metering infrastructure: A survey," *IEEE Commun. Surv. Tutor.*, 21(3), 2831–2848 (2019).

[29] Pliatsios, D., Sarigiannidis, P., Lagkas, T., and Sarigiannidis, A.G.: "A survey on SCADA systems: Secure protocols, incidents, threats and tactics," *IEEE Commun. Surv. Tutor.* 22(3), 1942–1976 (2020).

[30] Liu, J., Qin, C., and Yu, Y.: "A comprehensive resilience-oriented FLISR method for distribution systems," *IEEE Trans. Smart Grid*, 12(3), 2136–2152 (2021).

[31] REN21. *Renewables 2023 Global Status Report collection, Renewables in Energy Supply*. Paris: REN21 Secretariat (2023).

[32] Wen, L., Zhou, K., Feng, W., and Yang, S.: "Demand side management in Smart Grid: A dynamic-price-based demand response model," *IEEE Trans. Eng. Manage.* 71, 1439–1451 (2024).

[33] Nandkeolyar, S., and Ray, P.K.: "Multi objective demand side storage dispatch using hybrid extreme learning machine trained neural networks in a smart grid," *J. Energy Storage*, 51(104439), 1–12, (2022).

[34] Stadler, M., Palensky, P., Lorenz, B., Weihs, M., and Roesener, C.: "Integral resource optimization networks and their techno-economic constraints," *Int. J. Electr. Power Energy Syst.*, 1(4), 299–320 (2005).

[35] Machado, T.G., Mota, A.A., Mota, L.T.M., Carvalho, M.F.H., and Pezzuto, C.C.: "Methodology for identifying the cybersecurity maturity level of Smart Grids," *IEEE Latin Am. Trans.*, 14(11), 4512–4519 (2016).

Chapter 2

Renewable energy system integration in smart grids

Anindya Bharatee[1], Pragnyashree Ray[1], Pravat Kumar Ray[1] and Bidyadhar Subudhi[2]

2.1 Background

The era of growing environmental concerns and rising demand for energy consumption leads to the utilization of renewable energy resources (RERs). RERs are found to be substantial substitutes for fossil-fuel-based power generation systems, which are environment-friendly with zero greenhouse gas emissions [1]. The most commonly used sustainable technologies are solar PV systems and wind turbines. From this, the concept of microgrid technology arises which also provides more efficiency, resiliency, power balance, and some power quality features to the consumers [2]. Due to low maintenance, high efficiency, low cost, and high consistency, the solar PV generation system is adopted worldwide. However, the weather-dependent power generation from PV units affects the microgrid stability up to a large extent [3]. To reduce the instability in the power supply to the end users and to enhance the quality of power, one of the best choices is the addition of energy storage units [4].

In this smart era, the power system is reframed by incorporating smart technologies. These technologies involve power quality-sensitive components that demand quality power for their uninterrupted operation. There are custom power devices (CPD) to solve the power quality issue. Unified Power Quality Conditioner (UPQC) is the most suitable CPD to handle both current and voltage-related power quality concerns [5]. Also, penetration of RERs into the grid causes power quality issues. Therefore the integration of PV to the grid through UPQC ensures two-fold advantages to the smart grid, i.e., power quality improvement and renewable energy support. The control and operation of PV-fed UPQC with respect to PV intermittency are discussed in [6,7].

In microgrid applications, the utilization of battery energy storage is very popular. But, during sudden transients and high-power applications, the stress on

[1]Department of Electrical Engineering, National Institute of Technology Rourkela, India
[2]School of Electrical Sciences, Indian Institute of Technology Goa, India

the batteries will increase and it will cause premature failure because of low power density (10–400 W/L) characteristics [8]. Hence, in microgrid applications, it is advised to utilize two or more different kinds of energy buffer devices with complementary characteristics to avail both the facility of high energy and power density. In [9], the authors recommended the integration of battery units with double-layer supercapacitors (SC) to form hybrid energy storage units (HESUs). The high power density of SC (500–5,000 W/L) can support the transient power requirement in the system with fast charging and discharging capacity. Similarly, the high energy density of batteries (50–80 Wh/L) can help to deliver the average power requirement in the system for a longer period. The combined use of battery and SC helps to lessen the stress from each other and hence increases the lifetime and reduces the net present cost of the system.

Microgrids can operate autonomously or in coordination with the existing utility grids. In both cases, the stability can be decided by regulating the power flow and demand in the system. An efficient power management technique optimizes the utilization of energy sources and prevents overloads and blackouts. Hence, nowadays researchers mainly focus on different efficient power management techniques to implement in microgrid applications. A centralized control scheme using unified control and power management (CAPMS) is suggested in [10] for the smooth transfer of power between generation and loads. In [11], a fuzzy logic-based perturb & observe (P&O) maximum power point tracking (MPPT) algorithm is proposed for maximum utilization of the solar PV system with effective energy management among battery, PV, and diesel generators. However, in these papers, only the battery storage unit is considered, which can deteriorate the transient operations during variable generation. Hence, to improve the energy storage performances, in various literature both battery and supercapacitors are used with different techniques for power division among them. In [4,12], a low-pass filter (LPF) based frequency decomposition technique with proportional-integral (PI) controllers is proposed for the allocation of power among the battery and SC. Similarly, a state of charge (*SOC*) based power controlling algorithm is recommended for the control of power flow. By using a real-time digital simulator (RTDS) for verification, a centralized optimization technique is implemented at an advanced distribution level in [13] in an AC microgrid to track the voltage at the DC bus properly. Similarly, the DC bus signaling technique is hybridized with the fuzzy logic controller to control the power flow in a remote AC microgrid [14].

PI controllers are used in most of the literature discussed above. PI controllers are not robust and adaptable to any large and sudden uncertainties in the system. But, in microgrid applications, the integration of weather-dependent energy sources can create high surges and uncertainties in the system. Hence, to overcome this issue, some robust nonlinear controllers are suggested in other literature to adopt these variations. In [15], a sliding mode controller (SMC) is used to maintain the *SOC* of the storage units in the system. Also, for maintenance of DC link voltage, an SMC is utilized in [16] with feedback linearization. However, the discharging and charging of storage units are not addressed, which is an essential part of the power management scheme. A fixed frequency with variable pulse width modulation (PWM) carrier signal-based SMC was used in [17] for the switching signal generation of DC/DC converters connected to the hybrid energy storage units. In [18], with a large-signal stability analysis, a power-sharing scheme was recommended for a hybrid microgrid by using an SMC for nonlinear and unbalanced loads. A super-twisting sliding mode controller is implemented in [19] to regulate the power transfer between HESUs and RERs to advance the global flexibility of the microgrid system.

To ensure the effective functioning and regulation of grid-integrated systems, it is crucial to accurately estimate the fundamental frequency and phase information of the grid. However, challenges arise due to disturbances imposed on the grid voltage signal, either from the grid itself or from loads connected to it. One commonly used method for grid synchronization is the three-phase synchronous reference frame-based phase-locked loop (SRF-PLL). However, the performance of SRF-PLL can suffer in the presence of unbalanced or highly polluted grids [20,21]. To address these challenges, various enhancements have been proposed for conventional PLL methods. One such enhancement involves incorporating a pre-filtering step into the SRF-PLL [22–24]. This pre-filtering step aims to extract the fundamental positive sequence component (FPSC) from the grid signal before feeding it into the SRF-PLL, thus improving the accuracy of the estimated values. Several techniques exist for extracting the fundamental positive sequence component, including dual second-order generalized integrator-based (DSOGI), complex coefficient filters (CCFs), space vector discrete Fourier transform (SV-DFT), and cascaded delay signal cancellation-based (CDSC) methods. Researchers have explored different pre-filtering approaches in conjunction with PLL. In [25], authors have discussed CCF as a pre-filtering stage with respect to PLL. As a pre-filtering stage for accurate operation of SRF-PLL, the SV-DFT-based method is implemented in [26]. Among these approaches, CDSC-based filters have demonstrated superior performance in terms of both speed and accuracy, particularly in scenarios where the grid is highly disturbed. This indicates that CDSC-based pre-filtering can effectively enhance the performance of SRF-PLL in challenging grid conditions [27,28].

To ensure sinusoidal grid current and unity power factor (UPF) operation at the point of common coupling (PCC), compensating for reactive power and harmonics becomes imperative. This necessitates the decoupled control of active and reactive components of the load current. Various methods exist for achieving this decoupling, including Synchronous Reference Frame (SRF) based, Instantaneous Reactive Power Theory (IRPT) based, and Kalman Filter (KF) based approaches. However, both SRF and IRPT methods exhibit inadequate tracking performance under dynamic conditions, along with a sluggish rate of reference generation. Additionally, they often require low-pass filtering, which adds complexity to the control system [29,30]. Control based on KF introduces matrix operations that increase the computational burden [31]. In the context of the presented work, a CDSC filter-based method is employed for the decoupled control of Shunt Compensation (ShC). This approach offers advantages over other methods by providing improved tracking performance, faster reference generation, and reduced control complexity without the need for extensive filtering or computationally intensive matrix operations.

From the above study, it is found that for reliable, stable, efficient, and sustainable operation of renewable sources integrated microgrid system requires a proper power management algorithm along with a control strategy for power quality improvement. The power balance in the system ensures the system's operation by avoiding overloading, overcharging, and deep discharging issues. Similarly, enhanced power quality makes the grid current and load voltage distortion-free with UPF operation at the point of common coupling (PCC). In this chapter, an efficient active power management algorithm (APMA) and power quality improvement scheme corresponding to renewable integration is proposed with the following contributions. Therefore, this chapter is divided into two sections: (1) active power management algorithm for grid-connected hybrid AC/DC microgrid, and (2) efficient control algorithm for grid-integrated PV-UPQC system.

(1) A DC bus voltage regulation technique-based active power management algorithm is implemented for optimal distribution of power between generation and demand by reducing the difference between them.
(2) Battery and supercapacitor energy storage units are hybridized to enhance the transient operation of the proposed microgrid system.
(3) To overcome the external disturbances and effects of parameter deviations, a nonlinear sliding mode controller is utilized to govern the hybrid energy storage units. Also, by using Lyapunov's stability criteria, the stability conditions of the system are defined.
(4) The *SOCs* of the energy storage units are maintained within the defined limits to avoid overcharging and deep discharging.
(5) The proposed LPF-based power management technique with SM controller is verified in both MATLAB® simulation and with the developed test model in the laboratory.
(6) Accurate estimation of the fundamental frequency, FPSCs of load current, and FPSC-based grid voltage unit templates using a CDSC-based filter.
(7) Decoupled control of active and reactive power demand of load ensuring UPF operation at PCC and proper balance of power between PV, load, and grid.
(8) Reference signal generation for control of SeC and ShC of UPQC.
(9) Simulation and experimental analysis of developed laboratory scale grid-integrated PV-UPQC prototype.

2.2 Active power management algorithm for grid-connected hybrid AC/DC microgrid

2.2.1 Analysis of the proposed grid-integrated hybrid microgrid

2.2.1.1 Microgrid overview

Here, a grid-integrated microgrid system is considered for the research work as presented in Figure 2.1. The solar PV system is chosen for renewable energy sources and batteries and SCs are chosen for energy storage units. Individual power-generating

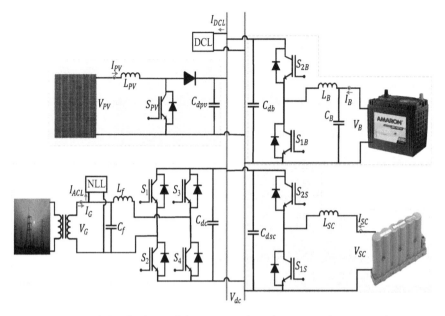

Figure 2.1   Outline of the proposed grid-connected microgrid

systems are connected together through their respective converters based on power electronics switches for the control of power transfer in the system. The PV units are coupled through a boost converter to step up the output voltage of the PV units to the voltage level of the DC link. Then, the switch of the boost converter (S_PV) is regulated through an MPPT algorithm. To maximize the energy production from PV units and to cost saving, the PV system is always operated at MPP. Here, the perturb & observe (P&O) algorithm is employed for the tracing of maximum power point. It ultimately reduces the grid dependency of the microgrid and increases the revenue. During low PV generation and high demand periods, the conventional grid will supply extra power demand to the system, and during high generation periods, the utility grid will take the extra generated power after supplying to the load. To avail this bidirectional control of power flow from and to the grid, one voltage source converter (VSC) is coupled at the grid side. This VSC will perform the AC/DC conversion and accordingly, the power flow will be done. The SC and battery units are coupled with their respective DC/DC bidirectional converters, which provide them with the flow of power in both directions for charging and discharging. During the charging period, the converters act as buck converters, and during the discharging period, they act as boost converters. The switching control of the bidirectional converters is done from the power management algorithm to avoid overcharging and deep discharging according to their available SOC levels. The detailed control mechanism of these converters is discussed in the next section of this chapter. Here, both AC and DC loads are considered for the testing of the proposed algorithm. Normal resistive DC loads are added to the DC bus. Nonlinear AC loads by combining the diode bridge rectifier in parallel with the resistive-inductive (RL) component are coupled at the AC side of the VSC.

2.2.1.2   Implementation of sliding mode controller for control of DC/DC bidirectional converters

The robustness of the sliding mode controller provides a reasonable option for the control of a nonlinear complex system with external interruptions and parameter variations [32]. In SMC, the main part of the design procedure is to choose a suitable sliding surface such that the controlling variable will stay on that surface [33].

A proportional function in corporation with an integral function is chosen for the design of the sliding surface where the control parameter is the current error of the energy storage units. The proportional function helps to bring the error in the current close to the surface when the deviation is high and after arriving near the surface, the integral function helps to slide the error in the current on the surface [34]. The current deviation (e) is defined as

$$e = I_x^* - I_x \tag{2.1}$$

In (2.1), I_x^* and I_x are the current references generated from the APMA and the real currents of the HESUs. Here, x will be denoted as B for battery units and SC for supercapacitor units.

Now, the PI-based sliding surface (S) is

$$S = \beta_1 e + \beta_2 \int e dt = \gamma e + \int e dt \tag{2.2}$$

In (2.2), $\gamma = \beta_1/\beta_2$ is the positive constant, which will regulate both the steady-state and transient performance of the HESUs. According to the requirement of the switching regulation of the bidirectional converters, the control laws can be defined as follows [35].

$$u = \frac{1}{2}[1 - \text{sign}(S)] = \begin{cases} 1 & S < 0 \\ 0 & S > 0 \end{cases} \tag{2.3}$$

where u is the required logical condition of the switches of the DC/DC bidirectional converters. The state of u can be defined as 0 or 1 to obey the control law given in (2.3) and also the Lyapunov stability criteria. The Lyapunov criterion can be expressed as

$$\dot{V}(S) = S\dot{S} < 0 \tag{2.4}$$

The sign of S and $\dot{S}$ should always be opposite to satisfy the condition in (2.4). $\dot{S}$ can be obtained from (2.2) as

$$\dot{S} = \gamma \dot{e} + e = \gamma \left(\frac{uV_{dc} - V_x}{L_x} \right) + (I_x^* - I_x) \tag{2.5}$$

V_x is the voltage output of HESUs (for battery V_B and for SC V_{SC}) and L_x is the inductor of the corresponding bidirectional converters (L_B and L_{SC}).

Condition I:
$S < 0$ and $\dot{S} > 0 \Rightarrow u = 1$

$$\gamma \left(\frac{uV_{dc} - V_x}{L_x} \right) + (I_x^* - I_x) > 0 \quad \Rightarrow \gamma \left(\frac{V_{dc} - V_x}{L_x} \right) > 0 \tag{2.6}$$

Condition II:

$S > 0$ and $\dot{S} < 0 \Rightarrow u = 0$

$$\gamma\left(\frac{uV_{dc} - V_x}{L_x}\right) + \left(I_x^* - I_x\right) < 0 \quad \Rightarrow \gamma\left(\frac{-V_x}{L_x}\right) < 0 \tag{2.7}$$

Both the inequalities obtained in (2.6) and (2.7) can be met only when the sliding coefficient (γ) is positive to make the system stable and improve performance.

2.2.2 Active power management algorithm

2.2.2.1 Generation of reference currents by utilizing the DC bus voltage regulation technique

Power management techniques mainly focus on maintaining the power equilibrium in all operating situations in the microgrid. The power balance ensures the system's stability by reducing the power difference between the generation and demand. An imbalance system operation can lead to voltage and frequency deviation which will affect both the microgrid and the connected utility grid. Hence, the power equilibrium expression for the proposed microgrid system can be defined as

$$P_{PV} \pm P_G \pm P_B \pm P_{SC} = P_{load} \tag{2.8}$$

$$P_{load} = P_{ACL} + P_{DCL} \tag{2.9}$$

In (2.8), P_{PV}, P_B, P_G, and P_{SC} are the PV, battery, grid, and SC powers respectively. The positive sign in (2.8) represents the supply of power from that particular device and negative power denotes the absorption of power from the microgrid. P_{load} is the power of total connected loads, which is a combination of both AC loads (P_{ACL}) and DC loads (P_{DCL}).

The balance of power in (2.8) is obtained by implementing a DC bus voltage control technique. Here, this method is adopted because any disturbance in the microgrid is reflected on the DC bus voltage. Hence, any deviation in DC link voltage will decide the operating condition of the system. This control technique is illustrated in Figure 2.2.

$$\Delta V_{dc}(s) = V_{dcr}(s) - V_{dc}(s) \tag{2.10}$$

$\Delta V_{dc}(s)$ is the variation in DC bus voltage between the reference bus voltage ($V_{dcr}(s)$) and actual bus voltage ($V_{dc}(s)$) in (2.10). The total current (I_T) required in the system is derived from the voltage error at the DC bus by using a PI controller.

$$I_T(s) = K_P \times \Delta V_{dc}(s) + \frac{K_I}{s} \times \Delta V_{dc}(s) \tag{2.11}$$

Here, K_P and K_I are the proportional and integral coefficients of the PI-based voltage controller. Now, this total current in the system will be decomposed into low-frequency (LF) and high-frequency (HF) current. The LF current part will be controlled by the battery and grid units and the high-frequency component will be

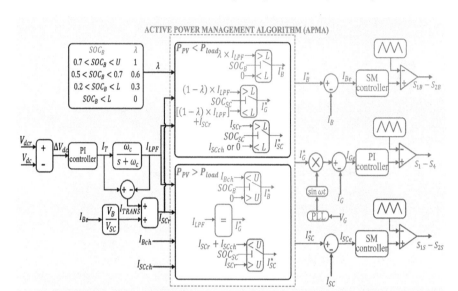

Figure 2.2   Control diagram of proposed active power control scheme

taken care of by the supercapacitors. The average low-frequency part ($I_{LPF}(s)$) is derived by using a low pass filter as follows.

$$I_{LPF}(s) = \frac{\omega_c}{s + \omega_c} \times I_T(s) \tag{2.12}$$

$$I_{TRANS}(s) = I_T(s) - I_{LPF}(s) \tag{2.13}$$

ω_c is the LPF cut-off frequency and $I_{TRANS}(s)$ is the high-frequency component. $I_{LPF}(s)$ will be supplied by both grid and battery according to the available battery SOC. $I_{TRANS}(s)$ will be supplied by the SC along with the battery current error compensation term as given in (2.14) to decrease the pressure from the batteries.

$$I_{SCr}(s) = I_{TRANS}(s) + I_{Be}(s) \times \frac{V_B}{V_{SC}} \tag{2.14}$$

In (2.14), $I_{SCr}(s)$, $I_{Be}(s)$, V_B, and V_{SC} are the SC reference current, battery error current, battery voltage, and SC voltage respectively. After generating these currents, the active power management algorithm is developed to create the individual current reference according to the PV generation, load demand, and $SOCs$ of energy storage units.

2.2.2.2   Active power management algorithm
The active power control technique mainly comprises two operating modes of the hybrid microgrid. The operating modes are mainly dependent on the amount of

power generated from the PV units and the total connected load power requirement in the system. The two operating situations are

1. Excess PV Power Mode (EPPM) ($P_{PV} > P_{load}$)
2. Deficit PV Power Mode (DPPM) ($P_{PV} < P_{load}$)

After deciding the operating modes, the reference currents for each component are generated by considering the *SOCs* of the battery and SC in the system. In the following sections, *L* denotes the lower limit of *SOC* and *U* denotes the upper limit of *SOC*.

2.2.2.3   Excess PV power mode

During this operation, the power generated from PV units is enough to fulfill the load connected. After delivering power to the loads, the additional available power is employed for the charging of the battery and supercapacitors. Four different operating cases are defined as follows.

Mode I and mode II ($SOC_B < U$)

Here, it is considered that the battery is not fully charged ($L < SOC_B < U$). So, the extra PV power available in the system is used for the charging of battery units. Similarly, for supercapacitors, when they are not fully charged, they will take power for charging along with supplying the HF power requirement to the microgrid.

$$I_B^* = I_{Bch} \quad L < SOC_B < U \tag{2.15}$$

$$I_{SC}^* = \begin{cases} I_{SCr} + I_{SCch} & L < SOC_{SC} < U \\ I_{SCr} & SOC_{SC} > U \end{cases} \tag{2.16}$$

$$I_G^* = I_{LPF} \tag{2.17}$$

In the above equations, I_B^*, I_{SC}^*, and I_G^* are the respective reference currents of the battery, SC, and grid generated from APMA. I_{Bch} and I_{SCch} are the charging currents of the battery and SC.

Mode III and mode IV ($SOC_B > U$)

These two modes are the extensions of the previous modes. The batteries are in charging conditions in modes I and II, here it is assumed that the batteries are fully charged ($SOC_B > U$). Hence, the battery becomes idle and the extra PV power is utilized for charging of SCs and also supplied to the utility grid to increase the revenue.

$$I_B^* \cong 0 \quad SOC_B > U \tag{2.18}$$

$$I_{SC}^* = \begin{cases} I_{SCr} + I_{SCch} & L < SOC_{SC} < U \\ I_{SCr} & SOC_{SC} > U \end{cases} \tag{2.19}$$

$$I_G^* = I_{LPF} \tag{2.20}$$

2.2.2.4   Deficit PV power mode

In DPPM, the produced PV power is not enough to provide the complete connected load power. Hence, the extra load power required in the system is taken from battery and grid units. By taking the *SOCs* of HESUs into consideration, the following four modes of operations are defined in this study.

Mode V and mode VI (SOC$_B$ > L)

In these two modes, it is found that there is some charge in the battery units. Hence, it can support the load till it becomes completely discharged. During these periods, the battery and grid combined supply the extra power to the loads.

$$I_B^* \cong \lambda \times I_{LPF} \quad SOC_B > L \tag{2.21}$$

$$I_{SC}^* = \begin{cases} I_{SCch} & SOC_{SC} < L \\ I_{SCr} & SOC_{SC} > L \end{cases} \tag{2.22}$$

$$I_G^* = \begin{cases} (1-\lambda) \times I_{LPF} + I_{SCr} & SOC_{SC} < L \\ (1-\lambda) \times I_{LPF} & SOC_{SC} > L \end{cases} \tag{2.23}$$

λ is the load-sharing constant and its value is decided according to the available battery *SOC* [4]. If the battery has high *SOC* values, the value of λ will be high and vice versa. The defined values of λ are given in Figure 2.2.

Mode VII and mode VIII (SOC$_B$ < L)

Here, the battery units are completely discharged (*SOC$_B$ < L*). Hence, they are not supplying any average power to the loads. The extra load power required during this situation is satisfied by the utility grid.

$$I_B^* = 0 \quad SOC_B < L \tag{2.24}$$

$$I_{SC}^* = \begin{cases} I_{SCr} & SOC_{SC} > L \\ 0 & SOC_{SC} < L \end{cases} \tag{2.25}$$

$$I_G^* = \begin{cases} I_{LPF} & SOC_{SC} > L \\ I_{LPF} + I_{SCr} & SOC_{SC} < L \end{cases} \tag{2.26}$$

2.2.3   *MATLAB simulation results and discussion*

The efficacy of the proposed APMA is tested by using MATLAB simulations. The obtained results from the simulation are discussed here. A single phase 230 V and 50 Hz grid is utilized and its voltage is stepped down to 50 V in this work. The values of different parameters taken during the research are given in Table 2.1.

2.2.3.1   Excess PV power mode

In EPPM, the system is operated with 380 W of PV generation and 250 W of load demand as illustrated in Figure 2.3. From 0 to 4 s, during modes I and II, the battery units are in charging condition by absorbing power from the microgrid. Hence, the battery current is −ve during this period as shown in Figure 2.4. Then, the batteries are completely charged during the period of 4–8 s and become idle as presented in Figure 2.3. Similarly, the supercapacitors are in a charging state in modes I and III and also provide the transient current in all four operating modes as given in Figures 2.3 and 2.4. Irrespective of the variations in the *SOCs* of energy storage units, the DC bus voltage is retained at 100 V as shown in Figure 2.4. The implementation of the DC bus voltage regulation method in active power management helps to maintain the constant bus voltage in all operating conditions. The behavior

Table 2.1   Values of different parameters

System	Parameters	Simulation values	Experimental values
PV Array	Open circuit voltage	60 V	70 V
	Short circuit current	20 A	2.5 A
Battery	Terminal voltage	12 V	12 V
	Ah capacity	42 Ah	42 Ah
	No. of batteries	4	4
Supercapacitor	Terminal voltage	16.2 V	16.2 V
	Capacitance	65 F	65 F
	No. of SCs	3	3
Utility grid	Voltage	230 V	230 V
	Frequency	50 Hz	50 Hz
DC bus	Voltage	100 V	100 V

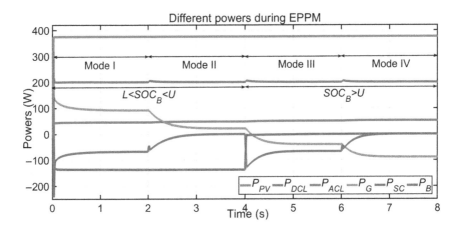

Figure 2.3   Different powers of the system during EPPM

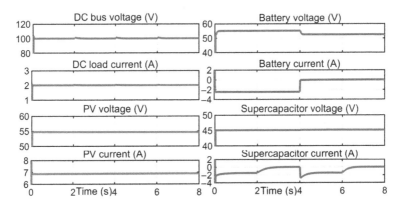

Figure 2.4   Voltages and currents during excess power operation

of the AC load current is shown in Figure 2.5. In modes III and IV after complete charging of the battery, the extra power from the PV system is supplied to the utility grid. Hence, the grid power is –ve during this period in Figure 2.3. As the grid is absorbing power from the system, the phase difference between the utility grid voltage and current is 180° as presented in Figure 2.5.

2.2.3.2   Deficit PV power mode
During DPPM operation, the power produced from the PV units is 380 W and the connected load is 550 W as presented in Figure 2.6. The additional power required by the load is supplied by the battery and grid from 0 s to 4 s as $SOC_B > L$. The discharging power of the battery is shown in Figure 2.6 and discharging current is presented in Figure 2.7. After 4 s, the battery is completely discharged and becomes

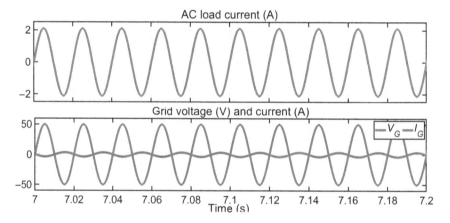

Figure 2.5   Plot of AC load current and phasor relation between grid voltage and current

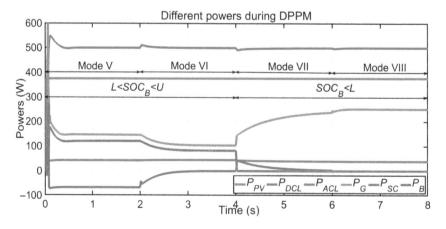

Figure 2.6   Different powers of the system during DPPM

idle. During this period, the grid units supply more power to fulfill the load demand as given in Figure 2.6. The supercapacitor in the system provides the transient power during any changes. The combined energy storage units in the microgrid make it possible to maintain the constant DC bus voltage at 100 V as shown in Figure 2.7 irrespective of all these operation and *SOCs* variations. During DPPM, the grid supplies power to the microgrid, and the voltage and current of the grid are operated at unity power factor as illustrated in Figure 2.8.

2.2.4   Experimental results and discussion

The proposed power management technique is also confirmed in a test model established in the laboratory as presented in Figure 2.9. The details of the equipment

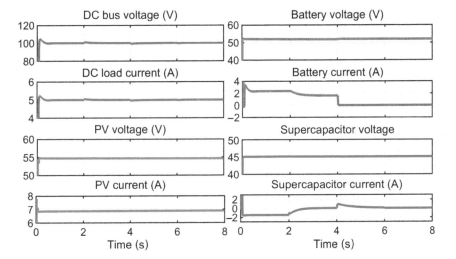

Figure 2.7   Voltages and currents during insufficient power operation

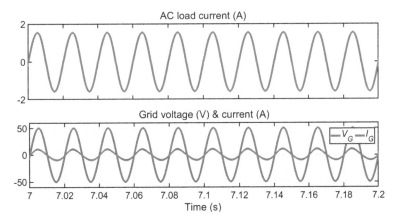

Figure 2.8   Plot of AC load current and phasor relation between grid voltage and current

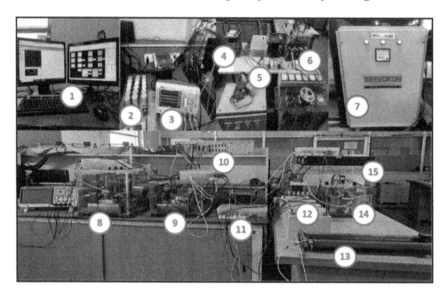

Figure 2.9   Developed model of the proposed microgrid (1. host PC, 2. SC, 3.
DSO, 4. Isolation transformer, 5. Nonlinear loads, 6. VSC, 7.
Autotransformer, 8. SC converter, 9. Battery converter, 10. ds1103,
11. Sensor board, 12. Batteries, 13. R load, 14. PV converter, 15. PV
simulator)

used for the development of the prototype are given in Table 2.1. It comprises a
utility grid, SCs (EATON XVM-16R2656-R), batteries (QUANTA 12AL042), and
PV simulator (Keysight E4360 PV simulator having 5 A and 130 V rating, 600 W).
A digital microcontroller board DS1103 is utilized to carry out the experiments and
for conversion of the power level voltage and current to signal level voltage and
current, LV 25-600 and LA 25-P are used. In experimental results, P_R represents the
dissimilarity between the load demand and PV generation ($P_{load} - P_{PV}$). The prototype id developed according to the design guidelines given in [36].

2.2.4.1   Performance during switching of operating conditions

Here, the system performance is tested during the switching of operating modes.
The produced PV power and the connected AC load power are considered as
constant. Only the power taken by the DC loads of the system is varied in this
operation. Initially, the system is operated in EPPM from t_0 to t_1 and the battery
units are in a charging state with −ve power as shown in Figure 2.10(a) as
$L < SOC_B < U$. At t_1 point, the DC load is increased intentionally to switch the
system operation to DPPM as given in Figure 2.10(b). Hence, after t_1, the battery is
started to discharge power to satisfy the extra load in the microgrid. The power
difference between load and PV generation (P_R) is reflected in Figure 2.10(b).
During these alterations, SC units support the high-frequency power as emphasized
in Figure 2.10(a) and the *SOCs* of supercapacitors are supposed to be within the

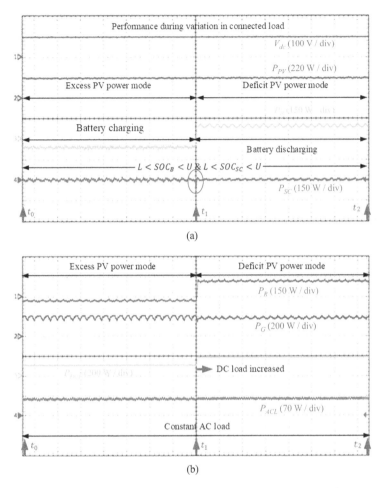

Figure 2.10   *(a) and (b) Behavior of microgrid during transition of operating conditions*

defined limits. During practical application also the voltage at the DC bus is kept constant at 100 V as given in Figure 2.10(a).

2.2.4.2 Unity power factor operation of the utility grid

One of the most important operations of a utility grid in a microgrid is to operate at unity power factor (UPF). Here, in Figure 2.11(a), before switching on VSC control, the grid current is not sinusoidal and also not in phase with the grid voltage. But in Figure 2.11(b), the voltage and current of the grid are working at UPF after switching on the VSC control. Overall, maintaining the unity power factor operation of grid voltage and current is essential for improving energy efficiency, ensuring equipment performance, reducing losses, obeying regulations, and preserving grid stability. Also, in Figure 2.11(b), the nonlinear AC load current is presented.

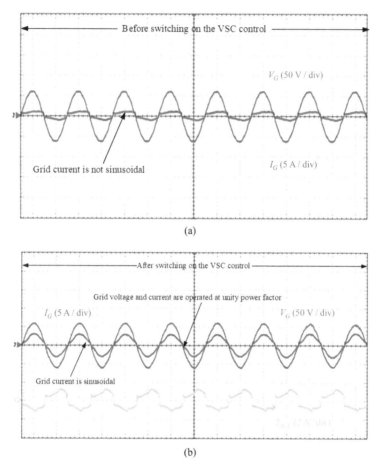

Figure 2.11   (a) Grid current and voltage before VSC control and (b) grid behavior after VSC control and nonlinear load current in grid-connected microgrid

2.3   Efficient control algorithm for grid-integrated PV-UPQC system

2.3.1   System architecture

The system setup outlined in Figure 2.12 comprises several key components: a three-phase three-wire utility grid, a non-linear load, PV panels, and a Unified Power Quality Conditioner (UPQC). Here is a breakdown of their roles and connections:

- **UPQC:** This device includes a series converter (SeC) and a shunt converter (ShC), both linked back-to-back via a dc capacitor (C_{dc}). The SeC is interconnected in series with the grid through a coupling inductor (L_r) and a series injection transformer (T_{inj}). Its purpose is to manage voltage-related power quality issues in the

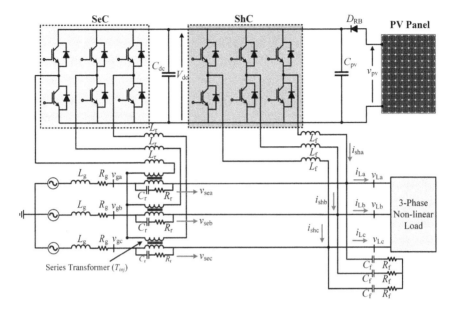

Figure 2.12 System configuration

grid, such as sag, swell, harmonics, and imbalance. Meanwhile, the ShC is parallel-connected to the load through a coupling inductor (L_f). It handles current-related power quality issues, including harmonics, imbalance, and reactive power compensation to maintain the unity power factor at PCC.

- **PV integration**: The PV system is directly linked to the DC-link of the UPQC. Here, the ShC plays a crucial role in regulating maximum power point tracking (MPPT) operations for the PV panels. It ensures optimal power extraction from the PV system while simultaneously controlling the voltage across the DC-link.
- **Non-linear load**: Here, a rectifier-based RL load is connected. Its characteristics introduce complexities into the system's behavior, necessitating effective power quality compensation strategies from the UPQC to mitigate any adverse effects on the grid and other connected components.

2.3.2 Sizing of the UPQC converters

In the proposed grid-integrated PV-fed UPQC, the efficient operation of the series converter (SeC) and shunt converter (ShC) relies on the conditions of both the grid and the load, both of which exhibit inherent uncertainty. Therefore, achieving optimal and effective operation of these converters necessitates appropriate sizing.

The sizing of the SeC is crucial and depends on two main factors: the magnitude of voltage sag or swell to be compensated and the maximum load current. This dependency on load current stems from the fact that the SeC is interconnected in series with the grid through an injection transformer, and the total current supplied by the grid passes through this transformer. The sizing calculation for the SeC can be expressed as follows:

$$S_{se} = 3v_{sag} \cdot i_{L\_max} = 3 \times 0.9 \times \frac{110}{\sqrt{3}} \times 10$$
$$= 1.7 \text{ kVA} \tag{2.27}$$

where, v_{sag} is the amount sag in the grid voltage. Here, the SeC is designed for 90% of sag. i_{L_max} is considered 20%–30% higher than the maximum load current. Equation (2.27) takes into account both the voltage-related disturbances that need to be compensated and the maximum load current, ensuring that the SeC is appropriately sized to handle the expected range of grid conditions and load demands. The design of ShC relies on maximum PV power at MPP (P_{MPP}), maximum load reactive power (Q_L), harmonic power (H), and maximum power consumed by SeC (P_{se}) from the DC-link during sag. The sizing of ShC is given as:

$$S_{Sh} = \sqrt{(P_{MPP} + P_{se})^2 + Q_L^2 + H^2}$$
$$= \sqrt{(1490 + 1714)^2 + 300^2 + (0.2 \times 1530)^2} = 3.2 \text{ kVA} \tag{2.28}$$

where, harmonic power is considered as 20% of the total load kVA rating. The rating of DC-link voltage is assessed as:

$$V_{dc} = \frac{2\sqrt{2}v_{L-L}}{\sqrt{3}} = \frac{2\sqrt{2} \times 110}{\sqrt{3}}$$
$$= 179.6 \text{ V} \tag{2.29}$$

where, V_{L-L} is the rms value of line voltage of grid voltage. In the proposed work, V_{dc} is taken higher than 179.6 V, i.e., 200 V for controlled operation of UPQC under dynamic conditions.

2.3.3   Control methodology

The control methodology of a grid-integrated PV-fed UPQC involves following the necessary steps for the efficient operation of the system.

(a) Generation of reference DC-link voltage for UPQC
(b) Estimation of fundamental positive sequence unit templates using cascaded delay signal cancellation filter.
(c) Reference signal generation for switching of SeC of UPQC
(d) Reference signal generation for the operation of ShC of UPQC

2.3.3.1   Generation of reference DC-link voltage for UPQC

The DC-link voltage (V_{dc}) holds paramount importance in controlling both UPQC converters, requiring it to consistently surpass the peak grid voltage for proper converter switch operation. With PV directly connected to the DC-link capacitor, its impact on V_{dc} is significant. Additionally, the PV array's output voltage (V_{PV}) must align with the DC-link's defined voltage thresholds ($180 \text{ V} \leq V_{dc} \leq 250 \text{ V}$) to ensure system accuracy. The minimum threshold is contingent upon the rated grid voltage, while the maximum threshold is determined by the rated voltage of the DC capacitor incorporated into the system. Meanwhile, the reference DC-link voltage (V_{dcref}) is dynamically regulated by the perturb and observe-based maximum power point tracking (P&O MPPT) algorithm. This algorithm constantly adjusts V_{dcref} to optimize PV system performance amidst changing environmental conditions and load requirements.

2.3.3.2 Estimation of fundamental positive sequence unit templates using CDSC-based filter

In this subsection, CDSC filter-based extraction of FPSC-based unit templates is discussed. In the presented work CDSC filter is used as a pre-filtering stage with respect to SRF-PLL to enhance the performance of PLL during highly non-ideal grid voltage. Firstly, the FPSC is extracted by using a CDSC filter and then the extracted FPSC is fed to SRF-PLL. As discussed earlier, SRF-PLL efficiently estimates grid information only under a balanced and pollution-free grid, therefore, the signal is first processed and filtered prior to feeding the signal to SRF-PLL. Hence, the processing is carried out by a CDSC filter. The elimination of a signal of frequency (f_n) is performed by adding that particular signal with its out-of-phase signal as given below [21]:

$$DSC_r = \frac{1}{2}[x_n(t) + x_n(t - T/4)] \qquad (2.30)$$

where, $x(t)$ refers to the current or voltage signal, T_f is the fundamental time period and $r \in N$. The delay time of the signal can be varied depending on the value of h. The minimum delay time required for the operation of DSC_r is considered from $t = 0$ to the time when the signal reaches negative zero crossing. For the elimination of nth harmonic frequency from the signal, the necessary delay time can be obtained as:

$$\frac{T_f}{r} = \frac{T_f}{2n} + K\frac{T_f}{n} \qquad (2.31)$$

where, $K < n - 0.5$ and $K \in N$. If delay time is known then the harmonic signal that can be eliminated is given as:

$$n = rK + \frac{r}{2} \qquad (2.32)$$

Therefore, harmonics attenuated by DSC_2, DSC_4, DSC_8, DSC_{16} and DSC_{32} are 2K+1, 4K+2, 8K+4, 16K+8 and 32K+16 respectively. By cascading these DSC filters, cascaded DSC (CDSC) is obtained where elimination of all the lower harmonics is possible.

Here, for a three-phase grid synchronization operation, $CDSC_{2,4,8,16}$ is considered. The three-grid voltage signal is transformed from abc-reference frame to $\alpha\beta$-frame of reference by using Clarke's transformation matrix as given below:

$$\begin{bmatrix} v_{g\alpha} \\ v_{g\beta} \end{bmatrix} = \frac{2}{3}\begin{bmatrix} 1 & -1\backslash2 & -1\backslash2 \\ 0 & \sqrt{3}\backslash2 & -\sqrt{3}\backslash2 \end{bmatrix}\begin{bmatrix} v_{ga} \\ v_{gb} \\ v_{gc} \end{bmatrix} \qquad (2.33)$$

where, v_{gabc} is the three-phase grid voltage. After transformation to a stationary frame of reference, the voltage signal ($v_{g\alpha\beta}$) is fed to CDSC-I filter to get the fundamental positive sequence component as shown in Figure 2.13. Once, positive sequence

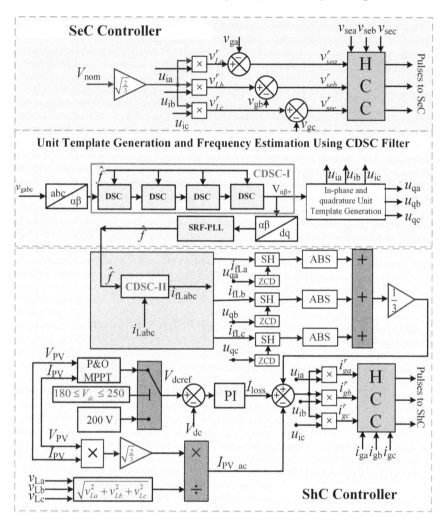

Figure 2.13   Overall control strategy of PV integration with grid through UPQC

components are obtained after processing through CDSC filter-based PLL, then, the FPSC of voltage in $\alpha\beta$-frame of reference ($v_{\alpha\beta+}$) is transformed to abc-reference frame by using inverse Clarke's matrix. Finally, fundamental positive sequence components based on three-phase *voltage* (v_{ga+}, v_{gb+}, and v_{gc+}) are extracted which is further utilized to generate in-phase and quadrature unit templates as given below:

$$u_{ia} = \frac{v_{ga+}}{V_{mag}}, \quad u_{ib} = \frac{v_{gb+}}{V_{mag}}, \quad u_{ic} = \frac{v_{gc+}}{V_{mag}} \tag{2.34}$$

$$u_{qa} = \frac{-u_{ib} + u_{ic}}{\sqrt{3}}, \quad u_{qb} = \frac{\sqrt{3}u_{ia}}{2} + \frac{(u_{ib} - u_{ic})}{2\sqrt{3}}, \quad u_{qc} = \frac{-\sqrt{3}u_{ia}}{2} + \frac{(u_{ib} - u_{ic})}{2\sqrt{3}} \tag{2.35}$$

where, V_{mag} is the magnitude of the grid voltage. The unit templates generated are utilized in the reference signal generation process for SeC and ShC.

2.3.3.3 Reference signal generation for switching of SeC of UPQC

Figure 2.13 illustrates the control algorithm employed for the series converter (SeC). The primary objective of this control algorithm is to generate a reference signal for SeC to compensate for grid voltage disturbances, ensuring that the load voltage remains purely sinusoidal and at its rated magnitude. The control algorithm operates by comparing the reference load voltage signal, which represents the desired sinusoidal voltage waveform, with the actual load voltage. Based on this comparison, the switching pattern of the SeC switches is determined. The switches are activated or deactivated as necessary to adjust the output of the SeC and align it with the reference load voltage signal, thereby mitigating grid voltage disturbances and maintaining the desired voltage waveform at the load. The reference load voltage is obtained by using in-phase unit templates and the magnitude of nominal grid voltage (v_{nom}) as mentioned below:

$$v_{La}^r = \sqrt{\frac{2}{3}}v_{nom}(u_{ia}), \quad v_{Lb}^r = \sqrt{\frac{2}{3}}v_{nom}(u_{ib}), \quad v_{Lc}^r = \sqrt{\frac{2}{3}}v_{nom}(u_{ic}) \qquad (2.36)$$

Then, reference SeC injected voltage is evaluated as

$$v_{sea}^r = v_{La}^r - v_{ga}, \quad v_{seb}^r = v_{Lb}^r - v_{gb}, \quad v_{sec}^r = v_{Lc}^r - v_{gc} \qquad (2.37)$$

Ultimately, the switching pulses are derived by employing a hysteresis current controller (HCC) to process both the reference injection voltage and the measured injection voltage for all three phases. This HCC ensures that the output voltage of the converter remains within a specified hysteresis band. By continuously comparing the reference and measured voltages, the HCC dynamically adjusts the switching signals to maintain stable and accurate compensation of grid voltage disturbances across all three phases.

2.3.3.4 Reference signal generation for operation of ShC of UPQC using CDSC filter

The control strategy for the shunt converter (ShC) is outlined in Figure 2.13. Its primary objective is to compensate for load current harmonics, meet the reactive power demands of the load, and address any unbalanced load conditions to ensure that the grid current at the point of common coupling (PCC) is purely sinusoidal, free of harmonics, and in phase with the grid voltage. Achieving this goal involves generating an appropriate reference signal for the ShC, which in turn improves the quality of the grid current. Here, a CDSC-based control approach is implemented [5]. To facilitate ShC control, a CDSC filter, denoted as CDSC-II, is utilized. Initially, the measured load current (i_{Labc}) is transformed into the $\alpha\beta$-frame and then fed into the CDSC-II filter to generate the Fundamental Positive Sequence Component (FPSC) of the load current (i_{fLabc}). The frequency input for CDSC-II is derived from the frequency estimated by CDSC-I. Using this frequency information, the time period delay for the operation of CDSC-II is determined, enabling effective filtering and generation of the FPSC of the load current. Once the Fundamental Positive Sequence Component (FPSC) of the load current is obtained, it is transformed back to the abc-reference frame. Subsequently, to achieve the decoupled control of the active and reactive components of the load current, the active and reactive components are separated using quadrature unit templates. These quadrature unit templates are then fed into a zero-crossing detector (ZCD) to detect the zero crossings of the quadrature components (u_{qabc}). At each zero crossing detected by the ZCD, it triggers a sample and hold block (SH). The SH block captures and retains the value of the load current corresponding to the zero crossing of uqabc until the next trigger by the ZCD occurs at the subsequent zero crossing of u_{qabc}. Using the absolute function, the amplitude of the active component of the load current for all three phases is determined. The average magnitude of the active component across all three phases is then calculated. This process facilitates the accurate separation and control of the active and reactive components of the load current

$$I_{avg} = \frac{I_{fLa} + I_{fLb} + I_{fLc}}{3} \tag{2.38}$$

For the improvement of current and voltage-related power quality issues, the regulation of DC-link voltage is very crucial. While the UPQC converters incur some losses during compensation. These losses are compensated by regulating DC-link voltage. It ensures the efficient operation of UPQC in compensating both voltage and current-related disturbances. The reference DC-link voltage (V_{dcref}) is generated from the PV MPPT algorithm. Then, the voltage across the C_{dc} is regulated to track V_{dcref} by compensating for the deviation in the DC-link voltage due to the losses. That loss current (I_{loss}) component is obtained by the using PI controller.

$$I_{loss}(t) = I_{loss}(t-1) + k_p[V_e(t) - V_e(t-1)] + k_i V_e(t) \tag{2.39}$$

In the proposed system, PV is connected across the DC link of the UPQC. The power produced by PV is handled by the ShC. PV delivers its power at PCC through the ShC. Therefore, the share of PV power towards load and grid is determined by (2.40).

$$I_{PVac} = \sqrt{\frac{2}{3}} \frac{I_{PV} V_{PV}}{\left(\sqrt{v_{La}^2 + v_{Lb}^2 + v_{Lc}^2}\right)} \tag{2.40}$$

Now, the final magnitude of the reference grid current is obtained as

$$\left|i_g^r\right| = I_{avg} + I_{loss} - I_{PVac} \tag{2.41}$$

The reference grid current is evaluated as

$$i_{ga}^r = \left|i_g^r\right| u_{ia}, \quad i_{gb}^r = \left|i_g^r\right| u_{ib}, \quad i_{gc}^r = \left|i_g^{re}\right| u_{ic} \tag{2.42}$$

The generated reference grid current and measured grid current are compared and processed through HCC to generate switching pulses for ShC.

2.3.4 Simulation results and discussion

The study on integrating PV with the grid via UPQC is comprehensively analyzed using MATLAB/Simulink® simulations. Various scenarios, such as grid voltage sag, swell, harmonics, load current harmonics, unbalanced conditions, and changing solar irradiance levels, are simulated to evaluate system performance. The parameters utilized for these simulations are listed in Table 2.2. This thorough simulation investigation allows for a detailed assessment of the proposed method's effectiveness across a wide range of operating conditions, enabling insightful analysis and potential optimizations.

2.3.4.1 Performance of SeC during grid voltage sag

Figure 2.14 illustrates the performance of the series converter (SeC) of the UPQC under grid voltage sag conditions. At $t = 0.3$ s, a voltage sag of 40% is introduced to the grid. However, the SeC adeptly compensates for this sag by injecting the necessary voltage through the injection transformer. Consequently, the load voltage obtained remains purely sinusoidal and at its rated magnitude (110 V L-L rms), unaffected by the grid voltage sag. This demonstrates the efficient and effective voltage compensation capability of the SeC, ensuring stable and reliable operation of the system despite adverse grid conditions.

2.3.4.2 Performance of SeC during grid voltage harmonics

Figure 2.15 depicts the operation of the PV-fed UPQC under grid voltage harmonic conditions. At $t = 0.2$ s, the grid voltage experiences significant pollution from harmonics. However, despite this disturbance, the load voltage remains purely sinusoidal and free from harmonics. This impeccable performance is attributed to the effective compensation provided by the series converter (SeC), which ensures that the grid voltage disturbance is properly mitigated. As a result, the UPQC maintains stable and high-quality voltage output to the load, demonstrating its robustness in adverse grid conditions.

2.3.4.3 Performance of ShC with balanced mode of PV power operation

In Figure 2.16, the performance of the shunt converter (ShC) during PV operation in balanced power mode is illustrated. The ShC effectively eliminates non-linearity

Table 2.2 System parameters

Parameters	Simulation values	Experimental values
Grid	110 V, 50 Hz	110 V, 50 Hz
Interfacing inductor (shunt VSC)	4 mH	5 mH
Interfacing inductor (series VSC)	4 mH	5 mH
DC-link capacitor	5,000 μF	4,700 μF
DC-link voltage	203 V	200 V
1ϕ series injection transformer	2 KVA, 110 V/110 V	2 KVA, 110 V/110 V
Three-phase non-linear load	1,400 W and 300 VAR	56 Ω/30 mH
V_{MPP}, I_{MPP}, P_{MPP}	203 V, 7.3 A, 1.49 kW	205 V, 3.90 A, 1.25 kW

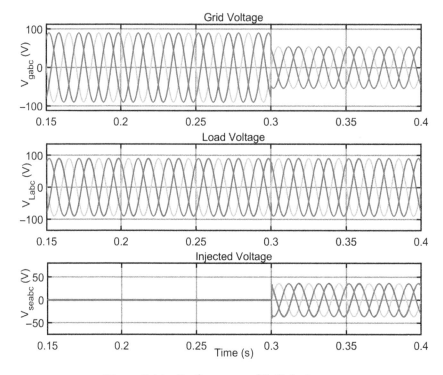

Figure 2.14   Performance of SeC during sag

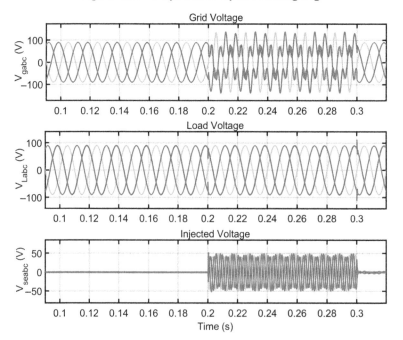

Figure 2.15   Performance of SeC during grid voltage harmonics

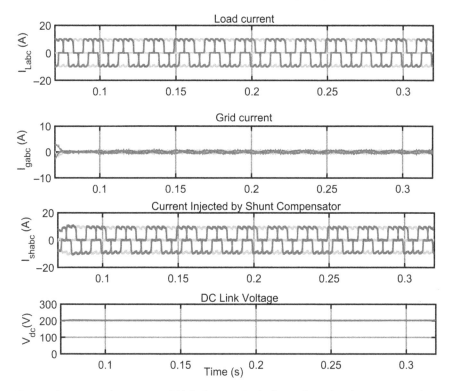

Figure 2.16 Performance of ShC during the balanced mode of operation of PV

from the grid current with respect to the load current by compensating for reactive and harmonic power. Additionally, the DC-link voltage of the UPQC is appropriately regulated at 200 V, ensuring stable operation of the system. Figure 2.17 further highlights the behavior of the grid power, which is notably reduced due to the PV system supplying the load's active power demand. Consequently, the grid only provides power to compensate for losses incurred in regulating the DC-link voltage. This demonstrates the efficient utilization of PV energy to meet the load demand while minimizing reliance on grid power, thereby optimizing system efficiency.

2.3.4.4 Performance of ShC under unbalanced non-linear loading

In Figure 2.18, the performance of the Shunt Compensator (ShC) under unbalanced load conditions is depicted. The load current across the three phases exhibits high levels of both imbalance and non-linearity. However, the ShC effectively addresses these issues by compensating for the non-linearity and imbalance in the load current. As a result, the grid current becomes harmonics-free and balanced. Additionally, it is noted that the DC-link voltage is precisely regulated at 200 volts, ensuring stable operation.

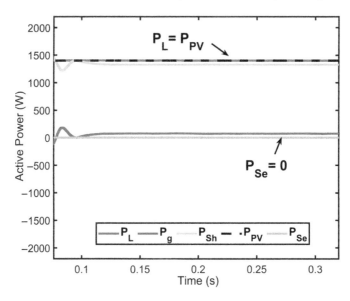

Figure 2.17   Active power distribution during the balanced mode of operation of PV

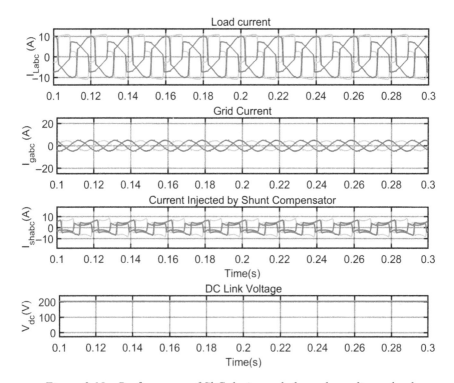

Figure 2.18   Performance of ShC during unbalanced non-linear load

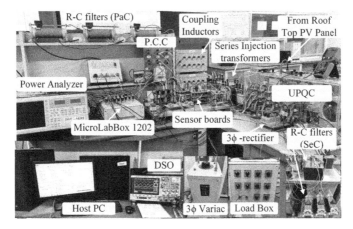

Figure 2.19   Experimental prototype of the proposed system

2.3.5   Experimental results and discussion

The proposed work on PV integration with the grid through UPQC is analyzed in the developed experimental prototype as shown in Figure 2.19. The experimental prototype consists of a three-phase variac as utility grid, UPQC, three-phase diode bridge rectifier connected to load box, and rooftop PV connected to DC-link of UPQC through a miniature circuit breaker. The system is experimental and studied under system scenarios such as grid voltage sag, PV operating in balanced power mode, and unbalanced non-linear loading conditions. The experimental parameters of the proposed system are given in Table 2.2. MicroLabBox 1202 is a digital controller used as an interface between the experimental prototype and the controller inside the host PC. For the study, all the voltage and current signals are taken from a digital storage oscilloscope, and the power curves are recorded from the plotter tool available on the control desk.

2.3.5.1   Performance of SeC during grid voltage sag

The proposed system is experimentally analyzed under a grid voltage sag of 25%. The sag in the grid voltage is compensated by SeC making the load voltage purely sinusoidal and of rated magnitude (110 V L-L rms) as shown in Figures 2.20 and 2.21. The corresponding series injection voltage of all the phases is presented in Figures 2.21 and 2.22. During sag the compensating voltage increases to compensate the sag in grid voltage. The DC-link voltage is regulated by ShC at the desired value of 200 V as shown in Figure 2.22.

The proposed system is experimentally analyzed under a grid voltage sag of 25%. The sag in the grid voltage is compensated by SeC making the load voltage purely sinusoidal and of rated magnitude (110 V L-L rms) as shown in Figures 2.20 and 2.21. The corresponding series injection voltage of all the phases is presented in Figures 2.21 and 2.22. During sag the compensating voltage increases to compensate the sag in grid voltage. The DC-link voltage is regulated by ShC at the desired value of 200 V as shown in Figure 2.22.

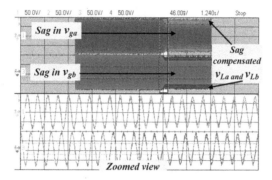

Figure 2.20   Grid voltage and load voltage during 25% of grid voltage sag of phases a and b

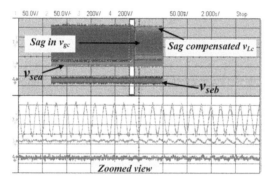

Figure 2.21   Grid voltage and load voltage of phase c and injection voltage of phases a and b during sag

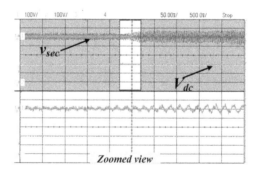

Figure 2.22   Injection voltage of phase c and DC-link voltage during sag

2.3.5.2   Performance of ShC during balanced PV power mode of operation

The proposed system is investigated under a balanced PV power mode of operation. The performance of ShC in maintaining current related power quality issues is depicted in Figure 2.23. As PV is feeding complete load power demand through ShC, therefore the grid current is almost negligible, a small amount of grid current is flowing to regulate DC-link voltage. The power feeding and balancing capability of ShC is depicted in Figure 2.24. Before $T = 14$ s, PV power is not connected to the system, hence, the load demand is fulfilled by the grid. At $t = 14$ s, PV is connected to the system and it feeds complete load power. However, the grid power is not zero because the grid provides the power to regulate DC-link and overcome losses.

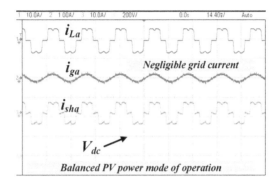

Figure 2.23   Performance of ShC during the balanced mode of PV power

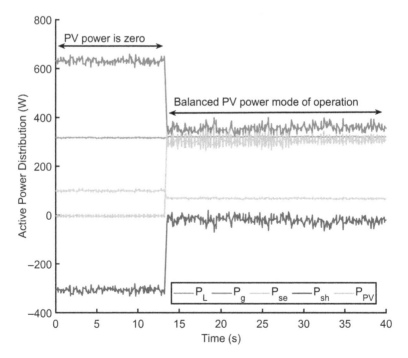

Figure 2.24   Active power balance between the system during balanced PV power mode of operation

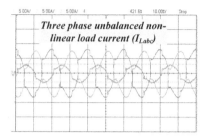

Figure 2.25   Unbalanced and non-linear load current

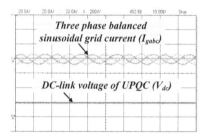

Figure 2.26   Compensated grid current and DC-link voltage

2.3.5.3   Performance of ShC during unbalanced non-linear loading

The proposed system is studied under unbalanced and non-linear load current as shown in Figure 2.25. The ShC operates to compensate for the reactive and harmonic power requirement of load and compensates for the unbalance in the load current making the grid current purely sinusoidal and balanced as in Figure 2.26. The DC-link voltage is also regulated at a desired value of 200 V.

2.4   Conclusion

The presented work delineates different approaches to renewable energy integration with smart grid. In this chapter, a novel active power management algorithm is implemented in a grid-integrated hybrid microgrid system. For the decomposition of power between the battery and SC, a LPF-based method is adopted and a sliding mode controller is utilized to control the switching of DC/DC bidirectional converters to overcome the external disturbances. The application of SC along with the battery units reduces the stress from the battery during the sudden transients in the system and ultimately, it enhances the life cycle of the batteries. The application of the DC bus voltage control technique in APMA helps to maintain the constant DC link voltage throughout the operations. Integration of PV with the grid through UPQC is studied and analyzed through simulation study as well as in experimental prototype. This method of integrating renewable energy with the grid offers numerous advantages, including support for green energy and improvement of power quality. The system demonstrates efficient performance in mitigating the impact of grid voltage sag, as observed in both simulation and experimental studies. Furthermore, it effectively regulates the DC-link voltage under dynamic conditions such as load unbalance and voltage sag. Additionally, proper power management among the grid, load, and PV systems is achieved through the shunt converter (ShC). These findings are validated through both MATLAB simulations and experimental testing using the developed test model.

References

[1]   A. Ehsan and Q. Yang, "Optimal Integration and Planning of Renewable Distributed Generation in the Power Distribution Networks: A Review of Analytical Techniques," *Applied Energy, Elsevier*, vol. 210, pp. 44–59, 2018.

[2]   Z. Tang, Y. Lin, M. Vosoogh, N. Parsa, A. Baziar, and B. Khan, "Securing Microgrid Optimal Energy Management Using Deep Generative Model," *IEEE Access*, vol. 9, pp. 63377–63387, 2021.

[3]   S. K. Kollimalla and M. K. Mishra, "A Novel Adaptive P&O MPPT Algorithm considering Sudden Changes in the Irradiance," *IEEE Transactions on Energy Conversion*, vol. 29, no. 3, pp. 602–610, 2014.

[4]   A. Bharatee, P. K. Ray, and A. Ghosh, "A Power Management Scheme for Grid-connected PV Integrated with Hybrid Energy Storage System," *Journal of Modern Power Systems and Clean Energy*, vol. 10, no. 4, pp. 954–963, 2022.

[5]   S. A. O. D. Silva, R. A. Modesto, L. P. Sampaio, and L. B. G. Campanhol, "Dynamic Improvement of a UPQC System Operating under Grid Voltage Sag/Swell Disturbances," *IEEE Transactions on Circuits and Systems II: Express Briefs*, vol. 71, no. 5, pp. 2844–2848, 2024.

[6]   L. B. G. Campanhol, S. A. O. da Silva, A. A. de Oliveira, and V. D. Bacon, "Power Flow and Stability Analyses of a Multifunctional Distributed Generation System Integrating a Photovoltaic System With Unified Power Quality Conditioner," *IEEE Transactions on Power Electronics*, vol. 34, no. 7, pp. 6241–6256, 2019.

[7]   P. Ray, P. K. Ray, and S. K. Dash, "Power Quality Enhancement and Power Flow Analysis of a PV Integrated UPQC System in a Distribution Network," *IEEE Transactions on Industry Applications*, vol. 58, no. 1, pp. 201–211, 2022.

[8]   A. Bharatee, P. K. Ray, A. Ghosh, and M. R. Jena, "Active Power Sharing Scheme in a PV Integrated DC Microgrid With Composite Energy Storage Devices," *IEEE Transactions on Power Systems*, vol. 39, no. 2, pp. 3497–3508, 2024.

[9]   A. Bharatee, P. K. Ray, A. Ghosh, and B. Subudhi, "Power Management Strategies in a Hybrid Energy Storage System Integrated AC/DC Microgrid: A Review," *Energies*, vol. 15, no. 19, pp. 1–18, 2022.

[10] Z. Yi, W. Dong, and A. H. Etemadi, "A Unified Control and Power Management Scheme for PV-Battery-based Hybrid Microgrids for both

Grid-connected and Isolated Modes", *IEEE Transactions on Smart Grid*, vol. 9, no. 6, pp. 5975–5985, 2018.

[11] A. B. Djilali, A. Yahdou, E. Bounadja, H. Benbouhenni, D. Zellouma, and I. Colak, "Energy Management of the Hybrid Power System based on Improved Intelligent Perturb and Observe Control using Battery Storage System," *Energy Reports, Elsevier*, vol. 11, pp. 1611–1626, 2024.

[12] P. Singh and J. S. Lather, "Power Management and Control of a Grid-Independent DC Microgrid with Hybrid Energy Storage System," *Sustainable Energy Technologies and Assessments, Elsevier*, vol. 43, pp. 1–11, 2021.

[13] S. Pannala, N. Patari, A. K. Srivastava, and N. P. Padhy, "Effective Control and Management Scheme for Isolated and Grid Connected DC Microgrid," *IEEE Transactions on Industry Applications*, vol. 56, no. 6, pp. 6767–6780, 2020.

[14] R. Al Badwawi, W. R. Issa, T. K. Mallick, and M. Abusara, "Supervisory Control for Power Management of an Islanded AC Microgrid Using a Frequency Signalling-Based Fuzzy Logic Controller," *IEEE Transactions on Sustainable Energy*, vol. 10, no. 1, pp. 94–104, 2019.

[15] C. Liang, Y. Zhang, X. Ji, X. Meng, Y. An, and Q. Yao, "DC Bus Voltage Sliding-mode Control for a DC Microgrid Based on Linearized Feedback," *2019 Chinese Automation Congress (CAC)*, Hangzhou, China, 2019, pp. 5380–5384, 2019.

[16] Y. A.-R. I. Mohamed and E. F. El-Saadany, "Adaptive Decentralized Droop Controller to Preserve Power Sharing Stability of Paralleled Inverters in Distributed Generation Microgrids," *IEEE Transactions on Power Electronics*, vol. 23, no. 6, pp. 2806–2816, 2008.

[17] D. B. W. Abeywardana, B. Hredzak, and V. G. Agelidis, "A Fixed-Frequency Sliding Mode Controller for a Boost-Inverter-Based Battery-Supercapacitor Hybrid Energy Storage System," *IEEE Transactions on Power Electronics*, vol. 32, no. 1, pp. 668–680, 2017.

[18] H. R. Baghaee, M. Mirsalim, G. B. Gharehpetian, and H. A. Talebi, "A Decentralized Power Management and Sliding Mode Control Strategy for Hybrid AC/DC Microgrids including Renewable Energy Resources", *IEEE Transactions on Industrial Informatics*, 2017.

[19] A. Raza, M. K. Azeem, M. S. Nazir, and I. Ahmad, "Robust Nonlinear Control of Regenerative Fuel Cell, Supercapacitor, Battery, and Wind based Direct Current Microgrid," *Journal of Energy Storage, Elsevier*, vol. 64, p. 107158, 2023.

[20] S. Pradhan, I. Hussain, B. Singh, and B. K. Panigrahi, "Modified vsslms- based adaptive control for improving the performance of a single stage pv-integrated grid system," *IET Science, Meas. & Tech.*, vol. 11, no. 4, pp. 388–399, 2017.

[21] N. Lokesh and M. K. Mishra, "A Comparative Performance Study of Advanced PLLs for Grid Synchronization," *2020 IEEE International Conference on Power Electronics, Smart Grid and Renewable Energy (PES-GRE2020)*, Cochin, India, 2020.

[22] P. Rodriguez, R. Teodorescu, I. Candela, A. V. Timbus, M. Liserre, and F. Blaabjerg, "New Positive-Sequence Voltage Detector for Grid Synchronization of Power Converters Under Faulty Grid Conditions," in *Proc. IEEE 37th Power Electron. Spec. Conf.*, pp. 1–7, 2006.

[23] S. Golestan, M. Monfared, F. D. Freijedo, and J. M. Guerrero, "Performance Improvement of a Prefiltered Synchronous Reference Frame PLL by Using a PID Type Loop Filter," *IEEE Transactions on Industrial Electronics*, vol. 61, no. 7, pp. 3469–3479, 2014.

[24] W. Li, X. Ruan, C. Bao, D. Pan, and X. Wang, "Grid Synchronization Systems of Three-Phase Grid-Connected Power Converters: A Complex Vector Filter Perspective," *IEEE Transactions on Industral Electronics*, vol. 61, no. 4, pp. 1855–1870, 2014.

[25] F. Neves, H. de Souza, F. Bradaschia, M. Cavalcanti, M. Rizo, and F. Rodriguez, "A Space-Vector Discrete Fourier Transform for Unbalanced and Distorted Three-Phase Signals," *IEEE Transactions on Industrial Electronics*, vol. 57, no. 8, pp. 2858–2867, 2010.

[26] K. K. C. Yu, N. R. Watson, and J. Arrillaga, "An Adaptive Kalman Filter for Dynamic Harmonic State Estimation and Harmonic Injection Tracking," *IEEE Transactions on Power Delivery*, vol. 20, no. 2, pp. 1577–1584, 2005.

[27] S. Golestan, M. Ramezani, J. M. Guerrero, and M. Monfared, "dq-Frame Cascaded Delayed Signal Cancellation-Based PLL: Analysis, Design, and Comparison with Moving Average Filter-Based PLL," *IEEE Transactions on Power Electronics*, vol. 30, no. 3, pp. 1618–1632, 2015.

[28] Y. F. Wang and Y. W. Li, "Grid Synchronization PLL Based on Cascaded Delayed Signal Cancellation," *IEEE Transactions on Power Electronics*, vol. 26, no. 7, pp. 1987–1997, 2011.

[29] K. Metin and O. Engin, "Synchronous-Reference-Frame-Based Control Method for UPQC Under Unbalanced and Distorted Load Conditions," *IEEE Transactions on Industrial Electronics*, vol. 58, no. 9, pp. 3967–3975, 2011.

[30] B. Singh and J. Solanki, "A Comparison of Control Algorithms for DSTATCOM," *IEEE Transactions on Industrial Electronics*, vol. 56, no. 7, pp. 2738–2745, 2009.

[31] N. Kumar, B. Singh, and B. K. Panigrahi, "ANOVA Kernel Kalman Filter for Multi-Objective Grid Integrated Solar Photovoltaic-Distribution Static Compensator," *IEEE Transactions on Circuits and Systems I: Regular Papers*, vol. 66, no. 11, pp. 4256–4264, 2019.

[32] M. R. Jena and K. B. Mohanty, "Robust Sliding Mode Current Controller for IPMSM Drives with MTPA Strategy," *2021 4th Biennial International Conference on Nascent Technologies in Engineering (INCTE)*, Navi Mumbai, India, pp. 1–6, July 2021.

[33] S. C. Tan, Y. M. Lai, and C. K. Tse, "General Design Issues of Sliding Mode Controllers in DC–DC Converters," *IEEE Transactions on Industrial Electronics*, vol. 55, no. 3, pp. 1160–1174, 2008.

[34] A. Verma, B. Singh, A. Chandra, and K. Al-Haddad, "An Implementation of Solar PV Array Based Multifunctional EV Charger," *IEEE Transactions on Industry Applications*, vol. 56, no. 4, pp. 4166–4178, 2020.

[35] L. Xu, J. Matas, B. Wei, Y. Yu, Y. Luo, J. C. Vasquez, and J. M. Guerrero, "Sliding Mode Control for Pulsed Load Power Supply Converters in DC Shipboard Microgrids," *International Journal of Electrical Power & Energy Systems,* vol. 151, p. 109118, 2023.

[36] A. Bharatee, P. K. Ray, and A. Ghosh, "Hardware Design for Implementation of Energy Management in a Solar-Interfaced DC Microgrid," *IEEE Transactions on Consumer Electronics*, vol. 69, no. 3, pp. 343–352, 2023.

Chapter 3

Adaptive control schemes for AC microgrid

Swagat Kumar Panda[1] and Bidyadhar Subudhi[1]

The integration of microgrids (MGs) with existing utility grids presents several challenges, including low inertia, intermittent nature of renewable energy sources (RES), sensor/actuator errors, the presence of imbalanced and nonlinear loads, supply-demand mismatches, uncertainties, and disturbances. Moreover, MG control relies on the communication network, which is prone to several types of failures such as communication noise, time delays, limited bandwidth, packet dropouts, and cyberattacks. So, this chapter provides a comprehensive analysis of the challenges encountered during MG integration with the existing grid. It also provides comprehensive knowledge of modern adaptive control approaches to address the aforementioned challenges in various MG topologies. A robust-adaptive distributed secondary control strategy for a photovoltaic (PV) based islanded AC MG is presented. The control objectives aim to restore voltage, frequency, and maintain active and reactive power sharing among distributed generations (DGs) in the presence of load perturbations, uncertainties, communication constraints, switching network topologies, and plug-and-play (PnP) functionalities.

3.1 Introduction

Nowadays, distributed generations (DGs) are preferred over traditional centralized generations because of economic challenges, technical improvements, and environmental impacts [1]. Figure 3.1 shows the world energy consumption in quadrillions from different sectors. As we can see the energy consumption from the industrial sector will rise magnificently in one decade. According to the aforementioned prediction, the anticipated rise in energy output by 2030 would result in higher usage of nonrenewable resources. The non-renewable resources (NRESs) would be considerably depleted as a result, raising the cost of such resources globally. The rising use of NRESs has a negative impact on the environment, leading to the release of greenhouse gases such as CO_2 and methane, which are considered to be the primary drivers of climate change and global warming [2].

Over the past few decades, there has been a worldwide increase in the pursuit of renewable energy resources (RESs) integration into the existing power system

[1]School of Electrical Sciences, Indian Institute of Technology Goa, India

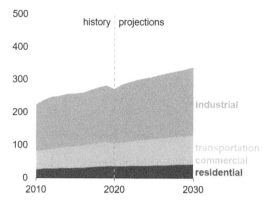

Figure 3.1   World energy consumption in quadrillion from different sectors

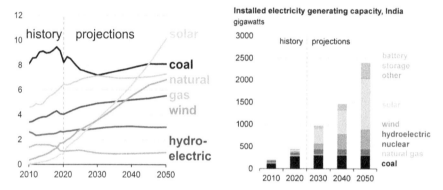

Figure 3.2   Potential of renewable resources compared to other resources with respect to world energy consumption

network. Figure 3.2 shows the potential of NRESs in the future worldwide and in India. The MG was introduced to address the inconsistent operation and growing penetration of distributed generation. The MG has started using DGs and appears to have a bright future. It can adapt to dynamic load changes, lowering feeder losses and enhancing local reliability at the same time, to meet local consumers' needs, and serves as an uninterruptible power source for critical loads [3].

The reliability of the MG may be increased if the right control techniques are used. The incorporation of MG into the existing power system has several advantages such as [4]:

- A decrease in greenhouse gas emissions,
- A consistent and independent supply of power to local and remote areas,
- An improvement in the quality of the entire power system,
- The availability of a backup power source during blackouts, operations,
- Effective bidirectional power flow.

The continual rise in power consumption, the constrained capacity of the power network to transmit power, and the overpriced reinforcement of the current transmission-distribution lines are all contributing to the stress on the traditional power grid [3]. Thus, the existing electrical power network is undergoing a significant shift from a traditional centralized architecture to a decentralized and distributed form. DGs, however, encounter technical challenges, when integrated with intermittent RES and feeble areas in the distribution network.

3.2   Microgrid control problems

Integration of DERs into the existing network has introduced several challenges. These control challenges need to be addressed while designing the controller in such a way that the potential benefit of the DG is fully realized as well as the reliability of the system is ensured. The control challenges may include incorrect assumptions that are frequently made about conventional distribution systems or may be due to the consequences of stability problems. MG integrates to the utility grid, the control action is divided into two sections, namely, input side control and grid side control [5]. Input side control is mainly concerned with extracting maximum power from the source, whereas grid side control focuses on active and reactive power control among DGs and the grid, ensuring power quality, synchronization with the grid, and protection from an islanding event. Figure 3.3 shows a generic MG model. Below are the lists of challenges in MG that encounters:

Bidirectional power flow: Although distribution feeders were initially designed for unidirectional power flow, the integration of DGs at low voltage levels can result in

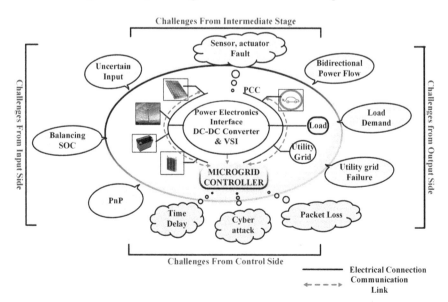

Figure 3.3   A generic model of MG

power flowing in the opposite direction [5]. This reverse power flow introduces complications in coordinating protection mechanisms and gives rise to undesirable power flow patterns, distribution of fault currents, and voltage control issues.

System modeling: Traditional assumptions are made when modeling power systems at the transmission level, such as balanced three-phase conditions, inductive transmission lines, and constant power loads. Consequently, it becomes necessary to remodel the system to accurately examine its behavior.

Low inertia: In contrast to large-scale power systems that benefit from multiple synchronous generators contributing to significant inertia, microgrids often exhibit low inertia characteristics. This can lead to notable frequency fluctuations during standalone operation. The low inertia of MG impacts both transient and steady-state stability [6]. Also, coordinating DGs within an MG introduces local oscillations, requiring transient stability analyses to ensure a smooth transition between grid-connected and islanded modes of operation.

Uncertainty: Effective coordination among multiple DERs is essential for the economical and reliable operation of MG. This coordination becomes more challenging in islanded MG, where maintaining a critical balance between demand and supply becomes crucial, especially considering the higher component failure rates and uncertainties associated with parameters like load profiles and weather forecasts. Due to lower loads and highly correlated variations among available energy suppliers, the uncertainty in microgrids surpasses that of bulk power systems, where the averaging effect is more pronounced. Consequently, microgrid control systems must address these challenges to ensure reliable and efficient operation.

Power balance: The various DERs output voltages and currents must match their reference values for stable operations. The components in the MG must be able to handle unforeseen active power imbalances while maintaining acceptable frequency and voltage variations. Appropriate demand-side management mechanisms must be devised to incorporate the capacity to control load [7].

Economic dispatch: The operational costs of an MG can be greatly decreased or the profit can be increased by properly dispatching the DER units involved in its operation. Reliability factors must also be taken into account when dispatching units, especially in stand-alone operations.

Seamless transition between operating modes: MG possesses a highly desirable feature of being able to function in both grid-connected and stand-alone modes, with a seamless transition capability between them. To ensure optimal control strategy adaptation, the inclusion of a high-speed islanding detection algorithm is of paramount importance [8]. This algorithm assumes a significant role in appropriately adjusting the control strategy based on the operational mode. As a result, distinct control strategies can be developed to cater to each mode of operation.

Limitations of power electronics components: The MG is made up of various electrical and electronic components. Semiconductors are the basis of these components that are more prone to different failures. In light of the current advancements in RESs and energy storage technologies, power-switching devices with high-voltage, high-frequency, low switching loss, and high-temperature operation capabilities have become crucial in MGs. The presence of unfavorable environmental factors, nearby electromagnetic fields, loose connections, and contacts lead to changes in the physical parameters [5].

Intermittency: A number of factors influence the production of RESs. In the case of PV arrays, the optimal energy generation is hindered by the availability of sunlight, altitude, day length, partial shading, solar insolation over PV arrays, and rain [2]. In wind generators, the wind velocity fluctuates significantly depending on the season and altitude. Therefore, voltage fluctuations, flickers, and system stability problems are the most frequent challenges for such resources, which have an impact on the dynamic and transient performance of MG [6].

Restoration of system parameters: The state parameters of the MG are to be properly tracked and ensured that the oscillations are properly damped out. The stability issues that commonly occur in MGs involve frequency and voltage stability. It indicates the ability of the MG system to maintain its operation parameters within an acceptable range. The voltage and frequency stability impacts a short-term and/or long-term stability problem. The high penetration of RES reduces system inertia, leads to a significant rise in the rate of change of frequency, increases the potential of unstable frequency in a power system, and increases the value of nadir frequency. A sudden change in loads, the presence of constant power load (CPL), penetration of DG units, and supply/demand imbalance are the major causes of voltage and frequency instability.

Maintain optimum power quality: The grid and the connected loads are designed to operate at specified frequencies with a rated voltage and current level [21]. Any power quality problem such as voltage sag and swell, flicker, current imbalance, and current and voltage harmonics can affect the synchronization of the PV system with the utility grid. So, the main power quality issues that affect the utility grid are increased reactive power demand, voltage imbalance, frequency deviations, and the presence of harmonic [22].

Mitigation of harmonics: The presence of nonlinear load and switching resonance in the MG distorts the voltage and current harmonics. Switching operation also often introduces harmonics in the system parameters. There are high-frequency harmonics in the voltage and the current [23]. Hence, in an MG system, the mitigation of PCC voltage harmonics, local load harmonics, voltage and current harmonics in the critical bus of multi-MG system, and DG line current are crucial issues to analyze [24].

3.2.1 Control challenges in cyber-physical layer of microgrid control

The degradation of communication in a network can be attributed to various factors, including additive noise, communication latency, packet loss, uncertainty, switching to unreliable communication networks, constrained communication bandwidth, and communication failure [9]. These factors collectively contribute to the decline in communication quality within the network.

Communication noise: The noise in the system is modeled as an additive constant term [9], a stochastic noise with a state dependence. The noise is added to the system due to sensors' non-fixed measurement precision, current errors, and time-varying stochastic disturbances. Also, the system noise can be classified as either multiplicative noise (disturbance present in internal dynamics, system states) or additive noise (additive Gaussian noise is added into the system as an external disturbance) [10].

Communication delay: One of the most important challenges with MG control is communication delay. However, the delay is rarely taken into account in the study. Signals that are sent and received experience some delay. The sample rate affects time delay. There are two types of delays: random time delays and continuous time delays [11]. The propagation delay, transmission delay, processing delay, and queuing delay are additional categories for the continuous time delay. Other factors that contribute to the communication delay include computation time for the generation of control inputs, slow communication speed, extra time for the reception of measurement messages, and execution time for the inputs. The impact of delay on system parameters is discussed in [12]. When a time delay is added to the system, the system poles move towards the right half of the s-plane, indicating that the system is not stable. Therefore, figuring out the delay margin is important for the time-delayed system. Any delay that exceeds the delay margin will cause instability in the system. The difference between the local feedback measurements in [13] may cause the Voltage Source Inverter (VSI)'s output voltage phase and magnitude to change, and it significantly lowers the output power. Therefore, it is difficult to take into account communication latency as another system parameter during the implementation of the control laws [9].

Packet dropouts: Packet dropouts prevent the information from being transmitted to the intended location despite the fact that it is present in the network bandwidth. As a result, the system's parameters lose their time synchronization, which lowers dynamic performance or potentially makes the system unstable.

Cyber-attacks: Cyberattacks also have an impact on the system during switching activities via the communication channel. The MG system is made more adaptable by the switching operation. The performance of the MG system can be significantly enhanced by the efficient use of switching operations. It can be done by lowering network losses, isolating fault areas, boosting resilience, and optimizing bus voltage profiles [14], but it becomes difficult when the system is fully loaded. Transient voltages and currents that occur when switching between MG operating modes have an impact on the system's relative stability [15]. The cyber system should ensure that the data are always available. Attackers obstruct or slow down data transmissions. Furthermore, due to the interconnected power converters between the hybrid AC/DC MG, any cyber assault on one side will have an impact on the other. It impairs the current sharing and state estimation capabilities of the MG [9]. Therefore, it is crucial to successfully fight against cyberattacks. Cyberattacks alter sensor data, which causes mistakes in the estimation of state variables like voltage and frequency [16]. The operations of MG are negatively impacted by this. In order to accomplish MG objectives, it is crucial to have access to specific accurate state information withstanding the occurrence of cyberattack [17].

All of the aforementioned criteria face additional difficulties with regard to the availability of measurements, high-speed communication, and computations.

Thus an advanced controller needs to be designed to handle the above challenges. The degree of intricacy and complexity of the solutions for the MG control requirements will much rely on whether it is intended to operate primarily in islanded or in grid-connected mode. Reliability concerns are more significant in stand-alone mode of operation than they are in grid-connected mode of operation, which places more focus on the interaction with the main grid.

3.3 Adaptive control for microgrid system

Adaptive control has evolved as one of the major research areas over decades, which involves improved implementation in both theoretical and practical fields. This covers improvements to system tracking, transient performance, and global stability. Controlling system uncertainty while in operation is one of the key features of adaptive control [18].

Figure 3.4 illustrates the fundamental structure of an adaptive control system. It consists of a control loop and an overlaying adapting mechanism, which are the primary components of adaptive control. To ensure the proper functioning of subsystems and the overall system, a supervisory system is also incorporated. This supervisory system is responsible for error detection, maintaining correct operation, and initiating appropriate responses when necessary. The adaptive control system performs three distinct tasks: identification, modification, and control. The identification task focuses on determining the characteristics of the system or its subsystems. The decision-making process involves comparing the collected data from system identification with desired attributes using predefined criteria. Based on this comparison, the controller is adjusted accordingly.

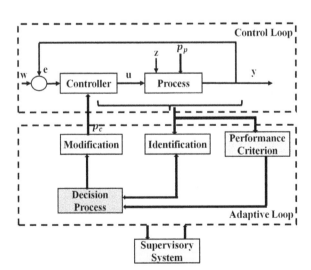

Figure 3.4 Block diagram of adaptive control

During adaptive control, the controller's gain parameters are automatically adjusted to optimize system performance under various circumstances. This adaptive behavior allows the controller to accommodate poorly understood processes and unexpected changes in environmental parameters. Adaptive control thus provides a means to effectively manage processes that involve uncertainties, such as nonlinearities and time-varying parameters [19].

3.3.1 History

Between the 1940s and 1950s, the initial concepts and applications of adaptive control were documented. During the 1950s to 1960s, various trial-and-error approaches were employed in building adaptive control systems. Notably, in 1951, a self-optimizing system that improved the performance of internal combustion engines was introduced. In 1958, a device utilizing the online least squares method to predict impulse response for self-optimizing control systems was developed. The model reference adaptive control law (MRAC), known as the M.I.T. rule, was proposed in the same year [20] (Figure 3.5).

Between the 1960s and 1970s, efforts were made to establish the theoretical foundations of adaptive control techniques. Lyapunov's second stability method-based design techniques were introduced for the first time. Simultaneously, the inclusion of random disturbances in adaptive controller design led to the formulation of stochastic approaches. Recursive least squares algorithms were used for process parameter estimation, assuming slow changes in system characteristics to accommodate non-stationary systems. However, the high cost and limited reliability of available process computers restricted the application of theoretically proposed adaptive control systems to a small number of prestigious projects. From the 1970s to the 1980s, significant developments took place in adaptive control. The first self-tuning regulator (STR) was developed by Aström and Wittenmark [21]. Also, a generalized self-tuning technique also emerged during this period. These controller types paved the way for updated versions of various STR strategies [22] with practical applications. The augmented error technique proposed by Monopoli [22] played a crucial role in constructing stable MRAC systems, enabling the development of globally robust adaptive control algorithms and leading to numerous important contributions. Landau [23] suggested a similar method utilizing Popov's hyperstability criterion for designing stable MRAC systems. The advancement of microprocessors during this time period significantly contributed to the progress of adaptive control, particularly in practical realizations across various fields.

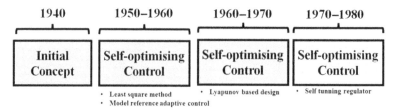

Figure 3.5   Chronology of adaptive control

3.4   Adaptive control, why?

Uncertainties arise owing to the inaccuracy in the modeling of a system. A suitable controller needs to be designed to achieve robustness. Traditional controllers cannot handle the nonlinearities and uncertainties of the MG system to achieve MG control objectives, i.e., voltage and frequency restoration and proportional active and reactive power sharing. An adaptive control system automatically adjusts the controller parameters to account for changes in the system dynamics while maintaining the overall performance of the system at its desired level. Any deterioration in plant performance over time is accounted for by an adaptive control mechanism. The adaptive control system estimates the dynamics of the process, and modify the controller's performances in accordance. Over many years, adaptive control approaches have been extensively investigated and given increased attention. Parametric uncertainties and disturbances are mostly addressed through adaptive control techniques. Thus, prolonged stability, resilience convergence, and tracking of the system dynamics may all be categorically attributed to adaptive control techniques. Important adaptive solutions are often applied to find close to ideal operating conditions.

3.4.1   Review of existing adaptive control technology

3.4.1.1   Adaptive sliding mode control

Sliding mode control (SMC) is a nonlinear efficient controller that shows robustness to external disturbances and internal perturbations. The fundamental idea behind the SMC approach is to use a reasonably high-frequency control law to force the trajectories of the system variables onto a sliding surface. The conventional SMC results in undesirable chattering caused by high-frequency switching. Chattering and a high level of control action activity are the two key factors impeding the application of SMC. If the magnitude is brought down to a minimum allowable level determined by the prerequisites for the SMC to exist, these two issues can be dealt with concurrently [24]. To reduce the chattering phenomenon many literatures proposed tunning of controllers [25]. Gains are frequently overestimated, which results in higher control magnitude and chattering. Controllers based on an intelligent approach have been used to adapt the gain. However, these methods do not ensure optimum tracking performance. Gain-dynamics control is an alternative method. However, this method's primary flaw is the gain's overestimation in relation to the bounds of uncertainty. Additionally, this method needs to be modified before it can be used with actual systems. As a result, the boundary layer width impacts accuracy and robustness, and the sign function is replaced by a saturation function. The strategy presented in [26] is based on the application of equivalent control. The occurrence of sliding mode makes it possible to evaluate the disturbance magnitude and properly tune the control gain. However, this strategy necessitates knowledge of uncertainty/perturbation bounds as well as the usage of a low-pass filter, when disturbances are present. Hall and Shtessel [27] propose a gain-adaptation technique that makes use of a sliding mode disturbance observer. The fundamental disadvantage is that designing an observer-based controller necessitates knowledge of uncertainty bounds.

SMC with a higher degree is additionally used to lessen chattering. However, it necessitates having knowledge of the upper bound of uncertainty. In actual situations, it can be challenging to predict the top bound of the uncertainty. Because the goal is to avoid having the uncertainty bound, an adaptive sliding mode controller (ASMC) is used, where the control gains can be adjusted dynamically without being aware of uncertainty or perturbation bounds. It makes sure that the dynamic control gain is adjusted to be as modest as feasible while yet being adequate to overcome uncertainties and perturbations. It is based on the idea of altering the control rule to create a system that demonstrates the same dynamic properties under uncertain circumstances by using available information. This results in an improvement in dynamical characteristics [28,29]. The ASMC's adaptive control technique estimates the system's ability to maintain the same dynamic characteristics in the presence of uncertainty based on the use of available data. When the parameters of the system being regulated are slowly changing over time or are unclear, it includes changing the control law. Additionally, adaptive control entails enhancing dynamic features when the characteristics of a regulated plant or environment are changing [28,29].

In the case of islanded microgrids, where nonlinear loads are prevalent, SMC offers an alternative approach to the control system of the inverter [30]. To address power distribution and tracking challenges in distributed energy storage systems, a cooperative adaptive terminal sliding mode controller based on a multi-agent network topology has been developed for MG. The controller incorporates a unique adaptive power allocation algorithm to maintain a constant State of Charge (SOC) for each battery in the distributed Energy Storage System (ESS). The control law is formulated as follows:

$$u_i = \frac{1}{\widehat{\delta}_{i2}} u_{dc} \left(\frac{1}{b_i} \left(\sum_{j=1}^{4} a_{ij} (\dot{e}_i - \dot{e}_j) - k_i sat(S_i) - \widehat{\delta}_{i1} x_i + \widehat{\delta}_{i2} u_{bi} - \dot{x}_{\text{iref}} + \beta_i \xi_i \left(\left(\int_0^t \xi_i dt \right)^{\left(\frac{p}{q} - 1 \right)} \right) \right) \right)$$

(3.1)

where $\widehat{\delta}_{i1}, \widehat{\delta}_{i2}$ are the adaptive variables, e_i, e_j are the errors of ith, j^{th} agent, $S_i = \xi_i + (\beta_i \int \xi_i dt)^{p/q}$ is the sliding surface, x_{ref} is the reference current, u_{bi}, u_{dc} is the voltage across the battery, DC bus. Another control strategy, discussed in [31], utilizes an adaptive three-order sliding mode controller to address load disturbances in microgrids during both the grid-connected mode and the transition to islanded mode.

Furthermore, an adaptive sliding-mode voltage control approach has been proposed to enhance the interference-rejection capabilities of the parallel-connected VSIs in microgrids. This adaptive algorithm considers both internal and external perturbations to ensure robustness in the microgrid system. Additionally, the switching strength of the control input is designed to be time-varying in order to effectively reduce undesired chattering in the control input signal [32].

3.4.1.2   Robust-adaptive control

The measured input and output of a real plant may be nonlinear, indefinitely dimensional, and distorted by noise and external disturbances. This impact on performance and stability. However, the real concern is how effectively adaptive schemes work in the presence of plant model uncertainties and bounded disruptions arising in the late 70s. The adaptive scheme is intended for a plant model that is free from disturbances, however, it may become unstable in the face of minor disturbances. In the early 1980s, the nonrobust behavior of adaptive schemes rose to the forefront of debate. The lack of robustness in the presence of unmodeled dynamics or bounded disturbances was demonstrated when further examples of instabilities were published. This inspired numerous scholars to investigate the causes of instabilities and devise solutions. By the middle of the 1980s, a number of fresh ideas and improvements, like robust-adaptive control, had been put forth. A robust nonlinear decentralized control scheme is proposed in [33] for islanded DC MG to achieve the desired voltage at the PCC and to maintain the power balance. The parametric uncertainties such as unknown control input are estimated through adaptive law, and the robustness against external disturbances is provided by the feedback linearization scheme.

3.4.1.3   Adaptive droop control

There are several problems with the typical droop control that need to be resolved, including inaccurate power sharing, line impedance dependency, and delayed transient response [34]. As a result, modifications to the traditional droop control i.e., adaptive droop controller have been suggested to deal with these issues.

To accurately share reactive power while maintaining the voltage amplitude, the adaptive droop control strategy was introduced by Kim *et al.* [35] (Figure 3.7). This method compares the reference value of reactive power Q_{ref} to the maximum reactive power Q_{max} from each unit. If the maximum reactive power is smaller than the reference value, the voltage amplitude follows the standard Q/E droop equation. Figure 3.6 depicts the fundamental idea of adaptive droop control. The difference between the reactive power (Q) and the reference value of reactive power is used to determine the target voltage amplitude. The voltage amplitude switches from lines 'a' and 'b' to lines 'e' and 'f' when $Q > Q_{ref}$. The maximum reactive power is stored, and the voltage amplitude is reduced by the constant value. As a result, the voltage amplitude shifts to lines 'c' and 'd' rather than lines 'a' and 'b' when Q drops below Q_{ref} once more.

Among the many variations of the conventional droop control, the adaptive droop control is one of the more intriguing tactics. Numerous studies have proposed various techniques for calculating the adaptive droop coefficient for adaptive droop control. To enhance reactive power sharing and transient response, Mohamed and El-Saadany [34] updated the voltage droop equation and used a second-order filter. The adaptive integral loop technique was used in [35–37] to enhance power-sharing dynamic response and address the impacts of line impedance on circulating current, however, it is impossible to identify the control parameters due to their complexity. By computing the microgrid impedance, Augustine *et al.* [38] predetermined the adaptive droop parameters.

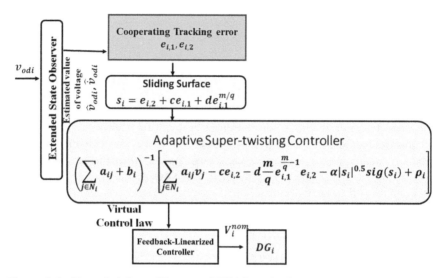

Figure 3.6   Extended State Observer (ESO)-based adaptive super twisting controller

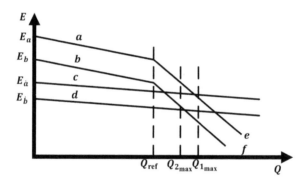

Figure 3.7   E–Q graph of adaptive droop control

Utilizing this method provides efficient power sharing and lowers line losses. Zhang *et al.* [39] proposed a derivative term in the adaptive droop control, which results in enhancement in the transient response of parallel-connected VSIs in DG systems.

$$\omega = \omega_{\text{ref}} + k_p \left(P_k - P_{k,\text{ref}} \right) + k_{p,d} \frac{dP_k}{dt} \tag{3.2}$$

$$V = V_{\text{ref}} + k_Q \left(Q_k - Q_{k,\text{ref}} \right) + k_{Q,d} \frac{dP_k}{dt} \tag{3.3}$$

In transient conditions, the power oscillation of an inverter is exacerbated by changes in output voltage. In order to reduce power oscillation, adaptive transient

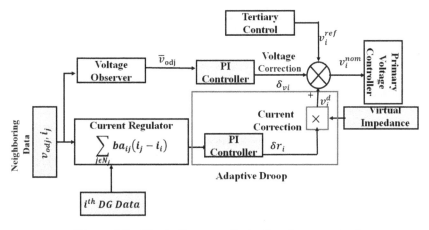

Figure 3.8 Block diagram of adaptive droop control

derivative droops are employed in [34]. A modified derivative controller-based control technique is provided in [35] to enhance the performance of the proposed controller in [34]. The following definitions describe the proposed control strategy equations. Another derivative-term-based method is provided in [36]. This method makes sure that power oscillation is actively dampened. Additionally, under various operating situations, this control technique produces the appropriate transient reaction to prevent the circulating current.

$$\omega = \omega_{\text{ref}} - m_p P - m_{dP}\frac{dP}{dt} + m_{dQ}\frac{dQ}{dt} \tag{3.4}$$

$$V = V_{\text{ref}} - n_q Q + n_{nP}\frac{dP}{dt} - n_{nQ}\frac{dQ}{dt} \tag{3.5}$$

In [37], an adaptive droop control scheme for DC microgrids integrating sliding mode voltage and current-controlled boost converters is proposed. The study suggests using an adaptive proportional-integral (PI) controller to mitigate current sharing errors by varying the droop resistance. Additionally, another adaptive PI controller is employed in the secondary control layer to regulate the DC bus voltage. Similarly, in another study focused on low-voltage standalone DC microgrids [38], an adaptive droop control strategy is presented. The objective is to address load current sharing and circulating current issues that arise in parallel-connected DC–DC converters. The suggested adaptive droop control approach reduces the current sharing disparity between the converters and minimizes circulating currents (Figure 3.8).

3.4.1.4 Adaptive backstepping controller
In the early 1990s, the "backstepping" approach emerged as a novel method for constructing adaptive controllers. Backstepping is a recursive Lyapunov-based technique specifically designed for strict feedback systems. This approach ensures

global or regional regulation and tracking by transforming the system into the parametric-strict feedback form. One of the key advantages of the backstepping design approach is its systematic and step-by-step methodology for designing stabilizing controllers, which avoids nonlinearity cancelations. However, there are certain practical challenges that have not been fully addressed using the backstepping approach. For example, dealing with nonsmooth nonlinearities remains an open issue [39]. The non-minimum phase characteristics of the boost converter in DC microgrids hinder the direct control of voltage and current using the backstepping algorithm. To overcome this, an adaptive backstepping control (ABSC) is proposed [40]. An observer is designed to estimate unknown parameters and integrated into the backstepping control to ensure that the system's output voltage remains unaffected by dynamic load perturbations. However, this method does not account for the impact of constant power load on the system. In [41], backstepping and integral links are combined for the boost converter with a constant power load. While backstepping guarantees system stability, the integration connection eliminates steady-state errors in the output voltage caused by faulty models and improper interference. However, the dynamic performance of the system is compromised. A technique called Nonlinear Disturbance Observer (NDO) is introduced in [42] as an efficient solution for compensating erroneous system models, uncertain parameters, and mismatched disturbances. NDO replaces conventional integral compensation disturbance and provides an effective method for the online estimation of nonlinear system disturbances. In a DC microgrid, the common DC-bus voltage is controlled by a nonlinear adaptive backstepping controller. This controller is developed using a recursive design process based on Lyapunov control theory, considering all parameters within the model as unknown. Adaptation laws are employed to estimate these unknown parameters, and the stability of the DC microgrid is ensured by formulating appropriate Control Lyapunov Functions (CLFs) at different design stages.

3.5 Distributed adaptive backstepping controller for microgrid

In this section, the design of a PV-integrated Voltage Source Inverter (VSI)-based islanded AC MG is presented. The control objectives are to manage undesirable conditions, support plug-and-play functionality, regulate voltage and frequency, and achieve proper active and reactive power sharing among Distributed Generators (DGs). Despite challenges such as unpredictable inverter switching pulses, intermittent input from PV sources, dynamic load variations, and disturbances, the control system aims to meet these objectives. To estimate disturbances, an Extended State Observer (ESO) is employed, while a Backstepping (BS) controller is utilized to track the nominal voltage and frequency. The ESO has the advantage of treating internal and external disturbances as a single disturbance. Previous works have focused on traditional DC supplies rather than renewable sources. Therefore, in this study, a robust-adaptive control strategy is suggested to restore both frequency and voltage accurately.

The topology of the DG model in the MG system is shown in Figure 3.9. It consists of a VSI, output connections, and a DC–DC boost converter that connects each DG's PV arrays to the Point of Common Coupling.

The dynamics of the $i\{th\}$ DG unit can be expressed as

$$\dot{v}_{pvi} = \frac{1}{C_{pvi}}\left(i_{pvci} - i_{pvi}\right) \tag{3.6}$$

$$\ddot{i}_{pvi} = \frac{1}{L_{pvi}}\left(v_{pvi} - i_{pvi}R_{pvi} - m_{pvi}i_{dci}\right) \tag{3.7}$$

$$\dot{v}_{dci} = \frac{1}{C_{dci}}\left(m_{pvi}i_{pvi} - m_{di}i_{ldi} - m_{qi}i_{lqi}\right) \tag{3.8}$$

where, the filtering capacitor, inductor, and internal resistor of the PV array are represented by C_{pvi}, L_{pvi}, and R_{pvi}, respectively. The output current and voltage of the PV array are indicated by i_{pvci}, v_{pvi}, and C_{dci} respectively. The input current to the converter is i_{pvi}, the filtered output current of the VSI is i_{ldi}, and the voltage across the DC link capacitor is v_{dci}. The DC-DC boost converter and VSI's duty cycles are indicated by m_{pvi}, m_{di}, and m_{qi} respectively.

Within the primary control layer of the MG that operates the VSI, there are three control loops: the inner voltage control loop, the inner current control loop, and the output power control loop. The droop controller is responsible for setting the voltage and frequency reference for the DG.

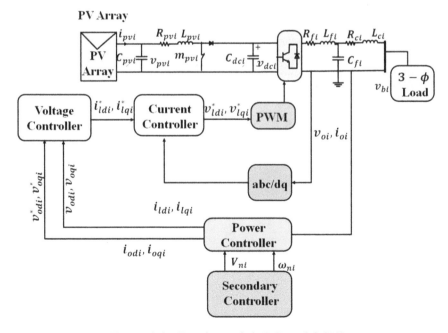

Figure 3.9   Topology of ith DG model [46]

In the voltage control loop, the reference voltage for the current control loop is generated using the output from the power controller. The current controller provides the reference signals for PWM-controlled operations. These inner loop controllers improve voltage performance and mitigate system disruptions. It is worth noting that the DG can be viewed as a regulated voltage source since the inner controller exhibits faster dynamics compared to the outer loop power controller [43]. A relationship exists between the active power frequency and reactive power voltage within the control system.

$$\omega_i = \omega_{ni} - m_{pi}P_i \tag{3.9}$$

$$v_{odi}^* = V_{ni} - n_{Qi}Q_i \tag{3.10}$$

$$v_{oqi}^* = 0 \tag{3.11}$$

where Q_i and P_i stand for instantaneous reactive and active powers. The nominal reference frequency and voltage for the ith DG are ω_{ni} and V_{ni}, respectively, while the dq components of v_{oi}^* are v_{odi}^* and v_{oqi}^*, respectively. The frequency and voltage droop coefficients of the primary droop controller are m_{pi} and n_{Qi}. The currents and voltages of the ith DG are denoted in the dq reference frame as i_{odi} and i_{oqi}, v_{odi} and v_{oqi}, respectively. To separate P_i and Q_i from the instantaneous power, two low-pass filters are utilized, each having a ω_{co} cut-off frequency.

$$\dot{P}_i = -\omega_{co}P_i + \omega_{co}\left(v_{odi}i_{odi} + v_{oqi}i_{oqi}\right) \tag{3.12}$$

$$\dot{Q}_i = -\omega_{co}Q_i + \omega_{co}\left(v_{oqi}i_{odi} + v_{odi}i_{oqi}\right) \tag{3.13}$$

In dq frame, the dynamics of DG are shown. Ignoring the fast dynamics of the inner loop controller, the dynamics of the three-phase filter and bus voltages can be represented by differential equations.

$$\dot{v}_{odi} = \frac{i_{ldi} - i_{odi}}{C_{fi}} + \omega_i v_{oqi} \tag{3.14}$$

$$\dot{v}_{oqi} = \frac{i_{lqi} - i_{oqi}}{C_{fi}} + \omega_i v_{odi} \tag{3.15}$$

$$\dot{i}_{odi} = -\frac{R_{ci}}{L_{ci}}i_{odi} + \omega_i i_{oqi} + \frac{1}{L_{ci}}\left(v_{odi} - v_{bdi}\right) \tag{3.16}$$

$$\dot{i}_{oqi} = -\frac{R_{ci}}{L_{ci}}i_{oqi} + \omega_i i_{odi} + \frac{1}{L_{ci}}\left(v_{oqi} - v_{bqi}\right) \tag{3.17}$$

$$\dot{i}_{ldi} = -\frac{R_{fi}}{L_{fi}}i_{ldi} + \omega_i i_{lqi} + \frac{1}{L_{fi}}\left(m_{di}v_{dci} + v_{ldi}^* - v_{odi}\right) \tag{3.18}$$

$$\dot{i}_{lqi} = -\frac{R_{fi}}{L_{fi}}i_{lqi} - \omega_i i_{ldi} + \frac{1}{L_{fi}}\left(m_{qi}v_{dci} - v_{oqi}\right) \tag{3.19}$$

where v_{bdi}, v_{bdi}, i_{ldi} and i_{lqi}, respectively, are the dq components of the bus voltage v_{bi} and the VSI output current i_{li}. The nonlinear model of the ith DG can be represented using Equations (3.1) to (3.12) as follows:

$$\dot{x}_i = f_i(x_i) + k_i(x_i)D_i + \theta_i^T(x_i)\theta_i + g_i(x_i)u_i \qquad (3.20)$$

$$y_{i,1} = h_{i,1}(x_i) = v_{odi}, y_{i,2} = h_{i,2}(x_i) = \omega_i \qquad (3.21)$$

where, $x_i = \left[\delta_i, P_i, Q_i, v_{odi}, v_{oqi}, i_{odi}, i_{oqi}, i_{ldi}, i_{lqi}, v_{pvi}, i_{pvi}, v_{dci}\right]^T$; $D_i = \left[i_{pvci}, v_{bdi}, v_{bqi}, \omega_{com}\right]^T$; $u_i = \left[V_{ni}, \omega_{ni}\right]^T$; $\theta_i = \left[m_{pvi}, m_{di}, m_{qi}\right], and y_i = \left[v_{odi}, \omega_i\right]^T$ denote the state variables, disturbances, inputs, uncertainties, and outputs respectively.

3.5.1 Problem formulation

The PWM switching capabilities of the VSI changes in the local loads, and communication failure all contribute to uncertainty and disruptions in the dynamics of the MG. The precise values of some parameters of an MG, such as $v_{dci}, i_{ldi}, i_{lqi}$ are unknown. This led to a change in the voltage across the bus bar. It is intended to control the MG in the presence of uncertainties and disturbances in order to restore the nominal frequency and voltage while achieving the appropriate active and reactive power sharing across DGs. The proposed ESO-BS-based resilient controller is made to choose $u_{vi}, u_{\omega i}$ in a way that v_{odi}, ω_i should monitor them. $v_{odi} \rightarrow v_{ref}$, $\omega_i \rightarrow \omega_{ref}$

We use multi-agent systems and graph theory to create a consensus-based distributed control system for MG. Below is a basic explanation of this. Each DG is seen as an agent in the distributed control design. A DG can communicate with other DGs via a sparse communication network. A directed graph $G_r = (V_G, E_G, A_G)$ depicting communication among DGs connected to MG can be created using the graph theory. The vertices and edges of the weighted graph are denoted as $V_G = \{1, .., 4\}, E_G \subset V_G \times V_G$. The adjacent matrix is represented by the notation $A_G = (a_{ij} \geq 0) \in R^{4 \times 4}.(i,j)\} \in E_G \Leftrightarrow a_{ij} = a_{ji} = 1$ otherwise $a_{ij} = a_{ji} = 0$, $a_{ii} = 0$ for all i_G. The in-degree matrix and adjacency matrix can be generated as follows for the communication topology under consideration (Figure 3.2).

$$D = \begin{bmatrix} 2 & 0 & 0 & 0 \\ 0 & 2 & 0 & 0 \\ 0 & 0 & 2 & 0 \\ 0 & 0 & 0 & 2 \end{bmatrix}; A = \begin{bmatrix} 0 & 1 & 0 & 1 \\ 1 & 0 & 1 & 0 \\ 0 & 1 & 0 & 1 \\ 1 & 0 & 1 & 0 \end{bmatrix} \qquad (3.22)$$

3.5.2 Controller design

This section describes how to create distributed resilient backstepping controllers for the AC MG. To synchronize all of the DG units' voltage to their nominal value, the control law chooses the appropriate control inputs $u_{vi}, u_{\omega i}$.

Differentiating (3.8) yields the relationship between the droop controller's input voltage and the VSI's output voltage.

$$\ddot{v}_{odi} = -\frac{1}{L_{fi}C_{fi}} \left[v_{odi} \left(-\omega_i^2 L_{fi}C_{fi} + 1 - \frac{L_{fi}}{C_{fi}} \right) -, R_{fi}, i_{ldi}, +2, \omega_i, L_{fi}, i_{lqi} \right.$$
$$\left. + \left(\frac{R_{ci}L_{fi}}{L_{ci}} \right) i_{odi} - 2\omega_i L_{fi} i_{oqi} + m_{di}v_{dci} - v_{ni} + n_{Qi}Q_i - \frac{v_{bdi}}{L_{fi}} \right] \tag{3.23}$$

It may be expressed as

$$\begin{cases} \dot{x}_{i1} = x_{i,2} \\ \dot{x}_{i2} = F_i(x_i) + f_i(x_i)\theta_i + g_i u_i \end{cases} \tag{3.24}$$

where the system states are represented by $x_{i,1} = v_{odi}$, $x_{i,2} = \dot{v}_{odi}$, and $F_i(x_i) = f_i(x_i) + k_i(x_i)D_i$. The estimated states must be obtained in order to track the reference signal. x_(i,2) is in an uncertain condition. m_{di}, v_{bdi}, v_{dci} stand for disturbance.

Equation (3.15) can be extended as

$$\begin{cases} \dot{x}_{i1} = x_{i,2} \\ \dot{x}_{i2} = x_{i,3} + g_i u_i \\ \dot{x}_{i3} = F_i(x_i) + f_i(x_i)\theta_i \\ y_i = x_{i,1} \end{cases} \tag{3.25}$$

The state vector of the system is denoted by $\left[x_{i,1}, x_{i,2} \right]^T$. The additional state, $x_{i,3}$ is known as the extended state since it represents the system's overall disturbances and uncertainty.

The following presumptions are made in order to design the ESO:

The desired trajectory, $y_d = x_d(t)$, is bounded, according to assumption 1.

Assumption 2: Input u_i and the term $F_i(x_i)$ are both locally Lipschitz in their arguments.

The ESO's dynamics can be described as

$$\begin{cases} \dot{\hat{x}}_{i1} = \hat{x}_{i,2} - G_{i1}e_i \\ \dot{\hat{x}}_{i2} = \hat{x}_{i,3} - G_{i2}e_i + g_i u_i \\ \dot{\hat{x}}_{i3} = -G_{i3}e_i \end{cases} \tag{3.26}$$

The estimated states are $\hat{x}_{i,1}, \hat{x}_{i,2}$ and $\hat{x}_{i,3}$. $\hat{x}_{i,3}$, which approximates the system's nonlinear dynamics and uncertainty, is used to depict the system's extended state. The estimation error is denoted by the equation $e_i = y_i - \hat{x}_{i,1}$. The observer's gains G_{i1}, G_{i2}, and G_{i3} are chosen so that the error e_i should be zero.

With the appropriate observer gain setting, the error will finally reach zero. It is possible to calculate the observer gains as follows: $G_{ij} = \delta_{n,j}\omega_0^j, \delta_{n,j} = \frac{n!}{j!.(n-j)!}$ $j = 1, 2 \ldots n, \omega_0 > 0$.

Equation (3.17) can be rewritten as

$$\dot{\hat{x}} = A_0\hat{x} + B_0 u + G(x_1 - \hat{x}_1) \tag{3.27}$$

where,

$$A_0 = \begin{bmatrix} 0 & 1 & 0 \\ 0 & 0 & 1 \\ 0 & 0 & 0 \end{bmatrix}; B_0 = \begin{bmatrix} 0 \\ 1 \\ 0 \end{bmatrix}; G = \begin{bmatrix} 4\omega_0 \\ 6\omega_0^2 \\ 4\omega_0^3 \end{bmatrix} \tag{3.28}$$

The dynamics of the state estimation error can be expressed as:

$$\dot{\tilde{x}} = A_0\tilde{x} + B_0u + G(x_1 - \tilde{x}_1) \tag{3.29}$$

Let $\varepsilon_i = \tilde{x}_i\omega_0^{i-1}, i = 1,2,3$ signify the scaled estimation error, be the expression for the dynamics of the state estimation error.

$\dot{\varepsilon} = \omega_0 A\varepsilon + B_1\frac{bu}{\omega_0} + B_2\frac{F(x)}{\omega_0^2}$, where $\varepsilon = [\varepsilon_1, \varepsilon_2, \varepsilon_3]^T$ and

$$A = \begin{bmatrix} -4 & 1 & 0 \\ -6 & 0 & 1 \\ -4 & 0 & 0 \end{bmatrix}; B_1 = \begin{bmatrix} 0 \\ 1 \\ 0 \end{bmatrix}; B_2 = \begin{bmatrix} 0 \\ 0 \\ 1 \end{bmatrix} \tag{3.30}$$

It is as an A in the Hurwitz matrix. Thus, there exists a positive definite matrix P that fulfills the following Lyapunov equation:

The local neighboring tracking errors are denoted as

$$z_{i1} = a_{ij}(x_{i,1} - x_{j,1}) + g_i(x_{i,1} - x_r) \tag{3.31}$$

$$z_{i2} = a_{ij}(x_{i,2} - x_{j,2}) + g_i(x_{i,2} - \dot{x}_r) - \alpha_{i,1} \tag{3.32}$$

$$z_{i3} = a_{ij}(x_{i,3} - x_{j,3}) + g_i(x_{i,3} - \ddot{x}_r) - \alpha_{i,2} \tag{3.33}$$

where x_r, $\dot{x}_r$, $\ddot{x}_r$ signify the reference values of v_{odi}, $\dot{v}_{odi}$, $\ddot{v}_{odi}$ respectively, and $\alpha_{i,1}$, $\alpha_{i,2}$ are the virtual control laws.

The result of z_{i1} rate of change is,
$\dot{z}_{i1} = (a_{ij} + g_i)x_{i,2} - a_{ij}x_{j,2} - g_i\dot{x}_r$, and the virtual control law α_{i1} can be expressed as:

$$\alpha_{i,1} = -k_1\left((a_{ij} + g_i)x_{i,1} - a_{ij}x_{j,1} - g_ix_r\right) \tag{3.34}$$

Thus, $\dot{z}_{i1}$ can be written as

$$\dot{z}_{i1} = z_{i2} - k_1z_{i1} \tag{3.35}$$

On differentiating z_{i2} with respect to t results in:

$$\dot{z}_{i2} = (a_{ij} + g_i)\dot{x}_{i,2} - a_{ij}\dot{x}_{j,2} - g_i\ddot{x}_r - \dot{\alpha}_{i1} \tag{3.36}$$

It can be simplified as:

$$\dot{z}_{i2} = z_{i3} + \alpha_{i,2} - (a_{ij} + g_i)g_iu_i - a_{ij}g_ju_j - k_1b_i\dot{x}_r + k_1(a_{ij} + g_i)x_{i,2}$$

$$- k_1a_{ij}x_{j,2} \tag{3.37}$$

The virtual control law $\alpha_{i,2}$ can be expressed as: $\alpha_{i,2} = \alpha_{i,2a} + \alpha_{i,2s}$

where

$$a_{i,2a} = -k_1\left((a_{ij} + g_i)\widehat{x}_{i,2} - a_{ij}\widehat{x}_{j,2} - g_i\dot{x}_r\right) \quad (3.38)$$

$$a_{i,2s} = \left((a_{ij} + g_i)\widehat{x}_{i,3} + a_{ij}\widehat{x}_{j,3} - \frac{a_{i1}}{k_1}\right) \quad (3.39)$$

$a_{i,2a}$ acts as a model-based compensation law via online state estimation, $a_{i,2s}$ serves as a robust control law to stabilize the system, and $k_1 > 0$ is a feedback gain. The control law for the voltage restoration is given as:

$$u_i = -\frac{1}{g_i\left(a_{\{ij\}} + b_i\right)}\left[a_{ij}g_ju_j - a_{ij}\left(\widehat{x}_{i,3} - \widehat{x}_{j,3}\right) + g_i\sum_{j=1}^{N}\widehat{x}_{i,3} + k_1z_{i1}\right] \quad (3.40)$$

On substituting the control law u_i in (3.23), $\dot{z}_{i,2}$, $\dot{z}_{i3}$ can be found as

$$\dot{z}_{i,2} = z_{i,3} + k_i\left(a_{ij} + g_i\right)\tilde{x}_{i,2} - a_{ij}\tilde{x}_{j,2} + (1 - k_i)z_{i,1} \quad (3.41)$$

$$\dot{z}_{i3} = \left(a_{ij} + g_i\right)\dot{\varepsilon}_{i3}\omega_0^2 - a_{ij}\dot{\varepsilon}_{j3}\omega_0^2 + z_{i,2} \quad (3.42)$$

To synchronize the frequency of each DG, ω_i, with ω_{ni}, we propose secondary frequency control inputs using the same distributed robust backstepping technique, where only information from neighboring units is required.

$$\dot{\omega}_i = \omega_{co}((\omega_{ni} + u_{\omega_i}) - \omega_i) - \omega\left(v_{odi}i_{odi} + v_{oqi}i_{oqi}\right) \quad (3.43)$$

The control law for the frequency restoration is given by

$$u_{\omega_i} = -\frac{1}{g_i\left(a_{\{ij\}} + b_i\right)}\left[a_{ij}b_ju_j - a_{ij}\left(\widehat{x}_{i,2} - \widehat{x}_{j,2}\right) - b_i\sum_{j=1}^{N}\widehat{x}_{i,2} + k_1z_{i1}\right] \quad (3.44)$$

The proposed distributed robust backstepping controller is shown in Figure 3.10. The suggested controller offers exponential convergence to produce strong transient performance and good tracking precision. Given that z and ε are both restricted, the actual and estimated states are both constrained. The appendix contains a detailed stability study of the proposed controller.

Theorem: A positive definite Lyapunov function V is defined as:

$$V = \frac{1}{2}z^Tz + \frac{1}{2}\varepsilon^TP\varepsilon \quad (3.45)$$

is bounded by

$$V(t) \leq V(0)\exp(-\tau T) + \frac{\varsigma}{\tau}[1 - \exp(-\tau T)] \quad (3.46)$$

where $\tau = 2\lambda_{\min}(\Lambda)\min\left\{\frac{1,1}{\lambda_{\max}(P)}\right\}$, in which $\lambda_{\min}(.)$, and $\lambda_{\max}(.)$ denote the minimum and maximum eigenvalues of a matrix, $z = [z_1, z_2, z_3]^T$ is the tracking error, and ε is the state estimation error bounded under.

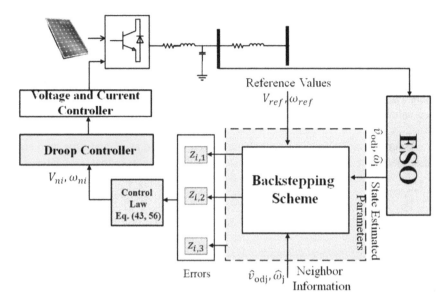

Figure 3.10   Block diagram of ESO-based backstepping controller [46]

Assumption 1: $F(x)$ is bounded. By properly choosing gains k_1, k_2, ω_0 in such a way that the matrix Λ defined below is positive definite:

$$\Lambda = \begin{bmatrix} \Lambda_1 & \Lambda_2 \\ \Lambda_2^T & \Lambda_3 \end{bmatrix} \tag{3.47}$$

where $\Lambda_1, \Lambda_2, \Lambda_3$ are defined as:

$$\begin{bmatrix} c_1 & -\dfrac{c_2}{2} & 0 \\ -\dfrac{c_2}{2} & 0 & -\dfrac{c_3}{2} \\ 0 & -\dfrac{c_3}{2} & 0 \end{bmatrix} \begin{bmatrix} 0 & 0 & -\dfrac{\gamma_2}{2} \\ 0 & -\dfrac{\gamma_1}{2} & 0 \\ -\dfrac{\gamma_2}{2} & 0 & 0 \end{bmatrix} \begin{bmatrix} \omega_0 & 0 & 0 \\ 0 & \omega_0 & 0 \\ 0 & 0 & \omega_0 \end{bmatrix} \tag{3.48}$$

Then the proposed control law guarantees the above Lyapunov candidate function.

Proof: Time derivative of V yields,

$$\dot{V} = z^T \dot{z} + \varepsilon^T P \dot{\varepsilon} \tag{3.49}$$

The above equation can be rewritten as:

$$\dot{V} = z_1(z_2 - k_1 z_1) + z_2(z_3 + k_2(L+B)\tilde{x}_2 + (1-k_1)z_1)$$

$$+ z_3\big((L+B)\dot{\varepsilon}_3\omega_0^2 + z_2\big) + \varepsilon^T P\left(\omega_0 A\varepsilon + \frac{B_1 bu}{\omega_0} + \frac{B_2 F(x)}{\omega_0^2}\right) \tag{3.50}$$

Lemma-1: In order to make the control law distributed, the ith agent depend upon its neighbor j, in the graph topology i.e.,

$$u_i = \sum_{j \in N_i} a_{ij}(x_{i1} - x_{j1}) + g_i(x_{i1} - x_r) = (L + B)x - g_{ix_r} \tag{3.51}$$

where x_i, x_j denote the state variable of ith, jth agent, and x_r represents the reference value of x.

Assumption 2: The nonlinear states x_i are estimated as $\hat{x}_i$, and estimated error is assumed to be bounded by some positive known function so that $|x_i - \hat{x}_i| \leq c_i \varepsilon_i$.

The set of known constants are defined as:$c_1 = k_1$, $c_2 = (2 + k_2)$, $c_3 = 2$, $\gamma_1 = (L + B)c_2 k_2$, $\gamma_2 = -4\omega_0^3(L + B)$, $\zeta = \left((L + B)z_3 + \frac{\varepsilon^T \delta_2}{\omega_0^2} \right) F(x)|_{\max} + \frac{\varepsilon^T \delta_1}{|\omega_0 u|_m}$,

where $\delta_i = ||PB_i||$ for $i = 1, 2, 3$, $|u|_{\max}$ represents the upper bound of the control law, and P is a positive definite matrix. It is defined as

From (3.36), the P is calculated as:

$$P = \begin{bmatrix} p_{11} & p_{12} & p_{13} \\ p_{21} & p_{22} & p_{23} \\ p_{31} & p_{32} & p_{33} \end{bmatrix} = \begin{bmatrix} 0.2 & 0.3 & 0.125 \\ 0.3 & 2.27 & 1.3 \\ 0.125 & 1.3 & 1.95 \end{bmatrix} \tag{3.52}$$

(3.26) can be simplified as:

$$\begin{aligned} \dot{V} &= c_1 z_1^2 + c_2 |z_1||z_2| + c_3 |z_2||z_3| - \omega_0 ||\varepsilon||^2 + \gamma_1 |z_2||\varepsilon_2| + \gamma_2 |z_3||\varepsilon_1| + \zeta \\ &= -\eta^T \Lambda \eta + \zeta \end{aligned} \tag{3.53}$$

where $\eta = \left[|z_1|, |z_2|, |z_3|, |\varepsilon_1|, |\varepsilon_2|, |\varepsilon_3| \right]^T$. Since the matrix Λ is a positive definite matrix; thus, we have

$$\dot{V} \leq -\lambda_{\min}(\Lambda) \left(||z||^2 + ||\varepsilon||^2 \right) + \zeta$$

$$\leq -\lambda_{\min}(\Lambda)(\Lambda) \left(||z||^2 + \frac{1}{\{\lambda_{\max}(P)\}\varepsilon^T P\varepsilon} \right) + \zeta \leq -\tau V + \zeta \tag{3.54}$$

Thus, z, ε are bounded, which means the state x, and its estimation are bounded.

3.6   Result and discussion

3.6.1   *Simulation results*

To validate the effectiveness of the proposed controller, simulations were run in MATLAB®/Simulink®.

3.6.2   *Change in irradiance*

Figure 3.11 depicts the changing irradiance on a PV array, where power changes with respect to irradiance dynamically. However, the DC link voltage remains stable within the specified range. The total harmonic distortion of the VSI output current is displayed in Figure 3.11(b). It can be observed that the total harmonics distortion (THD) is quite low i.e., 1.66 percent.

To optimize the output power of PV panels, an incremental conductance (InC) based Maximum Power Point Tracking (MPPT) approach is employed. This method involves summing the instantaneous conductance and capacitance of the PV panel and using the InC algorithm to control the operating point in the direction that maximizes power generation. The DC link capacitor plays a crucial role in the microgrid (MG) by providing virtual inertia when the frequency deviates from its nominal values. This helps maintain stability and regulate power flow within the MG. Figure 3.6 illustrates the *I–V* characteristic curves of the PV array under different irradiance and temperature conditions. It has been observed that for a given irradiance and temperature, there exists a specific DC voltage that enables the

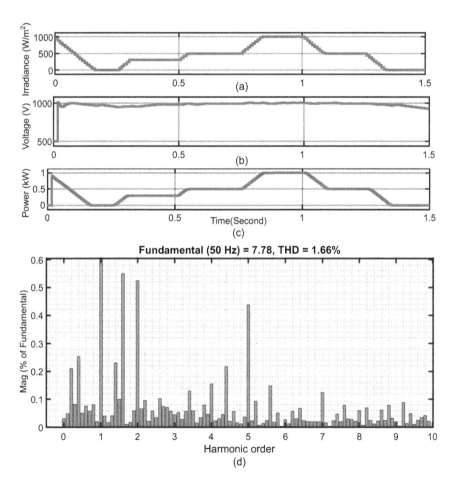

Figure 3.11   (a) Variable irradiance. (b) Voltage across DC link. (c) Power generated from PV array. (d) Inverter current THD [46].

extraction of the maximum solar energy. A MATLAB/Simulink model is utilized to generate the *P–V* characteristic curve at a temperature of 25 °C and an irradiance of 1000 W/m². By employing the InC-based MPPT approach and considering the optimal operating voltage based on the *I–V* characteristic, the PV panels can achieve maximum power output in various environmental conditions.

3.6.2.1   System performance evaluation

Case 1: Performance under steady state condition
At steady state conditions, the interaction among DGs is deemed to be ideal, and the MG functions at normal operating conditions. The responses of four DGs in terms of voltage, frequency, active, and reactive power are shown below. The voltage, frequency, active, and reactive power responses of DGs with voltage and frequency drops of 4% and 1% are shown in Figure 3.12(a)–(d). Initial primary droop control of MG causes the system parameters to deviate from their nominal values i.e., 415 V, 50 Hz. These nominal values of the state variables are restored by the proposed controller in fact because of the presence of the above challenges discussed. The power-sharing index is calculated using their droop coefficients.

$$n_{p_i} = \frac{\sum_i P_i}{\sum_i \frac{S_i}{m_{p_i}}}, \ n_{q_i} = \frac{\sum_i Q_i}{\sum_i \frac{S_i}{m_{q_i}}} \tag{3.55}$$

where proportional active and reactive power-sharing parameters, respectively, are denoted by n_{p_i}, n_{q_i}. Any deviation of P_i from n_{p_i} causes deviation from power distribution. The mean power-sharing index for real power (MPSI) and reactive

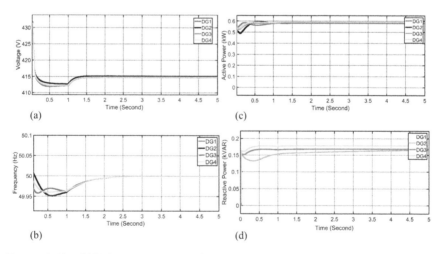

Figure 3.12   (a)Voltage responses. (b) Frequency responses. (c) Active power responses. (d) Reactive power responses of DGs [46].

power (MQSI) are as follows [11]:

$$MPSI = \frac{\frac{Droop\ coefficient*real\ power - n_{p_i}}{Rating\ of\ inverter}}{n_{p_i}}, MQSI = \frac{\frac{Droop\ coefficient*reactive\ power - n_{q_i}}{Rating\ of\ inverter}}{n_{q_i}},$$

Since MPSI, MQSI determines the average deviation of the inverter from equal power sharing. It is desired to be smaller values of these indicators, however, higher values of MQSI show a higher degree of circulating current. Tus at steady state condition these indicator values are desired to be 0, which indicates equal power sharing without interruption.

Case 2: Load perturbation

To assess the robustness of the controller against load perturbations, the following scenarios are considered. At time $t = 2$ s, Load 2 (15.3 kW + 7.6 kVar) is disconnected, and then reconnected at time $t = 3$ s. Initially, at $t = 0$ s, the primary droop control is responsible for controlling the microgrid (MG) system, while the secondary controller is turned off. Consequently, the voltage and frequency deviate from their nominal values. At $t = 1$ s, the secondary controller is activated, leading to the restoration of the voltage and frequency values.

When Load 2 is disconnected at $t = 2$ s, the voltage exhibits a slight increase and settles around the nominal values. At $t = 3$ s, when the load is reconnected, the load current also increases, resulting in equal active power sharing among the Distributed Generators (DGs) being restored. The performance of power sharing under load perturbations is depicted in Figure 3.13(a) and (b). It is evident from the results that the proposed controller achieves zero power-sharing performance in less than 1 second, even in the presence of load perturbations. Overall, the controller demonstrates its robustness and effectiveness in maintaining power sharing and system stability during load perturbations within the MG.

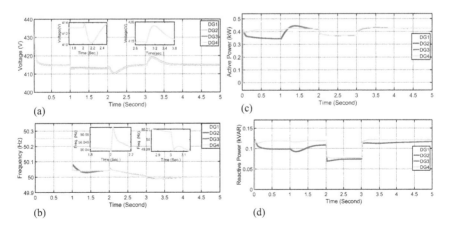

Figure 3.13 *(a) Voltage responses. (b) Frequency responses. (c) Active power responses. (d) Reactive power responses of DGs during load perturbations [46].*

Case 3: Plug and play capability

The robustness of the proposed controller is performed with respect to PnP capabilities. At $t = 2$ s, one of the DG is plugged out and the same is plugged into the existing network at $t = 3$ s. As one of the DG, i.e., DG_4 is plugged out from the corresponding network, as can be seen in Figure 3.10. The other DGs begin to produce more power to make up for the lost power from DG_4. As a result, while the $DG_4$is being powered by other DGs, the active and reactive power ratios of the other DGs rise. The plugged-out DG is reconnected at $t = 3$ s. It is obvious that less fluctuation is needed to enable smooth PnP activities. From Figure 3.14, it is observed proven that the proposed controller satisfies the requirement for plug-and-play functioning. Also, equal power-sharing performance is observed in Figure 3.14, demonstrating the rapid adaptability of the proposed controller.

Case 4: Comparison of the proposed controller with conventional controllers

In this evaluation, the performance of the proposed robust controller is compared to the distributed averaging-based controller [44], and the MG centralized controller [45]. It is observed that both the distributed averaging-based controller and the centralized controller exhibit asymptotic stability. However, the proposed robust controller not only restores the system's nominal frequency and voltage, and ensures active and reactive power sharing among DGs, but also demonstrates superior robustness in the presence of uncertainties. From the figures, it is evident that both the conventional controllers initially operate at values deviating from the nominal values at $t = 0$ s. In the case of [44,45], the voltage and frequency of the DG do not converge to their nominal values. However, the proposed distributed robust BS controller effectively maintains the voltage and frequency responses within the acceptable range. As mentioned in the introduction, the proposed

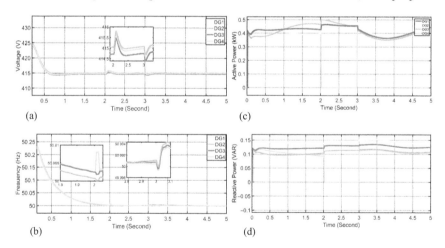

Figure 3.14   (a) Voltage responses. (b) Frequency responses. (c) Active power responses. (d) Reactive power responses of DGs during PnP operation [46].

controller effectively manages adverse circumstances without compromising system stability. Its robust performance ensures that the MG operates reliably and efficiently even in the face of uncertainties (Figure 3.15).

Case 5: Effects of communication degradation on the control performance

Communication deterioration in a network can be attributed to various factors such as communication noise, time delay, limited bandwidth, and latency. In this study, the performance of the proposed controller is evaluated based on communication delay and network switching topology. The responses of the DG under communication delays are depicted in Figure 3.16. It is observed that in the presence of both equal and unequal communication delays, the proposed controller successfully restores the nominal voltage and frequency within a finite convergence time. The convergence speed of the proposed controller is determined to be 0.075 s, demonstrating a faster convergence with reduced transient effects. These observations provide evidence of finite convergence times for voltage and frequency, indicating the effectiveness of the proposed controller in achieving stable and accurate voltage and frequency restoration even under communication delay conditions.

Case 6: Fault analysis

The following fault analysis is used to assess the proposed controller's fault tolerance capability. At the line intersection between two DG, a line-to-ground (L-G) fault occurs at $t = 0.2$ s, having a transition resistance of 0.1 ohm. As a result, phase A of the VSI provides a low resistance path for the current to flow along, which results in a very large current flow, as shown in Figure 3.17(a), and the voltage across phase A of the VSI is lowered to 0. In Figure 3.17(c), the voltage and frequency responses of the DG are displayed. A line-to-line (L-L) fault occurs at $t = 0.2$ s at the transition resistance of 1 ohm. As can be seen from Figure 3.17(b), this causes a strong circulation current to flow between phases A and B. Figure 3.17(d) shows the corresponding voltage and frequency responses of DG, where the voltage and frequency exhibit more transitory oscillation than in the prior case.

3.6.3   Real-time experimentation

The experimental set-up of two DGs is shown in Figure 3.18(a). A three-phase rectifier and a four-leg IGBT inverter power stack (3PINV85-R) are used to create three-phase voltages and currents for the PCC. The harmonics present in the voltage and current output of VSI are reduced by the LC filter. Resistive and inductive (RL) load is used as a local load at PCC. Six gate pulses are generated for the three-phase VSI through OPAL-RT (OP4510). Through an Ethernet cable, the OPAL-RT and host PC are linked, and the OPAL-RT and inverter are linked via the DAC port.

A scopecorder (DL950) is used to record the DG voltage, frequency responses, inverter output voltage, and regulated PWM pulses. Figure 3.16 shows a picture of the experimental setup along with a drawing of it. One DG is linked to another DG using an RS485 module, which communicates information between the two DG. The controller was built using the OPAL-RT platform, which incorporates MATLAB and Simulink. By collecting data from sensors and providing it to OPAL-RT through

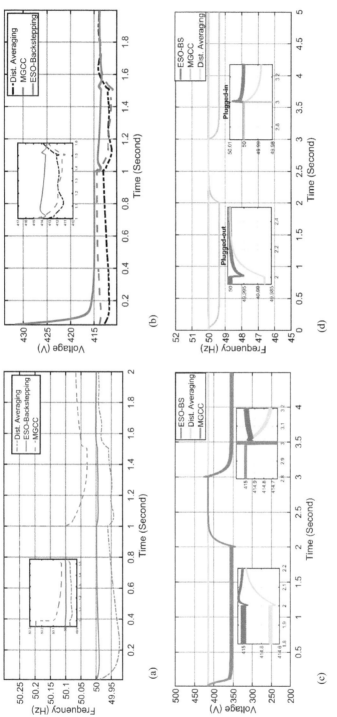

Figure 3.15   Comparison of the proposed controller with conventional controllers. (a and b) Voltage and frequency responses of DG under load changes. (c and d) Voltage and frequency responses of DG with PnP operation [46].

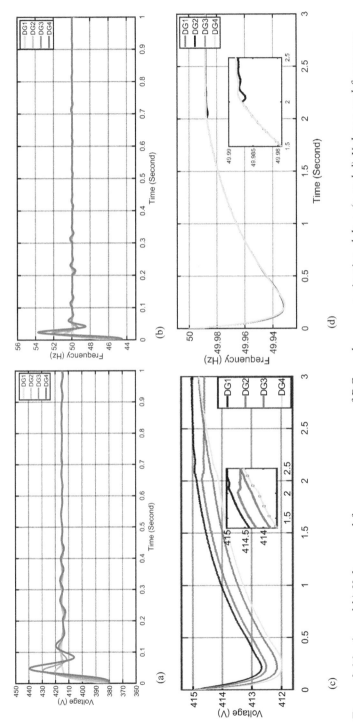

Figure 3.16 (a and b) Voltage and frequency response of DGs under communication delay. (c and d) Voltage and frequency responses of DG with switching network [46].

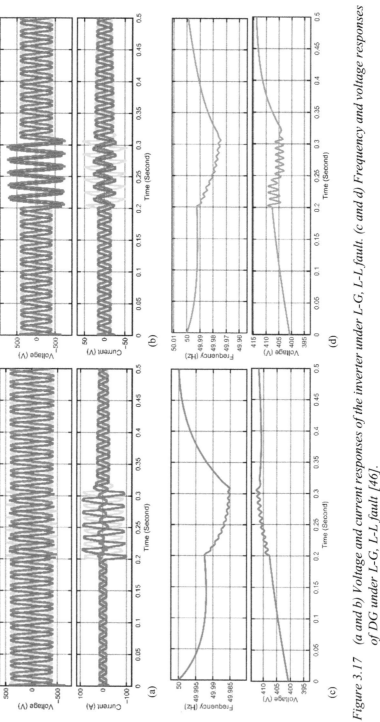

Figure 3.17 (a and b) Voltage and current responses of the inverter under L-G, L-L fault. (c and d) Frequency and voltage responses of DG under L-G, L-L fault [46].

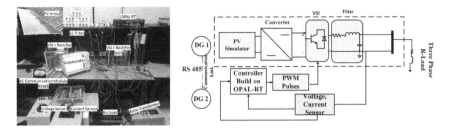

Figure 3.18   (a) Photo of the experimental set-up. (b) Sketch of the experimental set-up [46].

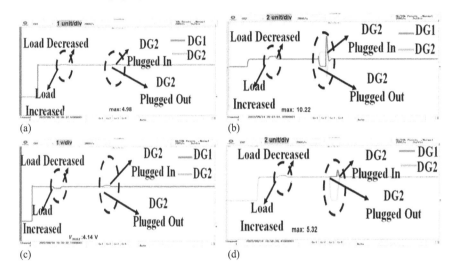

Figure 3.19   (a) Voltage responses. (b) Reactive power responses of DGs. (c) Frequency responses. (d) Active power responses of DGs under load variations and PnP operations [46].

analog input, the HIL simulation setup for distributed control of the MG is carried out. Analog output provides the PWM signal for the inverter. As an example, the load is varied, and one of the DGs was unplugged and then replugged into the MG. The real-time responses under these constraints are displayed in Figure 3.19.

3.7   Conclusion

This chapter provides a thorough analysis of the various challenges experienced when RESs were integrated into the existing grid. The control difficulties that need to be handled, as well as the most recent research developments in the area of adaptive control, are presented. Adaptive controllers perform significantly better than conventional controllers in terms of transient- and steady-state behavior and robustness in order to meet the control objectives of the MG and overcome the aforementioned problems. A design and implementation of a distributed extended state observer-based backstepping control system for an AC microgrid is done. The proposed controller is expected to manage voltage and frequency and achieve power-sharing among the DGs properly despite load fluctuations, communication delay, and other factors that were investigated through simulation and experiments.

References

[1]   S.K. Panda and B. Subudhi. 'A review on robust and adaptive control schemes for microgrid.' *Journal of Modern Power Systems and Clean Energy*, 2022;11 (4):1027–1040.

[2]   IEO. *International Energy Outlook 2021* (IEO2021). 2021.

[3]   V.A. Suryad, S. Doolla, and M. Chandorkar. 'Microgrids in India: Possibilities and challenges.' *IEEE Electrification Magazine*, 2017;5(2):47–55.

[4]   F.R. Badal, P. Das, S.K. Sarker, and S.K. Das. 'A survey on control issues in renewable energy integration and microgrid.' *Protection and Control of Modern Power Systems*, 2019;4(1):1–27.

[5]   D.E. Olivares, A. Mehrizi-Sani, A.H. Etemadi, *et al.* 'Trends in microgrid control.' *IEEE Transactions on Smart Grid*, 2014;5(4):1905–1919.

[6]   Y. Cheng, R. Azizipanah-Abarghooee, S. Azizi, L. Ding, and V. Terzija. 'Smart frequency control in low inertia energy systems based on frequency response techniques: A review.' *Applied Energy*, 2020;279:115798.

[7]   C. Wang and M.H. Nehrir. 'Power management of a stand-alone wind/photovoltaic/fuel cell energy system.' *IEEE Transactions on Energy Conversion*, 2008;23(3):957–967.

[8]   H. Karimi. 'Islanding detection and control of an islanded electronically-coupled distributed generation unit.' *Thesis*, University of Toronto; 2008.

[9]   X. Lu and J. Lai. 'Communication constraints for distributed secondary control of heterogeneous microgrids: A survey.' *IEEE Transactions on Industry Applications*, 2021;57(6):5636–5648.

[10]   N.M. Dehkordi, H.R. Baghaee, N. Sadati, and J.M. Guerrero. 'Distributed noise-resilient secondary voltage and frequency control for islanded microgrids.' *IEEE Transactions on Smart Grid*, 2019;10(4):3780–3790.

[11]   A. Afshari, M. Karrari, H.R. Baghaee, and G.B. Gharehpetian. 'Resilient cooperative control of AC microgrids considering relative state-dependent noises and communication time-delays.' *IET Renewable Power Generation*, 2020;14(8):1321–1331.

[12]   J. Lai, X. Lu, X. Yu, A. Monti, and H. Zhou. 'Distributed voltage regulation for cyber-physical microgrids with coupling delays and slow switching topologies.' *IEEE Transactions on Systems, Man, and Cybernetics: Systems*, 2020;50(1):100–110.

[13]   A.B. Shyam, S. Anand, and S.R. Sahoo. 'Effect of communication delay on consensus-based secondary controllers in DC microgrid.' *IEEE Transactions on Industrial Electronics*, 2021;68(4):3202–3212.

[14] Q. Zhou, M. Shahidehpour, A. Paaso, S. Bahramirad, A. Alabdulwahab, and A. Abusorrah. 'Distributed control and communication strategies in networked microgrids.' *IEEE Communications Surveys & Tutorials*, 2020;22(4):2586–2633.

[15] Y.A.R.I. Mohamed and A.A. Radwan. 'Hierarchical control system for robust microgrid operation and seamless mode transfer in active distribution systems.' *IEEE Transactions on Smart Grid*, 2011;2(2):352–362.

[16] F. Nejabatkhah, Y.W. Li, H. Liang, and R. Reza Ahrabi. 'Cyber-security of smart microgrids: A survey.' *Energies*, 2021;14(1):27.

[17] X. Liu, M. Shahidehpour, Y. Cao, L. Wu, W. Wei, and X. Liu. 'Microgrid risk analysis considering the impact of cyber attacks on solar PV and ESS control systems.' *IEEE Transactions on Smart Grid*, 2017;8(3):1330–1339.

[18] D. Popović. *Analysis and Control of Industrial Processes*. Wiesbaden: Springer Verlag; 1991.

[19] S.G. Santoso and M.A. Garratt. 'Progress in adaptive control systems: past, present, and future.' In *International Conference on Advanced Mechatronics, Intelligent Manufacture, and Industrial Automation (ICAMIMIA)*. 2015. IEEE.

[20] J.D. Day and H. Zimmermann. 'The OSI reference model.' *Proceedings of the IEEE*, 1983;71(12):1334–1340.

[21] D. Agarwal. *Fibre Optics in Communication*. New Delhi: AH Wheeler & Co. Pvt. Ltd.; 1988.

[22] D. Popović. 'Fibre optic communication systems in industrial automation.' In *Analysis and Control of Industrial Processes* (pp. 34–51). Wiesbaden: Springer Verlag; 1991.

[23] N. Kapany. *Fiber Optics: Principles and Applications*. New York: Academic Press; 1967.

[24] B. Bandyopadhyay, S. Janardhanan, and S.K. Spurgeon. *Advances in Sliding Mode Control*. Berlin, Heidelberg: Springer; 2013.

[25] I. Boiko and L. Fridman. 'Analysis of chattering in continuous sliding-mode controllers.' *IEEE Transactions on Automatic Control*, 2005;50(9):1442–1446.

[26] H. Lee and V.I. Utkin. 'Chattering suppression methods in sliding mode control systems.' *Annual Reviews in Control*, 2007;31(2):179–188.

[27] C.E. Hall and Y.B. Shtessel. 'Sliding mode disturbance observer-based control for a reusable launch vehicle.' *Journal of Guidance, Control, and Dynamics*, 2006;29(6):1315–1328.

[28] K.J. Åström and B. Wittenmark. *Adaptive Control*. 2nd edition. Reading, MA: Addison-Wesley Pub Co.; 1995.

[29] S. Sastry, M. Bodson, and J.F. Bartram. 'Adaptive control: stability, convergence, and robustness.' *Journal of the Acoustical Society of America*, 1990;88:588–589.

[30] Y. Yang, D. Xu, T. Ma, and X. Su. 'Adaptive cooperative terminal sliding mode control for distributed energy storage systems.' *IEEE Transactions on Circuits and Systems I: Regular Papers*, 2021;68(1):434–443.

[31] Y. Liu, Q. Zhang, C. Wang, and N. Wang. 'A control strategy for microgrid inverters based on adaptive three-order sliding mode and optimized droop controls.' *Electric Power Systems Research*, 2014;117:192–201.

[32] Z. Chen, A. Luo, H. Wang, Y. Chen, M. Li, and Y. Huang. 'Adaptive sliding-mode voltage control for inverter operating in islanded mode in microgrid.' *International Journal of Electrical Power & Energy Systems*, 2015;66:133–143.

[33]   M.A. Mahmud, T.K. Roy, S. Saha, M.E. Haque, and H.R. Pota. 'Robust nonlinear adaptive feedback linearizing decentralized controller design for islanded DC microgrids.' *IEEE Transactions on Industry Applications*, 2019;55(5):5343–5352.

[34]   Y.A.R.I. Mohamed and E.F. El-Saadany. 'Adaptive decentralized droop controller to preserve power sharing stability of paralleled inverters in distributed generation microgrids.' *IEEE Transactions on Power Electronics*, 2008;23(6):2806–2816.

[35]   J. Kim, J.M. Guerrero, P. Rodriguez, R. Teodorescu, and K. Nam. 'Mode adaptive droop control with virtual output impedances for an inverter-based flexible AC microgrid.' *IEEE Transactions on Power Electronics*, 2011;26(3):689–701.

[36]   M. Hassanzahraee and A. Bakhshai. 'Adaptive transient power control strategy for parallel-connected inverters in an islanded microgrid.' In *IECON 2012 – 38th Annual Conference on IEEE Industrial Electronics Society*. 2012. IEEE.

[37]   M. Mokhtar, M.I. Marei and A.A. El-Sattar. 'An adaptive droop control scheme for DC microgrids integrating sliding mode voltage and current controlled boost converters.' *IEEE Transactions on Smart Grid*, 2019;10 (2):1685–1693.

[38]   V. Nasirian, A. Davoudi, and F. Lewis. 'Distributed adaptive droop control for DC distribution systems.' *IEEE Transactions on Energy Conversion*, 2014;29(4):944–956.

[39]   Z. Zhang, G. Song, J. Zhou, *et al.* 'An adaptive backstepping control to ensure the stability and robustness for boost power converter in DC microgrids.' *Energy Reports*, 2022;8:1110–1124.

[40]   O. Boutebba, S. Semcheddine, F. Krim, and B. Talbi. 'Adaptive nonlinear controller design for DC-DC buck converter via backstepping methodology.' In *International Conference on Advanced Electrical Engineering (ICAEE)*. 2019. IEEE.

[41]   M.S. Khan, I. Ahmad, and F.Z. Ul Abideen. 'Output voltage regulation of FC-UC based hybrid electric vehicle using integral backstepping control.' *IEEE Access*, 2019;7:65693–65702.

[42]   W.-H. Chen, J. Yang, L. Guo, and S. Li. 'Disturbance-observer-based control and related methods—An overview.' *IEEE Transactions on Industrial Electronics*, 2016;63(2):1083–1095.

[43]   A. Bidram, A. Davoudi, F.L. Lewis, and Z. Qu. 'Secondary control of microgrids based on distributed cooperative control of multi-agent systems.' *IET Generation, Transmission & Distribution*, 2013;7(8):822–831.

[44]   J.W. Simpson-Porco, Q. Shafiee, F. Dörfler, J.C. Vasquez, J.M. Guerrero, and F. Bullo. 'Secondary frequency and voltage control of islanded microgrids via distributed averaging.' *IEEE Transactions on Industrial Electronics*, 2015;62(11):7025–7038.

[45]   Q. Shafiee, J.M. Guerrero, and J.C. Vasquez. 'Distributed secondary control for islanded microgrids—A novel approach.' *IEEE Transactions on Power Electronics*, 2014;29(2):1018–1031.

[46]   S.K. Panda and B. Subudhi. 'An extended state observer based adaptive backstepping controller for microgrid.' *IEEE Transactions on Smart Grid*, 2024;15 (1):171–178.

Chapter 4

Reinforcement learning-based inter-area damping controller considering communication channel latency and actuator saturation

Anirban Sengupta[1] and Dushmanta Kumar Das[2]

The power systems are increasingly integrating renewable energy sources. Therefore, the assurance of power system stability is important. Wide area damping controllers (WADCs) based on wide area measurement systems (WAMS) can boost power system stability and provide adequate damping, but their performance significantly deteriorates as a result of actuator saturation. This study develops a model-free control algorithm that takes actuator saturation and latency in the communication channel into account and increases power system stability through the use of reinforcement learning techniques. The Bellman equation is used to compute the controller gains. The Q-learning technique is used to build the state feedback and output feedback control techniques. The electromagnetic oscillation is mitigated by providing sufficient damping with the help of a supplementary damping controller (SDC) along with a unified power flow controller (UPFC), which acts as the actuator. The proposed deep learning-based control technique is validated on a modified 10-machine power system and a modified 16-machine test system.

4.1 Introduction

In order to meet the power demands of the modern power system, power system interconnection is becoming necessary [1]. Several kinds of disruptions can arise in an interconnected power system (IPS), while it is operating. Low frequency oscillation (LFO) issues in linked power systems could be brought on by these disruptions. LFO deteriorates the small signal stability of the IPS and as a result the reliable operation of the IPS also deteriorates [2–4]. There are two types of LFO: inter-area LFO (IALFO) (0.1–1 Hz) and local LFO (1–2 Hz) [5]. Unlike an IALFO, which involves a set of alternators, situated in a particular zone oscillating with some other

[1]Department of Electrical and Electronics Engineering, Sikkim Manipal Institute of Technology, Sikkim Manipal University, India
[2]Department of Electrical and Electronics Engineering, National Institute of Technology Nagaland, India

set of alternators, which are situated in a different zone, a local LFO involves a single generator oscillating independently inside a certain area. Local power system stabilizers (PSS) can effectively reduce local LFO, but they are unable to offset the IALFO [6]. WADC can effectively assist in lowering IALFO [7].

There are numerous research publications in the literature pertaining to WADC development. The advantages of a WADC in the operation of a power system are demonstrated in [8]. The development of WADC using a flexible AC transmission system (FACTS) is presented in [9,10]. In [11], a decentralized approach is demonstrated to design a control action to suppress inter-area oscillation. Although the WADC designed in [8–11] can effectively mitigate the IALFO, their performance degrades considerably if latency occurs during transmission. One of the very common contingencies in a communication channel is the latency of the communication channel. It can impair significantly the efficacy of any control action. Therefore, when building the damping controller, it is imperative to consider the impact of latency in the communication channel. A state feedback control technique based on WADC considering latency in communication channels is demonstrated in [12]. The effect of latency in communication channels and their mitigating technique is also demonstrated in [13]. A latency-resilient adaptive control technique is developed in [14]. A deep learning-based approach to mitigate the consequences of latency in data transmission lines is developed in [15]. A back-to-back converter-based control technique considering the latency in the data transmission channel is proposed in [16]. In [17], a controller is designed using an optimal control approach to suppress the rotor angle oscillation considering the latency in the data transmission channel. The consequences of having latency in the data transmission channel in the performance of a network control system are demonstrated in [18].

There are numerous literary works devoted to WADC design accessible. In [19], a decentralized WADC that takes operating uncertainty into account for IPS is given. In [20], for wind generator integrated power systems, a delay-dependent damping controller based on the Lyapunov-Krasovskii function is suggested. To determine the controller gains of the WADC for the integrated power system of a solar power plant, a gain scheduling technique is proposed in [21]. In [22], a proportional-derivative controller with a WADC that takes communication latency into account is designed. In [16], a WADC based on the data-driven approach is put into practice. In [23], a damping controller that takes load uncertainty into account is described as utilizing an output feedback controller. A damping controller is constructed in [24] utilizing the genetic algorithm (GA) and particle swarm optimization (PSO) algorithms while taking power system uncertainties into account. In [25], a method based on linear matrix inequality is used to create a coordinated WADC. In [26], an optimization algorithm-based WADC is created where the parameters of the WADC are calculated using the Jaya algorithm. A control law to enhance the system damping with the help of time scale theory is developed in [27].

The damping controllers discussed in [16,19–27] produce an ample amount of damping to reduce the IALFO, but most of the WADCs developed in [16,19–27] have not considered the actuator saturation. Saturation is an inherent property of any practical actuator and it can deteriorate the effectiveness of the WADC if it is not considered while designing the damping controller [28]. Several researchers are working towards the development of a damping controller considering saturation. The development of control law with the help of linear matrix inequality is demonstrated in [28–31] to mitigate the effect of saturation on WADC. A dynamic controller-based WADC considering the upper and lower limit of the actuator is developed in [32]. The design of a WADC including saturation is also presented in [15,33–35]. A feedback controller is developed considering saturation in [36].

While the controllers created in [15,28–36] offer enough dampening to reduce the IALFO, a significant drawback of the previously stated techniques is that the WADC must be designed with precise knowledge of the power system. A networked power system is extremely complicated, non-linear, and not necessarily observable or controllable. As a result, it is never possible to locate a realistic power system model [37]. This restriction can be addressed by the development of a controller that takes the help of artificial intelligence (AI) to develop the required control signal. This can also be helpful in creating a model-free controller that does away with IALFO [38]. In [39] and [40], the application of artificial neural networks (ANN) and deep neural networks (DNN) to eradicate IALFO is covered. In [41], a control law is developed using the Q-learning technique. The control law is created taking the unpredictable situations of a power system into account. Hadidi and Jayasurya [42] use a Q-learning method to create a WADC. In [43], a deep reinforcement learning-based technique to lower the IALFO is created.

Although the controllers developed in [39–43] have demonstrated the superiority of reinforcement learning techniques in power system applications, the aforementioned works do not take the effect of saturation and data transmission channel latency into account. Furthermore, to the best of the author's knowledge, no research has been done in the literature to construct a WADC utilizing the reinforcement learning technique while taking saturation and data transmission channel latency into account. In this work, a reinforcement learning method is created to reduce IALFO in light of the above given observations considering the effect of actuator saturation and data transmission channel latency. A Q-learning-based algorithm is developed in this chapter to ensure the stability of the power system considering the saturation effect of the actuator. A UPFC is considered as the actuator in this chapter because UPFC has the advantages of both series-connected and shunt-connected FACTS devices [44]. The objective of this chapter is to design a model-free controller using both state feedback and output feedback techniques to suppress the small signal stability problem of a power system taking saturation of the actuator and data transmission channel latency into consideration.

4.2 Novelty of the chapter

A Q-learning-based model free damping controller is designed in this chapter taking the saturation of the controller and data transmission channel latency into consideration. The formulated control technique is tested for both state feedback and output feedback-based controllers. The advantage of a Q-learning-based controller is that it is not required to know the information of all the states of the system to develop the control logic.

4.3   Problem formulation

The interconnected power system in state space representation can be written as:

$$x(k + 1) = Ax(k) + Bu(k - d) + B_\omega \omega(k), \tag{4.1}$$

$$y(k) = Cx(k), \tag{4.2}$$

where $x(k) \in R^n$ is the state vector; $u(k) \in R^m$ is the input vector; $y(k) \in R^p$ is the output; $\omega(k) \in R^{n_\omega}$ is the disturbance input and d is the latency in the data transmission channel.

Actuator saturation is inherent in any practical damping controller. The input vector $u(k)$ taking the upper and lower limit of the actuator can be represented as:

$$sat(u(k)) = \begin{cases} u_{\max}, & u \geq u_{\max} \\ u, & u_{\min} < u < u_{\max} \\ u_{\min}, & u \leq u_{\min} \end{cases} \tag{4.3}$$

Therefore, the power system model considering actuator saturation is given by:

$$x(k + 1) = Ax(k) + Bsat(u(k - d)) + B_\omega \omega(k), \tag{4.4}$$

$$y(k) = Cx(k), \tag{4.5}$$

4.4   Reinforcement learning

Reinforcement learning is a reward-based control technique. The basic strategy of reinforcement learning technique is demonstrated in this section. To use the reinforcement learning technique, at first a value function is chosen as:

$$F_V = \max(R(x, u_n)), \tag{4.6}$$

where $R(x, u_n)$ is a discounted reward function. The discount will be positive for fewer errors and will be negative for more errors in the control action. The discounted reward function is defined as:

$$R(x, u_n) = \sum_{n=0}^{\infty} \Upsilon^n r(x_n, u_n), \tag{4.7}$$

where Υ is a reward factor; u is the control signal.

The solution of the value function can be found from the solution of the Bellman equation [45] and is given by:

$$F_V = \max[r(x, u) + \Upsilon F_V(f(x, u))]. \tag{4.8}$$

It is possible to compute the control input's most optimum value as:

$$u_{\text{opt}} = \arg \max[r(x, u) + \Upsilon F_V(f(x, u))]. \tag{4.9}$$

The value function can also be written as:

$$F_V = \max(Q(x, u)), \tag{4.10}$$

where Q is expressed as:

$$Q = r(x, u) + \Upsilon F_V(f(x, u)), \tag{4.11}$$

and the optimum control signal can be calculated as:

$$u_{\text{opt}} = \arg \max(Q(x, u)). \tag{4.12}$$

From (4.12), it can be seen that the knowledge of Q is helpful in finding the optimal value of the control signal using the reinforcement learning technique.

4.5   Iterative Q-learning scheme

To develop the Q-learning-based control law, a quadratic utility function is chosen as:

$$r(x(k), u(k), \varepsilon) = \varepsilon x^T(k)x(k) + u^T(k)u(k), \tag{4.13}$$

where $\varepsilon \in (0, 1)$ is a low gain parameter [46].

A cost function is also chosen to develop the Q-learning-based control law, given by:

$$V(x(k), u(k)) = \sum_{i=k}^{\infty} r(i)(x(i), u(i), \varepsilon). \tag{4.14}$$

4.5.1   State feedback control law

The control signal in the case of a state feedback controller is given by:

$$u(k) = K(\varepsilon)x(k), \tag{4.15}$$

where $K(\varepsilon)$ is the gain of the controller. In the case of a feedback stabilization control policy, the cost function reduces when

$$V(K(x(k)) = x^T(k)P(\varepsilon)x(k), \tag{4.16}$$

where $P(\varepsilon)$ is a positive definite matrix.

Applying the Bellman optimally principle, (4.14) can be rewritten as:

$$V(Kx(k)) = r(x(k), Kx(k), \varepsilon) + V(Kx(k + 1)). \tag{4.17}$$

A Q-function can be defined similar to (4.17), which is given by:

$$Q(x(k), u(k), \varepsilon) = r(x(k), u(k), \varepsilon) + V(Kx(k + 1)). \tag{4.18}$$

(4.18) can be rewritten as:

$$Q(x(k), u(k), \varepsilon) = \varepsilon x^T(k)x(k) + u^T(k)u(k) + x^T(k + 1)P(\varepsilon)x(k + 1). \tag{4.19}$$

Applying (4.4), (4.19) can be modified as:

$$Q(x(k), u(k), \varepsilon) = \varepsilon x^T(k)x(k) + u^T(k)u(k) \\ + [Ax(k) + Bu(k) + B_\omega\omega(k)]^T P(\varepsilon) \\ \times [Ax(k) + Bu(k) + B_\omega\omega(k)]. \tag{4.20}$$

In matrix form, (4.20) can be written as:

$$Q(x(k), u(k), \varepsilon) = Z^T(k)M(\varepsilon)Z(k), \tag{4.21}$$

where $Z(k) = [x(k)u(k)w(k)]^T$ and

$$M(\varepsilon) = \begin{bmatrix} \varepsilon I + A^T P(\varepsilon)A & A^T P(\varepsilon)B & A^T P(\varepsilon)B_\omega \\ * & I + B^T P(\varepsilon)B & B^T P(\varepsilon)B_\omega \\ * & * & B_\omega^T P(\varepsilon)B_\omega \end{bmatrix}$$

Optimal stabilization condition can be determined using $\frac{\partial Q(x(k),u(k),\varepsilon)}{\partial u(k)} = 0$ and the corresponding control signal can be determined as:

$$u(k) = -[I + B^T P(\varepsilon)B]^{-1}[A^T P(\varepsilon)B]^T x(k), \tag{4.22}$$

which is similar to (4.15). Therefore, the controller gain can be written as:

$$K(\varepsilon) = -[I + B^T P(\varepsilon)B]^{-1}[A^T P(\varepsilon)B]^T \tag{4.23}$$

Using (4.17) and (4.18), it can be concluded that

$$V(Kx(k)) = Q(x(k), u(k), \varepsilon) \tag{4.24}$$

Therefore, (4.19) can be rewritten as:

$$Q(x(k), u(k), \varepsilon) = \varepsilon x^T(k)x(k) + u^T(k)u(k) \\ + Q(x(k+1), u(k+1), \varepsilon). \tag{4.25}$$

Solution of (4.25) requires estimation of $M(\varepsilon)$. This is possible by linearizing the Bellman equation

$$Q(x(k), u(k), \varepsilon) = \bar{M}(\varepsilon)^T \bar{Z}(k), \tag{4.26}$$

where $\bar{M}(\varepsilon) = [M_{11}, M_{12}, \ldots\ldots, M_{(i+j),(i+j)}]$. Also from Bellman equation, $\bar{M}(\varepsilon)^T \bar{Z}(k)$ can be expressed as:

$$\bar{M}(\varepsilon)^T \bar{Z}(k) = \varepsilon x^T(k)x(k) + u^T(k)u(k) + \bar{M}(\varepsilon)^T \bar{Z}(k+1). \tag{4.27}$$

Based on (4.27), a Q-learning technique can be used to estimate $M(\varepsilon)$.

Algorithm 1: Q-learning algorithm for state feedback control law

1. **Initialization:** Start with any random value of the control signal $u(k) = K_0 x(k) + v(k)$. Here $v(k)$ is the exploration signal and K_0 is the initial random control gain.

2. **Data collection:** Collect L number of datasets of $x(k)$ using the initial control signal for $k \in [0, L-1]$.

3. **Cost evaluation:** Calculate the value of $\bar{M}(\varepsilon)^T \bar{Z}(k)$ for ith iteration using (4.27), given by:

$$\bar{M}^i(\varepsilon)^T \bar{Z}(k) = \varepsilon x^T(k)x(k) + u^T(k)u(k) + \bar{M}^{i-1}(\varepsilon)^T \bar{Z}(k+1). \qquad (4.28)$$

4. **Update control gain:** Update the control gain for ith iteration using

$$K^{i+1}(\varepsilon) = -\left[\left[I + B^T P(\varepsilon)B\right]^{-1}\right]^i \left[\left[A^T P(\varepsilon)B\right]^T\right]^i. \qquad (4.29)$$

5. **Convergence criterion:** Repeat steps 3 and 4 until the convergence criterion is obtained. The convergence criterion is given by:

$$K^{i+1}(\varepsilon) - K^i(\varepsilon) < \delta, \qquad (4.30)$$

   where δ is a very small value.

6. **Saturation check:** For all iterations check for the saturation condition given by:

$$u_{\min} \leq K(\varepsilon)x(k) \leq u_{\max}. \qquad (4.31)$$

   If any iteration (4.43) is violated, then set $\varepsilon = \varepsilon \pm \Delta\varepsilon$, $i = 1$ and repeat steps 3, 4, and 5. Here $\Delta\varepsilon$ is a small change in ε.

7. **Termination criterion:** Stop the iteration when the control signal is not saturating for any iteration.

   Algorithm 1 can be used to determine a state feedback control signal considering saturation. Here L number of samples are used to develop the training datasets given by:

$$\alpha = \left[Z^1(k), Z^2(k), \ldots\ldots\ldots Z^n(k)\right], \qquad (4.32)$$

$$\begin{aligned} \beta = [r^1 + [M^{i-1}(\varepsilon)]^T Z^1(k+1), \ldots\ldots \\ \ldots\ldots, r^L + [M^{i-1}(\varepsilon)]^T Z^L(k+1)]^T. \end{aligned} \qquad (4.33)$$

Then, the solution of (4.40) is given by [47]:

$$M(\varepsilon) = \left[\alpha\alpha^T\right]^{-1}\alpha\beta. \qquad (4.34)$$

4.5.2 Output feedback control law

To find the control signal using Algorithm 1, the complete information of the system matrix is required. An interconnected power system is not always completely observable and therefore, it is not always possible to get the complete system matrix of the power system. Output feedback control law can be used to overcome this limitation of state feedback control law. Algorithm 1 can be extended into output feedback control law using state parameterizations. The following Lemma is used for state parameterizations:

Lemma 1 [48]: For any system, the states of the system can be represented in terms of input and output of the system given by:

$$x(k) = H_y y(k-1,k-N) + H_u u(k-1,k-N), \qquad (4.35)$$

where, $y(k-1,k-N)$ is the output vector; $u(k-1,k-N)$ is the input vector and $N \le n$ is the system observability index; $H_y = A^N (V_N^T V_N)^{-1} V_N^T$ and $H_u = U_N - A^N (V_N^T V_N)^{-1} V_N^T T_N$, where

$$V_N = \left[(CA^{N-1})^T \quad \ldots\ldots \quad (CA)^T \quad C^T \right]^T;$$

$$U_N = \left[B \quad AB \quad \ldots\ldots\ldots \quad A^{N-1}B \right];$$

$$T_N = \begin{bmatrix} 0 & CB & CAB & \ldots\ldots & CA^{N-2}B \\ 0 & 0 & CB & \ldots\ldots & CA^{N-2}B \\ \vdots & \vdots & \vdots & \ldots\ldots & \vdots \\ 0 & \ldots & \ldots & 0 & CAB \\ 0 & 0 & 0 & 0 & CB \end{bmatrix}.$$

Using the state parameterization process using Lemma 1, the state vector of the system can be written as

$$x(k) = \begin{bmatrix} H_u & H_y \end{bmatrix} \begin{bmatrix} u(k-1,k-N) \\ y(k-1,k-N) \end{bmatrix} \qquad (4.36)$$

Applying (4.36) in (4.21), a modified Q function can be defined as:

$$Q(\xi(k), \varepsilon) = \xi^T(k) \begin{bmatrix} S_{11} & S_{12} & S_{13} \\ * & S_{22} & S_{23} \\ * & * & S_{33} \end{bmatrix} \xi(k), \qquad (4.37)$$

where $\xi(k) = \left[u(k-1,k-N) \quad y(k-1,k-N) \quad u^T(k) \right]^T$; $S_{11} = M_u^T (\varepsilon C^T C + A^T P(\varepsilon) A) M_u$, $S_{12} = M_u^T (\varepsilon C^T C + A^T P(\varepsilon) A) M_y$, $S_{13} = M_u^T A^T P(\varepsilon) B$, $S_{22} = M_y^T (\varepsilon C^T C + A^T P(\varepsilon) A) M_y$, $S_{23} = M_y^T A^T P(\varepsilon) B$, $S_{33} = I + B^T P(\varepsilon) B$. The optimal control signal in output feedback control is given by:

$$u_o(k) = K_o(\varepsilon) \left[u_{k-1,k-N}^T \quad y_{k-1,k-N}^T \right]^T \qquad (4.38)$$

Using the modified Q-function described in (4.37), a modified Bellman equation can be written as:

$$S^T(\varepsilon) \xi(k) = \varepsilon y^T(k) y(k) + u^T(k) u(k) + S^T(\varepsilon) \xi(k+1) \qquad (4.39)$$

Based on (4.39), a Q-learning technique can be used to estimate $S(\varepsilon)$.

Algorithm 2: Q-learning algorithm for output feedback control law

1. **Initialization:** Start with any random value of the control signal $u(k) = K_o^0 \left[u_{k-1,k-N}^T y_{k-1,k-N}^T \right] + v(k)$. Here $v(k)$ is the exploration signal and K_o^0 the initial random control gain.

2. **Data collection:** Collect L the number of datasets of $x(k)$ using the initial control signal for $k \in [0,L-1]$.
3. **Cost evaluation:** Calculate the value of $S^T(\varepsilon)\xi(k)$ for ith iteration using (4.39), given by:

$$S_i^T(\varepsilon)\xi(k) = \varepsilon y_i^T(k)y_i(k) + u^T(k)u(k) + S^T(\varepsilon)\xi(k+1). \tag{4.40}$$

4. **Update control gain:** Update the control gain for ith iteration using

$$K^{i+1}(\varepsilon) = -\left[\left[I + B^T P(\varepsilon)B\right]^{-1}\right]^i \left[\left[A^T P(\varepsilon)B\right]^T\right]^i, \tag{4.41}$$

where $P(\varepsilon)$ is a positive definite matrix. The updated gain will be used to find the optimal value of u.
5. **Convergence criterion:** Repeat steps 3 and 4 until the convergence criterion is obtained. The convergence criterion is given by:

$$K^{i+1}(\varepsilon) - K^i(\varepsilon) < \delta, \tag{4.42}$$

where δ is a very small value.
6. **Saturation check:** For all iterations check for the saturation condition given by:

$$u_{\min} \leq K(\varepsilon)x(k) \leq u_{\max}. \tag{4.43}$$

If for any iteration (4.43) is violated, then set $\varepsilon = \varepsilon \pm \Delta\varepsilon$, $i = 1$ and repeat steps 3, 4, and 5. Here $\Delta\varepsilon$ is a small change in ε.
7. **Termination criterion:** Stop the iteration when the control signal is not saturating for any iteration.

4.6 Result analysis

In order to comprehend the performance of the created controller for various scenarios, the efficacy of the Q-learning-based controller for both state feedback and output feedback-based control signal is investigated using two distinct power system models.

4.6.1 Test system 1: Modified IEEE 39-bus New England power system

The purpose of this test case scenario is to evaluate the efficacy of the built controller using a modified IEEE 39-bus test system, as illustrated in Figure 4.1. Ten alternators make up the 39-bus system from the beginning. The original 39-bus system's variables can be found in [49,50]. To further understand the effect of the RES on inter-area oscillation, a 600 MW solar power plant and a 900 MW wind farm based on doubly fed induction generator (DFIG) are added to the test system at buses 9 and 36, respectively. The purpose of connecting the UPFC between buses 17 and 18 is to improve the power system's stability. The active power sensitivity technique is used to determine the UPFC's location [51].

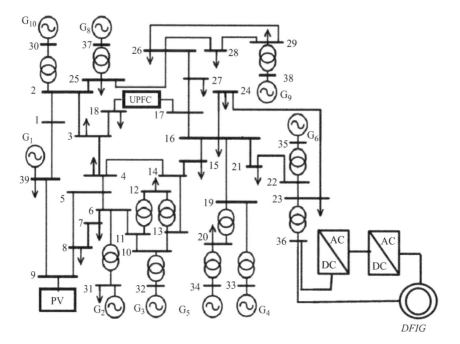

Figure 4.1   The first test system incorporating RES

4.6.1.1   Modal analysis

Eigenvalue analysis provides the details of the local and inter-area oscillation of the first test system. The redesigned 39-bus system is discovered to have three inter-area modes and many electromagnetic oscillation modes. Tables 4.1 and 4.2, respectively, illustrate the impact of solar PV and WECS integration on a 39-bus power system, where inter-area frequency (IAF) represents the inter-area frequency and IA ζ represents the inter-area damping ratio.

Tables 4.1 and 4.2 demonstrate that when solar PV plants or WECS are integrated with the IEEE 39-bus power system, the damping of the system is reduced. Furthermore, solar PV has a greater effect on the power system's damping than WECS. Table 4.3 illustrates the impact of overall dampening of the first test system when both renewable energy sources are added to the system.

It is seen from Table 4.3, that the effect of RES is more on the power system when both solar photovoltaic power plant and WECS are added with the first test system. The damping ratio of the RES-connected test system reduces considerably with the integration of RES on the IEEE 39-bus system. Table 4.4 displays the efficacy of the developed controller on the RES-integrated first test system.

It is seen from Table 4.4, that the overall system damping is increasing with the application of the designed controller. To choose the most effective signal as the input to the SDC, residue analysis [52] is used. The result of the analysis suggests that the most effective input signal will be the power signal between bus 1 and 16. It

Table 4.1   Impact of solar PV incorporation on first test system

Mode of oscillation	Excluding solar PV		Including solar PV	
	IAF	IA ζ	IAF	IA ζ
IA 1	0.1872	0.0489	0.2116	0.0396
IA 2	0.7104	0.0392	0.8132	0.0374
IA 3	0.9018	0.0326	1.1816	0.0267

Table 4.2   Impact of WECS incorporation on the first test system

Mode of oscillation	Excluding WECS		Including WECS	
	IAF	IA ζ	IAF	IA ζ
IA 1	0.1872	0.0489	0.1976	0.0454
IA 2	0.7104	0.0392	0.7216	0.05118
IA 3	0.9018	0.0326	1.0602	0.0307

Table 4.3   Impact of RES incorporation on first test system

Mode of oscillation	Excluding RES		Including RES	
	IAF	IA ζ	IAF	IA ζ
IA 1	0.2019	0.0496	0.2215	0.0319
IA 2	0.7018	0.0475	0.8212	0.0304
IA 3	0.9918	0.0424	1.0804	0.0208

Table 4.4   A comparative analysis of various controllers regarding inter-area frequency variation for the first test system

Controller	IA-1 ζ	IA-2 ζ	IA-3 ζ
Without control action	0.0391	0.0314	0.0252
Conventional H_∞ control [18]	0.0792	0.0718	0.0356
Dynamic control [30]	0.1018	0.0948	0.0721
Proposed technique	0.1516	0.1102	0.0961

was observed that the first test system considered in this study has 96 states. In the case of a small signal stability study, all the states are not required to be considered. Therefore, few states can be reduced for designing the controller. This reduction is possible using a balanced model truncation approach [53]. To finalize the number of states of the controller, the bode magnitude plot is compared between the original test system and the lesser-ordered system. From the comparison, it was observed that the

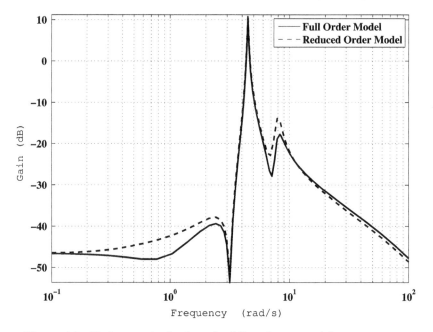

Figure 4.2   Bode magnitude plot of a full and truncated first test system

bode magnitude plot of the 8th ordered model is very similar to the original first test case model. Therefore, the 8th ordered model is selected as the required model to design the controller. The advantage of reducing the states of the system is it leads towards a lesser number of states of the controller (Figure 4.2).

4.6.1.2   Effect of latency in the communication channel

The unfavorable result of having latency in the communication channel can be seen in Figures 4.3 and 4.4. Figure 4.3 shows the deviation of speed between generator 1 and generator 2 and it is denoted by $\omega_{1,2}$. Figure 4.4 shows the deviation of speed between generator 1 and generator 5 and it is denoted by $\omega_{1,5}$. It is evident from Figures 4.3 and 4.4 that when the latency is not taken into account, the speed deviation sustains for a longer time and also the peak overshoot increases considerably. However, the control law proposed in this article can diminish the variation in speed and can lead toward more stability since latency is considered during the creation of the control action.

4.6.1.3   Effect of saturation

The problems that can occur of neglecting the upper and lower saturation limit of an actuator on the performance of the WADC are presented in Figures 4.5 and 4.6. An examination of Figures 4.5 and 4.6 reveals that the oscillation of the variation in speed increases visibly if saturation is neglected. However, the proposed Q-learning-based control technique has the ability to diminish the variation in speed very successfully as the saturation limit is taken into account in the design of the controller.

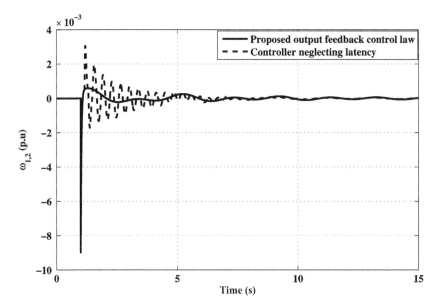

Figure 4.3 Effect of latency on $\omega_{1,2}$

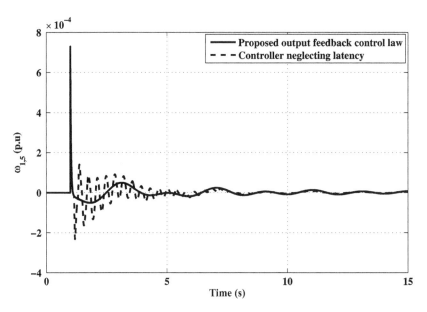

Figure 4.4 Effect of latency on $\omega_{1,5}$

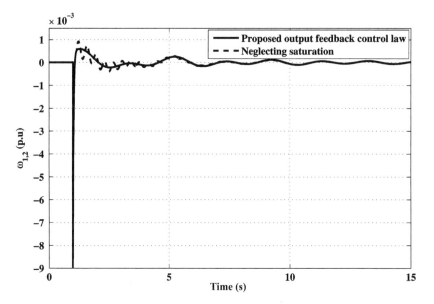

Figure 4.5   Effect of saturation on $\omega_{1,2}$

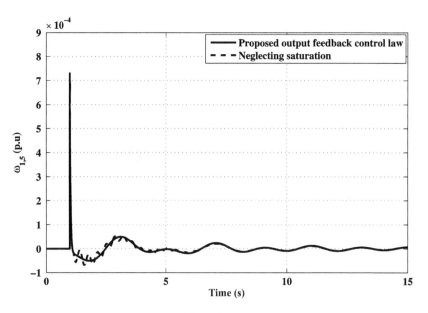

Figure 4.6   Effect of saturation on $\omega_{1,5}$

4.6.1.4 Fault analysis

The ability of the proposed Q learning-based control technique is examined using a balanced ground fault. The fault is assumed at bus 16. The power flow at the time of fault is assumed to be 600 MW. Figures 4.7 and 4.8 demonstrate the deviation in speed between generators of different areas due to different controllers. Variation of ω_{12} is shown in Figure 4.7 and that of ω_{15} is shown in Figure 4.8. ω_{12} is the

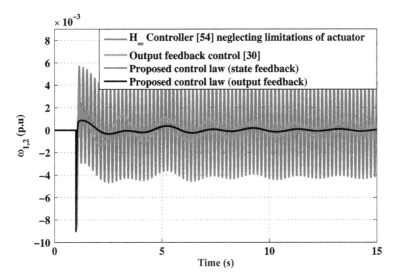

Figure 4.7   Variation of $\omega_{1,2}$ with different control law

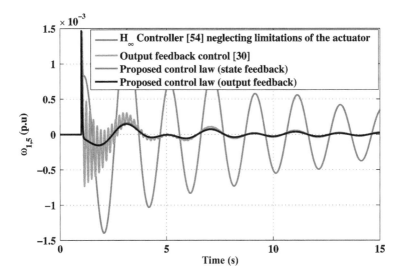

Figure 4.8   Variation of $\omega_{1,5}$ with different control law

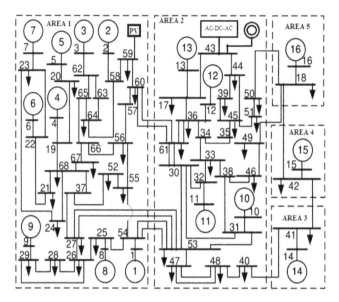

Figure 4.9   Modified IEEE 68-bus power system

difference in speed between generator-1 and generator-2, whereas ω_{15} is the difference in speed between generator 1 and 5. It can be observed from Figures 4.7 and 4.8 that the performance of conventional H_∞ controllers [54] deteriorates considerably if the effect of saturation is not considered. It can also be observed from Figures 4.7 and 4.8 that the controller designed in [30] can effectively eliminate the effect of saturation but the settling time of the controller is longer compared to the designed controller using a deep reinforcement learning algorithm. It is also seen that both state feedback and output feedback controllers designed using deep reinforcement learning strategy can mitigate the difference in speed between two generators effectively for the IEEE 39-bus power system.

4.6.2   Test system 2: Modified IEEE 68-bus power system

The performance of the designed controller is further verified using a more complex IEEE 68-bus power system. The original system has five areas and 16 generators. A solar power plant of 600 MW and a DFIG-based wind farm of 900 MW is added in bus 59 and 43 of the original power system as shown in Figure 4.9. The parameters of the 68-bus test system can be found in [54].

4.6.2.1   Modal analysis

Eigenvalue analysis of the test system reveals that the 68-bus power system has only four inter-area modes of oscillations. The effect of the integration of solar PV and WECS on the 68-bus test system is shown in Table 4.5 and 4.6 respectively.

Tables 4.5 and 4.6 show that the integration of solar PV plants has a more detrimental effect on the overall damping of the system as compared to WECS. The

Table 4.5   Impact of solar PV incorporation on the second test system

Mode of oscillation	Excluding Solar PV		Including Solar PV	
	IAF	IA ζ	IAF	IA ζ
IA 1	0.4001	0.0464	0.4102	0.0391
IA 2	0.5224	0.0396	0.5218	0.0321
IA 3	0.6104	0.0358	0.6168	0.0289
IA 4	0.8016	0.05	0.8194	0.0442

Table 4.6   Impact of WECS incorporation on the second test system

Mode of oscillation	Excluding WECS		Including WECS	
	IAF	IA ζ	IAF	IA ζ
IA 1	0.4001	0.0464	0.3994	0.0412
IA 2	0.5224	0.0396	0.5226	0.0362
IA 3	0.6104	0.0358	0.5954	0.0324
IA 4	0.8016	0.05	0.7996	0.0461

Table 4.7   Impact of RES incorporation on the second test system

Mode of oscillation	Excluding RES		Including RES	
	IAF	IA ζ	IAF	IA ζ
IA 1	0.4001	0.0464	0.3876	0.0372
IA 2	0.5224	0.0396	0.5189	0.0313
IA 3	0.6104	0.0358	0.6208	0.0264
IA 4	0.8016	0.05	0.8115	0.0392

effect of overall damping of the IEEE 68-bus power system with the integration of both renewable energy sources is shown in Table 4.7.

It is seen from Table 4.7, that the overall damping of the power system reduces considerably with integration of RES.

A comparison of different controllers for RES integrated IEEE 68-bus power station is shown in Table 4.8.

Table 4.8 shows that the designed controller can increase the overall damping of the PV-integrated 68-bus power system. The feedback signal of the damping controller is determined using residue analysis [52] and the feedback signal chosen in this case is the power flowing between buses 13 and 17. The PV-integrated IEEE 68-bus power system has a dynamic order of 136. Schur model reduction technique is used in this case as well to truncate the states of the second test system. The Bode magnitude plot

Table 4.8   *A comparative analysis of various controllers regarding inter-area*
            frequency variation for the second test system

Controller	IA 1 ζ	IA 2 ζ	IA 3 ζ	IA 4 ζ
Without control action	0.0391	0.0321	0.0289	0.0442
Conventional H_∞ control [18]	0.0672	0.0771	0.0682	0.0718
Dynamic control [30]	0.0819	0.0719	0.0714	0.0882
Proposed technique	0.1126	0.0998	0.0868	0.0918

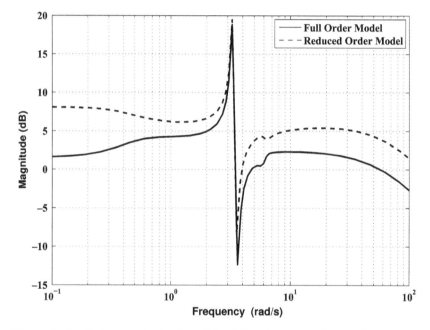

Figure 4.10   Bode magnitude plot of the full and truncated second test system

of the original second test system and truncated system is shown in Figure 4.10. It is
seen that the full-order model and the 8th order model are somehow similar. Therefore,
the reduced order of the RES integrated power system is chosen as 8.

4.6.2.2   Effect of latency in the communication channel

Latency in communication channels plays a vital role in the performance of a
WADC. The effect of latency can be visualized from Figure 4.11 and 4.12. An
examination of Figures 4.11 and 4.12 reveals that when latency is not considered in
the design of the WADC, the peak overshoot and settling time increases tremendously. However, the speed deviation can be effectively suppressed with the
help of a developed controller, which takes latency in the communication channel
into consideration.

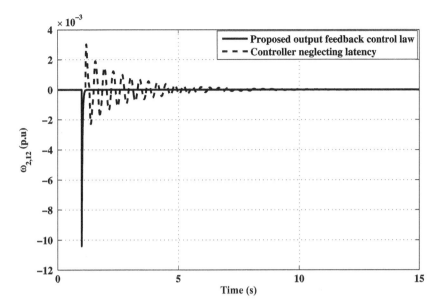

Figure 4.11   Effect of latency on $\omega_{2,12}$

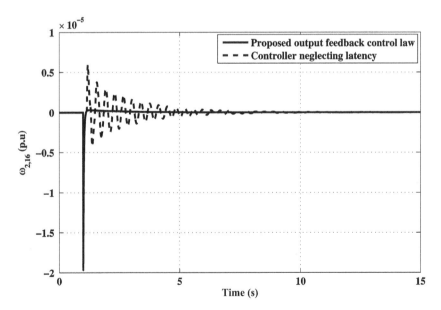

Figure 4.12   Effect of latency on $\omega_{2,16}$

4.6.2.3   Effect of saturation

The repercussions of neglecting the limitations of the actuator output level are presented in Figures 4.13 and 4.14. A careful observation of Figures 4.13 and 4.14

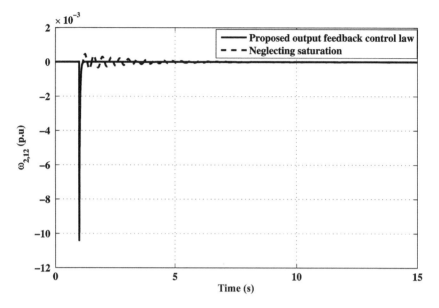

Figure 4.13   Effect of saturation on $\omega_{2,12}$

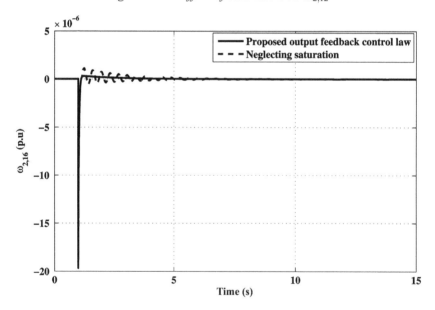

Figure 4.14   Effect of saturation on $\omega_{2,16}$

reveals that the effectuality of the WADC deteriorates tremendously if limitations of actuator output level are neglected in the design stage of the control signal. However, if the limitation is taken into consideration, then the speed deviation can be mitigated effectively under a saturation environment. Also, it can be observed from Figures 4.13 and 4.14 that after taking limitation values into account oscillation and peak overshoot can be reduced.

4.6.2.4 Fault analysis

The potential of the reinforcement learning-based proposed control law is observed for a 68-bus power system when a balanced three-phase to-ground fault takes place at bus 41 with 1,500 MW of power flowing. The superiority of the designed controller can be visualized using Figures 4.15 and 4.16. $\omega_{2,12}$ is the speed deviation between alternator 2

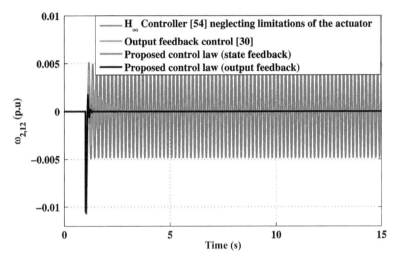

Figure 4.15   Variation of $\omega_{2,12}$ with different control law

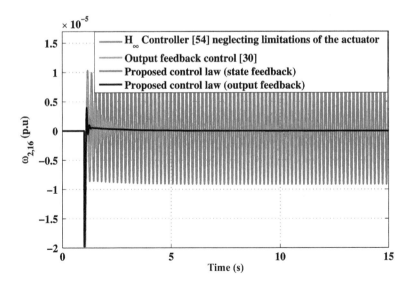

Figure 4.16   Variation of $\omega_{2,16}$ with different control law

and 12, whereas $\omega_{2,16}$ is the speed deviation between alternator 2 and 16. It is seen that the designed state and output feedback-based controller using reinforcement learning technique can effectively mitigate the speed deviation of the 68-bus power system.

4.7 Real-time simulation result

Using OPAL-RT, the proposed controller's efficacy is confirmed in real-time. The simulation model is split for this reason into two subsections. The name of the first subsection is master. In this subsection, only the simulation model of the test system is placed. The name of the second subsection is console. The work of the second subsection is to accumulate all the outputs. The solver used in this study is a fixed-time one. The step time is chosen as 0.001 seconds.

4.7.1   Test system 1

Figure 4.17 illustrates the real-time efficacy of the developed controller for the first test system. The variation of ω_{12} and ω_{15} for the balanced three-phase to-ground fault with the previously mentioned amount of power flow for state and output feedback control law using deep reinforcement learning is shown in Figures 4.17 and 4.18 respectively. The required amount of time for ω_{12} and ω_{15} to settle down within 5% of the final steady-state value is found to be 1.2 and 1.8 seconds for $\omega_{1,2}$ and $\omega_{1,5}$ respectively in real time for both state and output feedback controller.

4.7.2   Test system 2

The productivity of the proposed control law in real-time is also tested for the second test system. The test result is shown in Figures 4.19 and 4.20 respectively

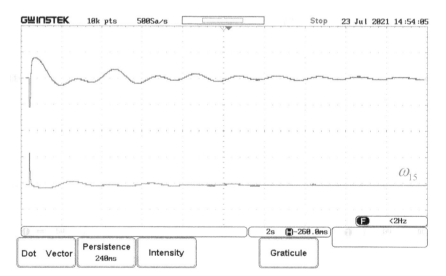

Figure 4.17   Real-time performance of the designed controller on the first test system for state feedback

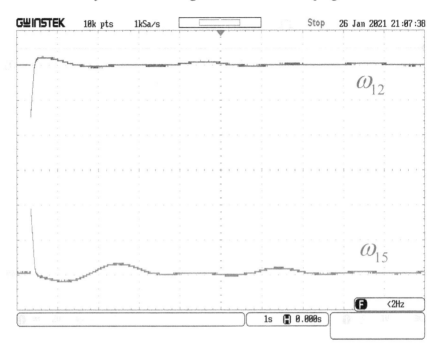

Figure 4.18   *Real-time performance of the designed controller on the first test system for output feedback*

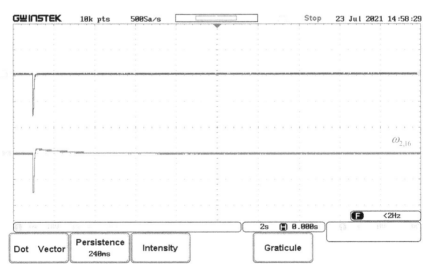

Figure 4.19   *Real-time performance of the designed controller on the second test system for state feedback*

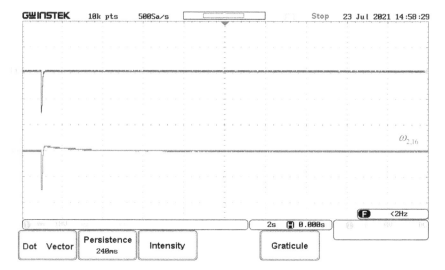

Figure 4.20   Real-time performance of the designed controller on the second test system for output feedback

for state and output feedback control law. The time required for $\omega_{2,12}$ and $\omega_{2,16}$ to settle down within 5% of the final steady-state value is found to be 1.1 and 0.8 seconds respectively in real-time for both state and output feedback control law using deep reinforcement learning.

The usefulness of the proposed control law is examined for both state feedback and output feedback controllers. The result shows the control law can be used for both cases. The advantage of output feedback control law is it eliminates the requirement of full state feedback.

4.8   Conclusion

In this research, taking actuator saturation into consideration, a control signal to remove inter-area oscillation is devised. Effective reduction of inter-area oscillation can be achieved by the proposed control technique, as it has the ability to significantly boost the damping of the interconnected power system. It is compared to the controller designed in [30] to see how effective the designed controller is. The control law is created in this study utilizing deep reinforcement learning techniques. In order to confirm that the reinforcement learning method is effective, both output feedback and state feedback controllers are used. When taking into account the effect of actuator saturation, it is found that the developed controller can stabilize the frequency variation rather successfully when compared to [30]. When constructing the controller, the communication channel's delay is disregarded. As a future project, delay and saturation can be taken into account when constructing the controller.

References

[1] C. Lu, Y. Zhao, K. Men, L. Tu, and Y. Han, Wide-area power system stabiliser based on model-free adaptive control, *IET Control Theory & Applications* 9(13) (2015) 1996–2007.

[2] I. Zenelis, and X. Wang, Wide-area damping control for interarea oscillations in power grids based on PMU measurements, *IEEE control systems letters* 2(4) (2018) 719–724.

[3] S. P. Azad, R. Iravani, and J. E. Tate, Damping inter-area oscillations based on a model predictive control (MPC) HVDC supplementary controller, *IEEE Transactions on Power Systems* 28(3) (2013) 3174–3183.

[4] M. Klein, G. J. Rogers, and P. Kundur, A fundamental study of inter-area oscillations in power systems, *IEEE Transactions on power systems* 6(3) (1991) 914–921.

[5] P. Kundur, N. J. Balu, and M. G. Lauby, *Power System Stability and Control*, Vol. 7, McGraw-Hill, New York, 1994.

[6] A. Patel, S. Ghosh, and K. A. Folly, Inter-area oscillation damping with non-synchronised wide-area power system stabiliser, *IET Generation, Transmission & Distribution* 12(12) (2018) 3070–3078.

[7] M. Zahid, Y. Li, J. Chen, J. Zuo, and A. Waqar, Inter-area oscillation damping and voltage regulation by using UPFC for 500 kv transmission network, in: 2017 2nd International Conference on Control and Robotics Engineering (ICCRE), IEEE, Piscataway, NJ, 2017, pp. 165–169.

[8] A. Chakrabortty, and P. P. Khargonekar, Introduction to wide-area control of power systems, in: 2013 American Control Conference, IEEE, Piscataway, NJ, 2013, pp. 6758–6770.

[9] M. Weiss, B. N. Abu-Jaradeh, A. Chakrabortty, A. Jamehbozorg, F. Habibi-Ashrafi, and A. Salazar, A wide-area SVC controller design for inter-area oscillation damping in wecc based on a structured dynamic equivalent model, *Electric Power Systems Research* 133 (2016) 1–11.

[10] A. Chakrabortty, Wide-area damping control of power systems using dynamic clustering and tcsc-based redesigns, *IEEE Transactions on Smart Grid* 3(3) (2012) 1503–1514.

[11] I. Kamwa, R. Grondin, and Y. Hébert, Wide-area measurement based stabilizing control of large power systems-a decentralized/hierarchical approach, *IEEE Transactions on Power Systems* 16(1) (2001) 136–153.

[12] A. Sengupta, S. K. Pradhan, and D. K. Das, Mitigating inter-area oscillation of an interconnected power system considering time varying delay, in: 2020 IEEE Applied Signal Processing Conference (ASPCON), IEEE, Piscataway, NJ, 2020, pp. 152–157.

[13] H. Wu, K. S. Tsakalis, and G. T. Heydt, Evaluation of time delay effects to wide-area power system stabilizer design, *IEEE Transactions on Power Systems* 19(4) (2004) 1935–1941.

[14]   S. Ghosh, M. S. El Moursi, E. El-Saadany, and K. Al Hosani, Online coherency based adaptive wide area damping controller for transient stability enhancement, *IEEE Transactions on Power Systems* 35(4) (2019) 3100–3113.

[15]   A. Sengupta, and D. K. Das, Delay dependent wide area damping controller with actuator saturation and communication failure for wind integrated power system, *Sustainable Energy Technologies and Assessments* 52(2022) 102123.

[16]   L. Zeng, W. Yao, Q. Zeng, *et al.*, Design and real-time implementation of data-driven adaptive wide-area damping controller for back-to-back VSC-HVDC, *International Journal of Electrical Power & Energy Systems* 109 (2019) 558–574.

[17]   Y. Nie, P. Zhang, G. Cai, Y. Zhao, and M. Xu, Unified smith predictor compensation and optimal damping control for time-delay power system, *International Journal of Electrical Power & Energy Systems* 117 (2020) 105670.

[18]   B. P. Padhy, S. C. Srivastava, and N. K. Verma, A wide-area damping controller considering network input and output delays and packet drop, *IEEE Transactions on Power Systems* 32(1) (2016) 166–176.

[19]   M. E. Bento, Fixed low-order wide-area damping controller considering time delays and power system operation uncertainties, *IEEE Transactions on Power Systems* 35(5) (2020) 3918–3926.

[20]   A. Noori, M. J. Shahbazadeh, and M. Eslami, Designing of wide-area damping controller for stability improvement in a large-scale power system in presence of wind farms and SMES compensator, *International Journal of Electrical Power & Energy Systems* 119(2020) 105936.

[21]   Y. Zhou, J. Liu, Y. Li, C. Gan, H. Li, and Y. Liu, A gain scheduling wide-area damping controller for the efficient integration of photovoltaic plant, *IEEE Transactions on Power Systems* 34(3) (2018) 1703–1715.

[22]   S. Roy, A. Patel, and I. N. Kar, Analysis and design of a wide-area damping controller for inter-area oscillation with artificially induced time delay, *IEEE Transactions on Smart Grid* 10(4) (2018) 3654–3663.

[23]   M. Maherani, I. Erlich, and G. Krost, Fixed order non-smooth robust h_∞ wide area damping controller considering load uncertainties, *International Journal of Electrical Power & Energy Systems* 115(2020) 105423.

[24]   M. E. Bento, A hybrid particle swarm optimization algorithm for the wide-area damping control design, *IEEE Transactions on Industrial Informatics* 18 (1) (2021) 592–599.

[25]   P. Gupta, A. Pal, and V. Vittal, Coordinated wide-area control of multiple controllers in a power system embedded with HVDC lines, *IEEE Transactions on Power Systems* 36(1) (2020) 648–658.

[26]   T. Prakash, V. P. Singh, and S. R. Mohanty, A synchrophasor measurement based wide-area power system stabilizer design for inter-area oscillation damping considering variable time-delays, *International Journal of Electrical Power & Energy Systems* 105 (2019) 131–141.

[27]   F. Z. Taousser, M. E. Raoufat, K. Tomsovic, and S. M. Djouadi, Stability of wide-area power system controls with intermittent information transmission, *IEEE Transactions on Power Systems* 34(5) (2019) 3494–3503.

[28] C. O. Maddela, and B. Subudhi, Delay-dependent supplementary damping controller of tcsc for interconnected power system with time-delays and actuator saturation, *Electric Power Systems Research* 164 (2018) 39–46.

[29] A. Sengupta, and D. K. Das, Mitigating inter-area oscillation of an interconnected power system considering time-varying delay and actuator saturation, *Sustainable Energy, Grids and Networks* 27 (2021) 100484.

[30] C. O. Maddela, and B. Subudhi, Robust wide-area TCSC controller for damping enhancement of inter-area oscillations in an interconnected power system with actuator saturation, *International Journal of Electrical Power & Energy Systems* 105 (2019) 478–487.

[31] M. C. Obaiah, and B. Subudhi, Anti-windup compensator design for power system subjected to time-delay and actuator saturation, *IET Smart Grid* 2(1) (2019) 106–114.

[32] M. C. Obaiah, and B. Subudhi, A delay-dependent anti-windup compensator for wide-area power systems with time-varying delays and actuator saturation, *IEEE/CAA Journal of Automatica Sinica* 7(1) (2019) 106–117.

[33] J. Fang, W. Yao, Z. Chen, J. Wen, and S. Cheng, Design of anti-windup compensator for energy storage-based damping controller to enhance power system stability, *IEEE Transactions on Power Systems* 29(3) (2013) 1175–1185.

[34] H. M. Soliman, and H. A. Yousef, Saturated robust power system stabilizers, *International Journal of Electrical Power & Energy Systems* 73 (2015) 608–614.

[35] M. E. Raoufat, K. Tomsovic, and S. M. Djouadi, Power system supplementary damping controllers in the presence of saturation, in: 2017 IEEE Power and Energy Conference at Illinois (PECI), IEEE, Piscataway, NJ, 2017, pp. 1–6.

[36] D. Roberson, and J. F. O'Brien, Loop shaping of a wide-area damping controller using HVDC, *IEEE Transactions on Power Systems* 32(3) (2016) 2354–2361.

[37] Y. Hashmy, Z. Yu, D. Shi, and Y. Weng, Wide-area measurement system-based low frequency oscillation damping control through reinforcement learning, *IEEE Transactions on Smart Grid* 11(6) (2020) 5072–5083.

[38] M. Glavic, (deep) reinforcement learning for electric power system control and related problems: A short review and perspectives, *Annual Reviews in Control* 48 (2019) 22–35.

[39] T. K. Chau, S. S. Yu, T. Fernando, H. H.-C. Iu, and M. Small, A load-forecasting-based adaptive parameter optimization strategy of STATCOM using ANNs for enhancement of LFOD in power systems, *IEEE Transactions on Industrial Informatics* 14(6) (2017) 2463–2472.

[40] S. Gurung, S. Naetiladdanon, and A. Sangswang, A surrogate based computationally efficient method to coordinate damping controllers for enhancement of probabilistic small-signal stability, *IEEE Access* 9 (2021) 32882–32896.

[41] J. Duan, H. Xu, and W. Liu, Q-learning-based damping control of wide-area power systems under cyber uncertainties, *IEEE Transactions on Smart Grid* 9(6) (2017) 6408–6418.

[42] R. Hadidi, and B. Jeyasurya, Reinforcement learning based real-time wide-area stabilizing control agents to enhance power system stability, *IEEE Transactions on Smart Grid* 4(1) (2013) 489–497.

[43] G. Zhang, W. Hu, D. Cao, *et al.*, Deep reinforcement learning-based approach for proportional resonance power system stabilizer to prevent ultra-low-frequency oscillations, *IEEE Transactions on Smart Grid* 11(6) (2020) 5260–5272.

[44] N. G. Hingorani, and L. Gyugyi, *Understanding FACTS: Concepts and Technology of Flexible Transmission AC Systems*, Wiley-IEEE Press, New York, 2000.

[45] D. Bertsekas, *Dynamic Programming and Optimal Control*, Vol. 1, Athena Scientific, 2012.

[46] Z. Lin, *Low Gain Feedback*, Springer, London, 1999.

[47] S. A. A. Rizvi, and Z. Lin, An iterative Q-learning scheme for the global stabilization of discrete-time linear systems subject to actuator saturation, *International Journal of Robust and Nonlinear Control* 29(9) (2019) 2660–2672.

[48] S. A. A. Rizvi, and Z. Lin, Output feedback reinforcement Q-learning control for the discrete-time linear quadratic regulator problem, in: 2017 IEEE 56th Annual Conference on Decision and Control (CDC), IEEE, Piscataway, NJ, 2017, pp. 1311–1316.

[49] M. Pai, *Energy Function Analysis for Power System Stability*, Springer Science & Business Media, Berlin, 2012.

[50] R. Jabr, B. Pal, N. Martins, and J. Ferraz, Robust and coordinated tuning of power system stabiliser gains using sequential linear programming, *IET Generation, Transmission & Distribution* 4(8) (2010) 893–904.

[51] H. Hasanvand, M. R. Arvan, B. Mozafari, and T. Amraee, Coordinated design of PSS and TCSC to mitigate interarea oscillations, *International Journal of Electrical Power & Energy Systems* 78 (2016) 194–206.

[52] M. Sun, X. Nian, L. Dai, and H. Guo, The design of delay-dependent wide-area DOFC with prescribed degree of stability α for damping inter-area low-frequency oscillations in power system, *ISA transactions* 68 (2017) 82–89.

[53] M. G. Safonov, and R. Chiang, A Schur method for balanced-truncation model reduction, *IEEE Transactions on Automatic Control* 34(7) (1989) 729–733.

[54] B. Pal, and B. Chaudhuri, *Robust Control in Power Systems*, Springer Science & Business Media, Berlin, 2006.

Chapter 5

Application of advanced and robust control schemes with cyber resiliency in microgrid network

Vivek Kumar[1], Pratyush Prateek[2], Soumya R. Mohanty[3] and Nand Kishor[4]

The contemporary electricity grid is currently exploring the potential of dc microgrids to expand their operational capabilities and ranges. One of the primary challenges faced by dc microgrids is achieving optimal current distribution among converters, a task accomplished through droop control mechanisms. The process of current distribution in a dc microgrid gives rise to a voltage disparity, necessitating compensation alongside the current allocation. This chapter delves into the examination of secondary control methods aimed at rectifying voltage fluctuations resulting from converter current distribution. The approach investigated here involves consensus-based sliding mode control (SMC). Fault-tolerant control is established through the creation of an optimal control computation aligned with event-triggering regulations. Additionally, Islanded DC microgrids are progressing into expansive and intricate cyber-physical systems (CPS), encompassing sophisticated intelligent controllers and communication networks layered over the underlying physical structures of power electronic converters and circuit components. These microgrids are susceptible to various forms of cyber-attacks, with the intent of disrupting their regular grid operations. Consequently, the identification and alleviation of such threats emerge as a significant research domain within the realm of cybersecurity for intelligent DC microgrids. This chapter delves into the exploration of an AI (Artificial Intelligence)-driven approach to identify instances of false data injection (FDI) within the CPS environment of microgrids. It also distinguishes the grid in the presence of disturbances for example incremental delays in communication and variation in the load from FDIs. An LSTM auto-encoder network to detect the FDI is utilized. The outlined approach involves the operation and management of microgrids through a consensus control framework,

[1]Electrical and Electronics Department, National Institute of Technology Sikkim, India
[2]Software Engineer-II at Microsoft, Hyderabad, India
[3]Department of Electrical Engineering, Indian Institute of Technology (BHU), India
[4]Faculty of Engineering, Department of Electrical Engineering, Østfold University College, Norway

ensuring smooth functioning under standard conditions, free from any FDIs. To demonstrate the efficacy of the strategy, simulations are performed within the MATLAB®/Simulink® environment, employing a test DC microgrid system.

5.1   Introduction

The DC microgrid has gained substantial attention in recent times as a response to evolving power needs, owing to its dependable performance and capacity to meet power demands [1–2]. Fundamentally equipped with dependable DC distributed sources, for example, photovoltaic and fuel cells, the utilization of power electronic converters is to connect various loads to the microgrid network. Consequently, DC microgrids exhibit reduced harmonic distortion and losses [3–4] in comparison to AC microgrids. These characteristics translate to an elevated quality of power as well as enhanced control abilities [5–7]. Droop control stands out as a remarkably resilient method widely employed to ensure effective current distribution for achieving proper power allocation within DC microgrids [8–10]. Within the realm of droop mechanisms, a virtual resistance is meticulously devised to facilitate optimal current sharing among various DC sources, incorporating well-suited gains. As detailed in [11], the application of high gains as a primary control mechanism can lead to changes in the DC bus values by its designated value. Relying solely on primary control within microgrids, as illustrated in Figure 5.1, can impact both the precision of current sharing and the accuracy of bus voltage. This necessitates the emergence of secondary control principles in DC microgrids, aiming to rectify any breaches in current and voltage distribution. A configuration for secondary control and centralized proportionality is outlined in [12]. While it presents the advantage of robust centralized control performance, it does expose itself to the vulnerability of potential single-point failures [13]. A secondary control scheme elaborated in [14–15], was introduced to guarantee the voltage of DC bus regulation with constrained transient improvement. However, the necessity of transmitting the information to converters introduces potential limitations, possibly hindering a common agreement point if certain conditions are unmet.

A novel secondary control approach, proposed in [16], revolves around averaging all DC source currents, analogous to the concept of average voltage presented in [11]. Nevertheless, the specifics of the averaging techniques in both feed-forward controller designs remain ambiguously outlined. For voltage correction and load sharing, separate voltage and current regulators are employed in [17–18]. While effective, this approach does not assure the exact restoration of the bus voltage. In [19], a secondary control strategy catering to discrete and continuous time components is introduced. Its focus primarily centers on minimizing communication overhead, current sharing, and restoration of bus voltage. However, it does not offer insights into handling actuator faults or reducing control computation efforts. In [20], a control scheme with feedback incorporates a secondary control strategy with consensus laws. It employs a PI controller to achieve consensus for secondary control to attain voltage regulation, but notably the event trigger mechanism is absent. While

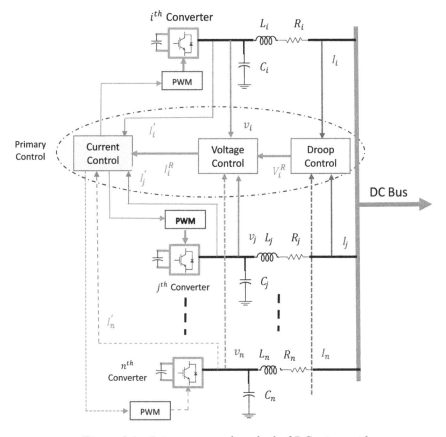

Figure 5.1 Primary control method of DC microgrid

Xing *et al.*'s work [21] briefly touches upon event trigger control strategy, it neglects to address scenarios involving control actuation faults.

DC microgrids offer superior efficiency, reliability, and cost-effectiveness compared to AC microgrids [22–23]. The control structure of DC microgrids involves hierarchical control and droop control-based techniques [24]. However, these methods encounter scalability issues and exhibit suboptimal voltage regulation, attributed to significant transmission line impedances [25–28]. As a remedy, distributed cooperative control emerges as an appealing solution, enhancing voltage regulation while demanding less extensive communication networks. Nevertheless, the inclusion of communication networks in the consensus-based control framework exposes DC microgrids to vulnerabilities associated with cyber-attacks [29]. In the event that a DC microgrid falls victim to an undetected and inadequately mitigated cyber-attack, the entirety of its control and power management mechanisms can be thrown into disarray. Consequently, devising solutions for the recognition of cyber-attacks and establishment of attack-resilient operations in the DC microgrid becomes paramount, ensuring its efficient and dependable functionality [30].

Cyber-attacks manifest in various forms. For instance, intermittent denial of service (DoS) attacks strive to render the network infrastructure inaccessible within the DC microgrid [31]. False Data Injection Attacks (FDIs) aim to manipulate the system's state by injecting fabricated data into sensors or communication channels [32–34]. A subset of DoS attacks, jamming, entails disrupting communication [35]. Replay cyber-attacks deceive operators by recording sensor readings for a specific timeframe and then replaying these readings within the system [36]. The presented study narrows its focus to FDIs, as they constitute the most frequently reported form of cyber-attacks. FDIs disrupt the steady-state operation of DC microgrids through the introduction of a signal, referred to as an attack vector. The creation of this attack vector is orchestrated by an intruder with access to either knowledge about the communication infrastructure or the sensors themselves. Prior research on detecting FDIs in DC microgrids has employed various techniques. Some studies, such as [37] and [38], utilized Kalman Filters for state estimation, measuring discrepancies between real measurements and estimates to uncover FDIs. A method utilizing the Kullback-Leibler distance was introduced in [39] to estimate differences in sensor data distributions between healthy and attacked systems. In [40], the implementation of a Chi-square detector was demonstrated for FDI detection. Works like [41] and [42] proposed forecasting and machine-learning-based approaches, while a generalized likelihood ratio was applied for state estimation in [43] to identify FDIs.

However, certain strategies, for instance, the one presented in [40], encountered challenges in detecting FDIs on specific state variables. Furthermore, some methods failed to consider all state variables within an integrated framework. For example, Sahoo *et al.* [44] focused on detecting attacks on voltage sensors, while [45] addressed attacks on current sensors. Machine learning and deep learning methodologies were also introduced in the detection of FDIs, as seen in [46] and [47], where behavior patterns were recognized based on historical data for future identification. This chapter introduces a novel approach, showcasing how particular types of neural networks can discern between FDIs and normal grid conditions without necessitating extensive statistical or mathematical feature analysis. The proposed framework leverages long short-term memory (LSTM) autoencoder networks' learning-reconstruction capabilities to detect FDIs in DC microgrids. The framework relies on a robust DC microgrid model, incorporating a distributed cooperative control topology, to extract training data. Since LSTM autoencoders inherently capture crucial features by transforming high-dimensional data into lower dimensions, separate feature extraction, and dimensionality reduction techniques are unnecessary. Notably, the method does not demand any *a priori* knowledge of system parameters or mathematical models, effectively sidestepping additional implementation and computational complexity.

Within this chapter, a comprehensive methodology is introduced for formulating a distributed secondary control strategy tailored for DC microgrids. The approach is meticulously crafted to guarantee the restoration of DC bus voltage while simultaneously establishing stable and equitable current distribution among power converters. The innovation extends further by alleviating the communication and control computation load, accomplished through strategic utilization of triggering action. A resilient SMC-based consensus scheme is considered to fulfill the requirements of secondary control design.

The key contributions highlighted in the chapter primarily center around the following focal points:

1. *Enhanced voltage regulation and adequate current sharing:* The chapter introduces a novel approach where the average virtual voltage droop, derived from the combination of line resistance and droop gain, is employed. This innovative approach employs the sliding mode technique to ensure effective voltage restoration while maintaining a balanced current distribution.
2. *Fault-tolerant distributed secondary control with event triggering:* The chapter presents a robust fault-tolerant distributed secondary control mechanism. This control strategy is seamlessly integrated with an event-triggering system, operating under a sliding mode action framework. Notably, this control configuration effectively handles actuation faults within the DC microgrid system.
3. *FDI attack detection and mitigation via LSTM-based autoencoder and decoder:* The chapter introduces an approach for the detection and mitigation of FDI attacks. It is obtained through the utilization of LSTM autoencoder and decoder architectures. These neural network structures play a crucial role in identifying anomalous data patterns indicative of FDI attacks and subsequently mitigating their effects.

5.2   Dynamics of DC microgrid

5.2.1   Modeling the cyber layer

Each agent (converter) can exchange data between neighbors according to the cyber layer. Let $V_g = \{v_1^g, v_2^g, v_3^g \cdots v_N^g\}$ represent nodes connected by edges $E_g \subset V_g \times V_g$. Let $A_g = [a_{ij}]$ be the time-invariant adjacency matrix of the given graph containing the communication weights. a_{ij} is the allocated weight to the data exchange from node j & i. Also, $a_{ij} > 0$ if $\left(v_j^g, v_i^g\right) \in E_g$ and $a_{ij} = 0$ otherwise. $N_i = \left\{ j \mid \left(v_j^g, v_i^g\right) \in E_g \right\}$ denotes the node ith neighbors, i.e., if $j \in N_i$, then ith node will receive information from jth node. The cyber network spanned across the DC microgrid is a digraph, i.e., the links are not necessarily reciprocal. $\left(v_j^g, v_i^g\right) \in E_g$ does not imply that $\left(v_i^g, v_j^g\right) \in E_g$. Therefore, we define the in-degree matrix of E_g denoted by $D_G^{in} = diag\{d_i^{in}\}$, where $d_i^{in} = \sum_{j \in N_i} a_{ij}$. Likewise, the out-degree matrix is denoted by $D_G^{out} = diag\{d_i^{out}\}$, where $d_i^{out} = \sum_{i \in N_j} a_{ji}$. Converters can transmit their measured control variables $\varphi_i = [\bar{v}_i, v_i, i_i^{pu}]$ to its neighbors. Here $\bar{v}_i$ denotes the averaged voltage value of the microgrid, v_i is the local voltage of the concerned converter connected to the bus that is the output of ith converter. i_i^{pu}

refers to the converter current divided by its rated current. The communication links are simulated as low-pass filters to replicate the inherent delays present in the data exchange process, analogous to the approach employed in [34].

The method for the detection of FDIs has two main steps:

1. **Offline training and analysis of the reconstruction errors:** In this phase, we gather enough training data for the offline training of the LSTM autoencoder. It is essential to select appropriate variables for efficient training and implementation of these networks. Two error norms are defined for this purpose, for each distributed energy resources (DER) unit. They are based on the weighted difference between the states of a DER unit and its neighbors. The mismatch of these norms is used to train the LSTM autoencoder networks. After training and fine-tuning, the networks are used to reconstruct the mismatch between the two norms. In the process, the distribution of the reconstruction errors of the mismatches is analyzed and a threshold is selected for each DER unit.
2. **Online detection:** After the first step, the trained networks are used to estimate the mismatch of norms for each DER unit. The estimation error is scored under the distribution for the given DER unit. If this score is greater than the threshold, the system is under FDI.

5.2.2   *Droop scheme for DC microgrid operation*

Figure 5.1 depicts the primary droop method setup, featuring 'N' DC-DC converters organized in the parallel connection alongside the Bus. The primary role of the droop method is to facilitate the decentralized sharing of current in all converters within the islanded DC microgrid. This approach allows for effective current distribution. However, it is important to note that droop control has inherent limitations, particularly in terms of voltage deviation. These limitations regarding voltage variations are succinctly highlighted within this section.

In Figure 5.1 primary control strategy is explained to achieve proper current sharing among converters. Where the control scheme is properly designed with the desired reference V_i^R of DC voltage having V_i as output of the converter's voltage [19]. Under droop scheme, reference voltage is mathematically represented as $V_i^R = V^* - K_i I_i$, with I_i, K_i and V^* respectively denoting current of the ith output converter, gain of the droop, and the reference open-loop DC voltage. Here V_{bus} represents the DC bus voltage as $V_{bus} = V_i - I_i R_i$ or $V_{bus} = V^* - I_i(K_i + R_i)$. Consequently, the DC voltage control across each converters will satisfy: $I_i(K_i + R_i) = I_j(K_j + R_j) = \cdots = I_n(K_n + R_n)$. The aforementioned equation establishes accurate sharing of current between the converters through the significant gain in the droop, i.e., $(K_i \gg R_i)$. Consequently, the ratio $I_i/I_j = K_j/K_i$ holds, and the droop gain serves as a "virtual resistance." This configuration ensures precise current sharing among the converters.

5.2.3   *DC microgrid control essentials*

The control objective encompasses the simultaneous realization of voltage bus regulation and adequate current sharing, which experiences deviation due to a substantial droop gain. Therefore, the primary aim of this chapter is to formulate a robust and fault-tolerant secondary control strategy that optimally utilizes control computation resources across multiple agent converters. This is realized by addressing the fundamental requirements of the DC microgrid:

1. Attaining proper current sharing within a finite timeframe, characterized by $\lim_{t \to t_f} I_i/I_j = K_j/K_i$.

2. Ensuring precise DC bus voltage regulation, expressed as $\lim_{t \to t_f} (V_B - V^*) = 0$.

One of the chapter's core focuses revolves around developing a control approach that fulfills these essential DC microgrid criteria while maintaining fault tolerance and robustness and optimizing computational resources for multiagent converters.

5.3   Secondary control of DC microgrid

To meet the primary scheme objectives (1–2), a control signal U_i is incorporated into the primary control scheme, considering that the converter's sharing of current is headed by droop control (Figure 5.2). Specifically, $V_i^R = V^* - K_i I_i + U_i$ leading to $V_{bus} = V^* - I_i(K_i + R_i) + U_i$. The design of the secondary control U_i becomes feasible when global information from all converters is available. However, an alternate approach is required to design scheme for secondary control that remains effective in the absence of comprehensive converter information. Hence, the secondary control is structured around the notion of "virtual voltage drop" $\{\overline{V}_i = I_i(K_i + R_i)\}$ for each converter. This approach defines $U_i = \overline{V}_a = \frac{1}{N}\sum_{i=1}^{N} \overline{V}_i$ where $\overline{V}_a$ is the mean value of $\overline{V}_i$. This way, the secondary scheme is designed with

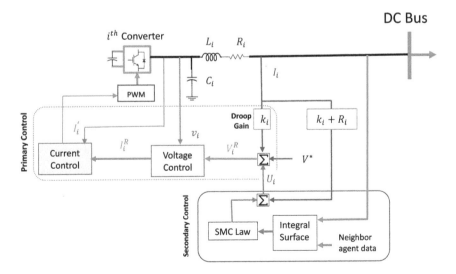

Figure 5.2   Multiagent DG-based secondary control of DC microgrid

the consideration of time-varying global parameter $\overline{V}_i$ availability [21]. However, this method necessitates global information of $\overline{V}_i$ for computation of $\overline{V}_a$. While this approach suits small-scale system configurations, it becomes problematic for larger-scale systems featuring multiple converters, where the availability of such global information $\overline{V}_i$ is often limited [21]. This challenge serves as the motivation behind the development of a consensus-based secondary control technique. This technique aims to address the complexities arising from the lack of global signal information, providing a more practical solution for large-scale systems.

5.3.1   Advance secondary control design using SMC scheme

The essence of the secondary control Equation (5.9) hinges on the availability of global information $\overline{V}_i$. However, in scenarios where this global information is inaccessible, the realization of Equation (5.9) becomes unattainable. This predicament can be effectively addressed by devising a secondary control strategy, granting the converter's capability in order to compute $\overline{V}_i$ through data exchange solely among neighboring converters within a multiagent context. This innovative approach empowers the calculation of $\overline{V}_i$ as:

$$
\begin{aligned}
\overline{V}_i(t) - V_i(t) &= -\int_0^t \sum_{j \in N_i} a_{ij}\left(\overline{V}_i(\tau) - \overline{V}_j(\tau)\right)d\tau \\
\dot{\chi}_i^v(t) &= -6 \sum_{j \in N_i} a_{ij}\left(\overline{V}_i(\tau) - \overline{V}_j(\tau)\right)
\end{aligned}
\tag{5.1}
$$

$$
U_{ci} + d_i(t) = \chi_i^v(t) + V_i(t)
\tag{5.2}
$$

Here $6 > 0$ and 'χ_i^v' is a transitional variable. Now consensus error 'e_i' can be obtained as [21],

$$
\tilde{e}_i = U_{ci} + d_i(t) - V_i(t)
\tag{5.3}
$$

and

$$
\dot{\tilde{e}}_i = \dot{V}_i - \dot{V}_i^{avg} - 6 \sum_{j \in N_i} a_{ij}\left(\tilde{e}_i - \tilde{e}_j\right)
$$

or

$$
\dot{\tilde{e}}_i = -6\pounds\tilde{e} + Л_N \dot{V}_i
\tag{5.4}
$$

Here, $Л_N = I_N - \frac{1}{N} 1_N 1_N^T$.

Assumption 1: The $K_i(t), i = 1, \dots N$, as droop gain derivative is bounded. Surface manifold for secondary control application can be considered as:

$$
s_i(t) = \tilde{e}_i(t) + \int \tilde{e}_i(t)dt
\tag{5.5}
$$

$$
\dot{s}_i(t) = \dot{\tilde{e}}_i(t) + \tilde{e}_i(t) = -6\pounds\tilde{e} + Л_N \dot{V}_i + \tilde{e} = (I_N - 6\pounds)\tilde{e} + Л_N \dot{V}_i.
\tag{5.6}
$$

Here $s_i(t) = [s_1(t), s_2(t), \dots s_n(t)]^T$. Taking the value of $\tilde{e}_i$ from (5.3),

$$
\dot{s}_i(t) = (I_N - 6\pounds)(U_{ci} - V_i + d_i(t)) + Л_N \dot{V}_i
$$

Or

$$\dot{s}_i(t) = (I_N - 6\pounds)U_{ci} - (I_N - 6\pounds)V_i + (I_N - 6\pounds)d_i(t) + \text{Л}_N\dot{V}_i$$
$$\dot{s}_i(t) = \text{Л}_N\dot{V}_i + \Theta^{-1}V_i - \Theta^{-1}U_{ic} - \Theta^{-1}d_i(t) \tag{5.7}$$

Here, $\Theta = (6\pounds - I_N)^{-1}$. Hence, the consensus-based control input can be achieved as:

$$U_{ic} = \Theta\varkappa_i(s_i) + \Theta\text{Л}_N\dot{V}_i + V_i - d_i(t) \tag{5.8}$$

with

$$\varkappa_i(s_i) = \beta_i\text{sign}(s_i(t)), \quad \beta_i = \text{diag}\{\beta_{11}, \beta_{12}, \ldots \beta_{1n}\}, \quad \text{sign}(s_i(t)) = [\text{sign}(s_1(t)),$$
$$\text{sign}(s_2(t)) \ldots \text{sign}(s_n(t))]^T.$$

Adaptive law, $\dot{\beta}_i = \delta(\|s_i\| - k\beta_i)$ with discontinuous tracking protocol is considered as [24].

$$\varkappa_i(s_i) = \begin{cases} -\beta_i\dfrac{s_i}{\|s_i\|}, & if\,\beta_i\|s_i\| \geq \xi \\ -\beta_i^2\dfrac{s_i}{\xi} & if\,\beta_i\|s_i\| < \xi \end{cases} \tag{5.9}$$

where $\delta > 0, k > 0, \beta_i(0) > 0$ & $\xi > 0$.

5.3.2 *Event trigger clause derivation*

For the control effort optimization event triggering condition in stable control operation is designed by through following Lyapunov function:

$$V_{1i}(t) = \frac{1}{2}s_i^T(t)s_i(t) \tag{5.10}$$

$$\dot{V}_{1i}(t) = s_i^T(t)\dot{s}_i(t) \tag{5.11}$$

Now $\dot{s}_i(t)$ can be formulated in terms of triggering error $\varsigma_i(t)$. Here $\varsigma_i(t) = (\varkappa_i(s_i(t)) + \Theta^{-1}V_i(t) + \text{Л}_N\dot{V}_i(t)) - (\varkappa_i(s_i(t_k)) + \Theta^{-1}V_i(t_k) + \text{Л}_N\dot{V}_i(t_k))$. Hence $\dot{s}_i(t) = \varsigma_i(t) - \varkappa_i(s_i(t)) + \tilde{D}_i$. Here $\tilde{d}_i(t) = d_i(t) - d_i(t_k)$. $\|\Theta^{-1}\tilde{d}_i(t)\| \leq \tilde{D}_i$, and $\tilde{D}_i$ is considered as an unmodeled disturbance of *i*th DC-DC Converter.

Putting $\dot{s}_i(t)$ in (5.11):

$$\dot{V}_{1i}(t) = s_i^T(t)[\varsigma_i(t) - \beta_i\text{sign}(s_i(t)) + \tilde{D}_i] \tag{5.12}$$

Or

$$\dot{V}_{1i}(t) \leq \|s_i(t)\|\|\varsigma_i(t)\| - \lambda_{\min}(\beta_i)\text{sign}(s_i(t))s_i^T(t) + \|s_i(t)\|\tilde{D}_i$$

Here, $\lambda_{\min}(\beta_i)$ represents the minimum eigenvalue of β_i matrix. Therefore, the event trigger condition is designed as:

$$\|\varsigma_i(t)\| \leq \beta_0 \tag{5.13}$$

where, $\beta_0 < (\lambda_{\min}(\beta_i) - \tilde{D}_i)$ & $\dot{V}_{1i}(t) = -\beta_0\|s_i(t)\|$.

5.3.3   Fault tolerance control and event trigger clause

The fault tolerant control scheme consists of the uncertainties accumulated with the control input and affects the system performance. To overcome such faults in the system the following analysis is required [32].

A secondary control input $U_c(t)$ denotes the actuator's actual output as below:

$$U_c(t) = (1 - \varphi_i(t))U_i + \omega_i(t) \tag{5.14}$$

In the presence of faults within the ith Distributed Generation (DG) unit, the real output $U_c(t)$ encompasses the formulated U_i control, which includes $\varphi_i(t)$ as a loss of partial effectiveness in ith DG and $\omega_i(t)$ as a biased component. It is worth noting that for real-world systems, disturbances and faults are generally confined within certain boundaries.

Assumption 2: $\varphi_i(t)$ & $\omega_i(t)$ are bounded as $\|\varphi_i(t)\| \leq \overline{\varphi}$ and $\|\omega_i(t)\| \leq \overline{\omega}$.

This section focuses on establishing stability constraints during actuation faults using the Lyapunov stability criteria. By adhering to these stability conditions, control triggering rules are derived. Furthermore, the section proceeds to achieve finite-time control under the sliding mode law (5.8), augmented by an adaptive arrangement (5.9) to accommodate actuation faults. The validity of the approach is substantiated using the following Lyapunov function:

$$V_{3i} = \frac{1}{2}s_i^T s_i + \frac{1}{2\delta}\sum_{i=1}^{n}(\beta_i - \overline{\beta})^2 \tag{5.15}$$

here $\overline{\beta}$ denotes the upper bound of β_i. Now event trigger condition during actuation fault condition can be obtained as:

$$\|\varsigma_i(t)\| < (\overline{\beta} - \tilde{D}_i) \tag{5.16}$$

5.3.4   Even trigger calculation with inter-execution time estimation

The comprehensive triggering scheme is elucidated through the flowchart depicted in Figure 5.3. Within the triggering condition, the "inter-execution time" signifies

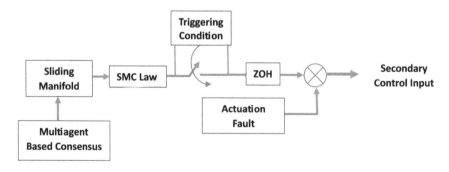

Figure 5.3   Event trigger action

the duration of two triggering moments successively. Regulation of the bus voltage is accomplished via the application of input (5.8), which is renewed instantaneously based on the event-trigger condition deduced in (5.13). Consequently, the value of T_i is expressed in accordance with (5.17).

$$T_i = \frac{1}{\bar{\beta}(n\gamma)} \ln\left(\frac{(\bar{\beta} - \tilde{D}_i)\bar{\beta}n\gamma + 1}{G}\right) \qquad (5.17)$$

Proof: Inter execution can be derived using error differentiation with respect to time.

$$\frac{d}{dt}\|\varsigma_i(t)\| \leq \left\|\frac{d}{dt}\varsigma_i(t)\right\| \leq \left\|\frac{d}{dt}\left(\Theta^{-1}V_i(t) + \beta_i \mathrm{sign}(s_i(t)) + \mathrm{J}_N \dot{V}_i(t)\right)\right\|$$

As $V_i(t)$ & $\dot{V}_i(t)$ are bounded.

$$\frac{d}{dt}\|\varsigma_i(t)\| \leq \bar{\beta}\|[1 - \tanh^2(\gamma s_i(t))]\gamma \dot{s}_i(t)\|$$

Here $\mathrm{sign}(s_i(t))$ is equivalent to $\tanh(\gamma s_i(t))$ with $(\gamma \gg 1)$, and $\|1_{n\times n} - \tanh^2(\gamma s_i(t))\| \leq \|1_{n\times n}\| = n$.

$$\frac{d}{dt}\|\varsigma_i(t)\| \leq (\bar{\beta}n\gamma)\|\dot{s}_i(t)\|$$

$$\frac{d}{dt}\|\varsigma_i(t)\| \leq (\bar{\beta}n\gamma)\|\varsigma_i(t) - \beta_i \mathrm{sign}(s_i(t)) + \tilde{D}_i\|$$

$$\|\dot{\varsigma}_i(t)\| - (\bar{\beta}n\gamma)\|\varsigma_i(t)\| \leq (\bar{\beta}n\gamma)\|(-\beta_i \mathrm{sign}(s_i(t)) + \tilde{D}_i)\|$$

For $t \in [t, t_k]$, a solution to the above inequality can be obtained by Lemma (3.4) described in the [Khalil, 2002 book] as:

$$\|\varsigma_i(t)\| \leq \frac{G\left(\exp^{\bar{\beta}n\gamma(t-t_k)} - 1\right)}{\bar{\beta}n\gamma} \qquad (5.18)$$

Here, $G = \|(\tilde{D}_i - \beta_i \mathrm{sign}(s_i(t)))\|$
From (5.16) and (5.18).

$$(\bar{\beta} - \tilde{D}_i) \leq \frac{G\left(\exp^{\bar{\beta}n\gamma(t-t_k)} - 1\right)}{\bar{\beta}n\gamma} \qquad (5.19)$$

Inter-execution time is therefore obtained as in (5.17).

5.3.5   Stability condition under actuation fault

Stability proof is provided by taking the derivative of Lyapunov candidate $\mathcal{V}_{3i}$ given in (5.15), as:

$$\dot{\mathcal{V}}_{3i} = s_i^T \dot{s}_i + \sum_{i=1}^{n} \frac{1}{\delta}(\beta_i - \bar{\beta})\dot{\beta}_i$$

$$\dot{\mathcal{V}}_{3i} \leq (\varsigma_i(t) + \tilde{D}_i)\|s_i\| - s_i^T G_c(s_i) + \sum_{i=1}^{n} (\beta_i - \bar{\beta})(\|s_i\| - k\beta_i)$$

Case 1: if $\beta_i\|s_i\| \geq \xi \; \forall i = 1, \ldots n$, $s_i^T G_c(s_i) = \sum_{i=1}^{n} \beta_i\|s_i\|$

Then

$$\dot{V}_{3i} \leq \left(\varsigma_i(t) + \tilde{D}_i\right)\|s_i\| - \sum_{i=1}^{n}\overline{\beta}\|s_i\| - k\sum_{i=1}^{n}(\beta_i - \overline{\beta})\beta_i$$

$$\dot{V}_{3i} \leq -(\overline{\beta} - \varsigma_i(t) - \tilde{D}_i)\|s_i\| + \frac{k}{4}\sum_{i=1}^{n}\overline{\beta}^2$$

As $(\beta_i - \overline{\beta})\beta_i\big|_{\max} = -\frac{\overline{\beta}^2}{4}$, when $\beta_i = \frac{\overline{\beta}}{2}$.

$$\dot{V}_{3i} \leq -\beta_0\|s_i\| + \psi_1$$

where $\psi_1 = \frac{nk}{4}\overline{\beta}^2$

Case 2: if $\beta_i\|s_i\| < \xi \; \forall i = 1, \ldots n$, $s_i^T G_c(s_i) = -\sum_{i=1}^{n}\frac{(\beta_i^2)}{\xi}\|s_i\|^2$.

$$\dot{V}_{3i} \leq \left(\varsigma_i(t) + \tilde{D}_i\right)\|s_i\| - \sum_{i=1}^{n}\frac{(\beta_i^2)}{\xi}\|s_i\|^2 + \sum_{i=1}^{n}(\beta_i - \overline{\beta})(\|s_i\| - k\beta_i)$$

$$\dot{V}_{3i} \leq -\beta_{01}\|s_i\| + \psi_1 + \sum_{i=1}^{n}\left(-\frac{(\beta_i^2)}{\xi}\|s_i\|^2 + \beta_i\|s_i\|\right)$$

Since $\beta_i\|s_i\| < \xi$, $\quad -\frac{(\beta_i^2)}{\xi}\|s_i\|^2 + \beta_i\|s_i\|\big|_{\min} = \frac{\xi}{4}$ when $\beta_i\|s_i\| = \frac{\xi}{2}$ for $\forall i = 1, \ldots n$.

$$\dot{V}_{3i} \leq -\beta_0\|s_i\| + \psi_2$$

where $\psi_2 = \psi_1 + \frac{nk}{4}\overline{\beta}^2$

Case 3: if $\beta_i\|s_i\| \geq \xi$ for a few converters, whereas $\beta_i\|s_i\| < \xi$ for the rest of the converters.

$$\dot{V}_{3i} \leq -\beta_0\|s_i\| + \psi_3$$

where $\psi_1 \leq \psi_3 \leq \psi_2$

By combining cases 1, 2, and 3 the $(\dot{V}_{3i})$ can meet the following norms:

$$\dot{V}_{3i} \leq -\beta_0\|s_i\| + \jmath$$

$$\dot{V}_{3i} \leq -(1 - \partial_0)\beta_0\|s_i\|$$

Here, $\|s_i\| \geq \frac{\jmath}{\partial_0\beta_0}$.

Also, $\jmath = \max\left(\psi_1, \frac{n\delta}{4}\overline{\beta}^2, n\frac{\xi}{4}\right)$, $0 < \partial_0 < 1$ so as to s_i will attain Ψ_i in the presence of actuation disturbance of $\Psi_i = \|s_i\| \leq \frac{\jmath}{\partial_0\beta_0}$ in finite time (Figure 5.4).

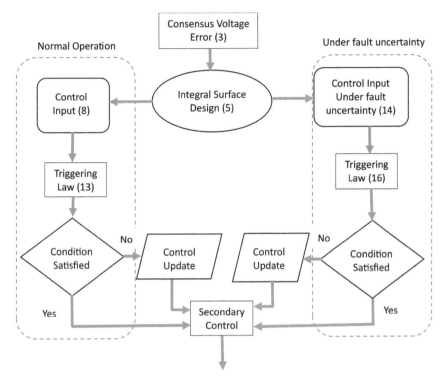

Figure 5.4 Flow chart describing triggering effect

5.4 LSTM autoencoder networks

Artificial neural networks (ANNs) and deep learning techniques are used in smart grids for a variety of applications. For instance, in [33] the ANN is applied to forecast the PV unit's output. In [34], a standalone DC microgrid is controlled by a scheme of ANN integrated with a droop mechanism. In [35], RNN is used to determine FDIs in microgrid systems. As shown in the above techniques, ANNs can be used where the output depends only on the input, hence the relationship is static. On the other hand, when the relationship between input and output is non-static, i.e., the output depends upon input and the historic value of outputs, the recurrent neural network and its variants become a good choice as they have an in-built memory state embedded within them.

The LSTM autoencoders are an encoder-decoder architecture that is highly capable of learning and reconstructing sequences. The LSTM encoder realizes the input sequence vector depictions and in a similar fashion LSTM decoder applies these depictions to estimate the sequence. In this chapter, as stated earlier, the first step is the offline training of the LSTM autoencoder networks. In the later sections, it will be discussed how a well-trained LSTM autoencoder is used to distinguish the cyber-attack conditions from the normal grid conditions including change in load conditions and increased communication delays. It also described that the trained architecture can detect cyber-attacks with minimal weights, i.e., the attacks that are supposed to bypass the FDI detection framework. The architecture and components of the LSTM autoencoder are discussed next.

5.4.1 LSTM network

The LSTM is basically an RNN alternative that efficiently prevails over the vanishing gradients in the naïve RNN [36]. The network of LSTM includes memory cells which consist of the sequence summary of past input. It also consists of the gating control mechanism through which the information flows among the cell memory, input, and output. The construct of LSTM is provided in Figure 5.5. The working of LSTM can be described by the following recursive equations:

$$f_t = \sigma\left(W_{uf}x_t + W_{hf}h_{t-1} + b_f\right) \tag{5.20}$$

$$i_t = \sigma(W_{ui}x_t + W_{hi}h_{t-1} + b_i) \tag{5.21}$$

$$o_t = \sigma(W_{uo}x_t + W_{ho}h_{t-1} + b_o) \tag{5.22}$$

$$c_t = f_t \odot c_{t-1} + i_t \odot \tanh\left(W_{uc}x_t + W_{hc}h_{t-1} + b_c\right) \tag{5.23}$$

$$h_t = o_t \odot \tanh\left(c_t\right) \tag{5.24}$$

where

- $\sigma(x) = 1/(1 + e^{-ax})$: Sigmoid function (elementwise).
- $x \odot y$: Element-wise product.
- $W_{hf}, W_{uf}, W_{ui}, W_{hi}, W_{uo}, W_{ho}, W_{uc}, W_{hc}$: Linear transformation matrices.
- b_i, b_f, b_o, b_c : Bias vectors.
- i_t, f_t, o_t : Gating vectors.
- c_t : Vector states of cell memory.
- h_t : Output vector states.
- u_t : Input vectors.

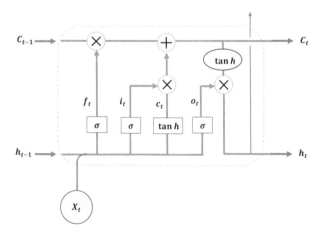

Figure 5.5   Structure of an LSTM cell

Functions for the memory cell which are updating of state, forgetting the state, and returning the state are obtained by the vector gates provided in (5.20), (5.21), and (5.22). The states of output and cell are renewed as per (5.23) and (5.24). The state of the cell is restored or reset as per the gating sequence of the forget vector f_t. Rest of other two vector gates are i_t & o_t.

5.4.2 LSTM autoencoder networks

The structure of an LSTM autoencoder network is shown in Figure 5.6. The construction works with two LSTM configurations encoder and decoder. Let input to the encoder network be $U = \{u_1, u_2, u_3 \cdots u_T\}$. The encoder realizes a fixed length of depiction sequence and the decoder applies these depiction sequences to rebuild the sequence of input as $S = \{s_1, s_2, s_3 \cdots s_T\}$ in reverse order. Both encoder and decoder are commonly trained to rebuild the output. During training, the input u_t and similarly the states of hidden layer h_{t-1} of the earlier LSTM layer are used to compute the state of hidden layer h_t of next LSTM layer using (5.21), (5.22), (5.23), and (5.24). The last hidden state c_T from encoder is applied as the first state c_0' for the initial LSTM layer of the decoder. A linear layer with a weight matrix $w \in \mathbb{R}^{c \times 1}$, where c denotes the LSTM units in LSTM layers of the decoder circuit, and biasing vector $b \in \mathbb{R}$ is applied to compute s_t given by:

$$s_t = w^t h_t' + b \tag{5.25}$$

The next decoder's LSTM layer uses the present hidden state h_t' and the output s_t to achieve the next state of hidden layer h_{t-1}' (output sequence is constructed in reverse order) using (5.21), (5.22), (5.23), and (5.24). The training of the model is done in order to minimize c, an objective function given by:

$$C = \sum_{U \in X} \sum_{t=1}^{T} |s_t - u_t| \tag{5.26}$$

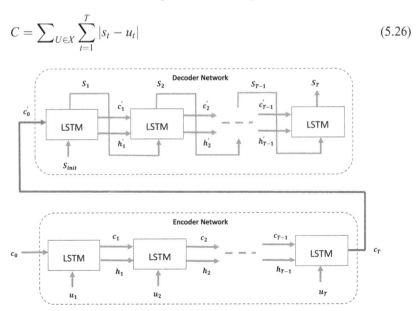

Figure 5.6   Architecture of a LSTM autoencoder

where $u_t \in U$, the set of training sequences for a given time-series X. This is achieved by optimizing the weights and the bias matrices in (5.21), (5.22), (5.23), and (5.24) for each LSTM and output layer.

5.5 Method and modeling of FDIs

5.5.1 FDI scenario formulation

The converters in a DC microgrid share their local states with their neighbors and update their local states by processing their own and neighbors' states, as we have seen in previous sections. These states are vulnerable to FDI. FDI can be targeted in two ways. In a Link FDI scenario, the communication link between the two converters is targeted, affecting either both the converters in the process (if the link is bidirectional) or one among the two converters. Link attack can be formulated as:

$$X_j^k = X_j^k + \mu_{jk}\chi \tag{5.27}$$

where X_j^k is the state variable of converter k received by converter j and $k \in N_j$. χ is the corrupted or false data added to modify the state variable. $\mu_{jk} \in \{0, 1\}$, and $\mu_{jk} = 1$ signifies the occurrence of an attack. In a node FDI scenario, an individual converter is targeted, thus affecting all its neighbors. Node FDI can be modeled in two ways. Initially, corrupted data is intruded in the attacked state variables as:

$$X_j = X_j + \mu_j\chi \tag{5.28}$$

Secondly, the corrupted data entirely replaces the attacked state variables as:

$$X_j = \left(1 - \mu_j\right)X_j + \mu_j\chi \tag{5.29}$$

where X_j is the attacked state variable of the converter j. $\mu_j \in \{0, 1\}$, and $\mu_j = 1$ signifies the occurrence of an attack. The second attack model is known as controller hijacking [29].

5.5.2 FDI detection methodology

For each converter j, the weighted norm of the state mismatch is defined as:

$$\delta_j(t) = C_j \left\| \sum_{k \in N_j} a_{jk}\left(x_k - x_j\right) \right\| \tag{5.30}$$

Another weighted norm is defined as:

$$\sigma_j(t) = C_j \sum_{k \in N_j} a_{jk}\left\| x_k - x_j \right\| \tag{5.31}$$

In (5.31), x_j and x_k are state variables and C_j denotes the local confidence factor [29] of the converter j given by:

$$\dot{C}_j(t) = \alpha_j\left(d_j - C_j(t)\right) \tag{5.32}$$

where, $\alpha_j > 0$ is a factor that permits the normal grid transients like load changes and d_j is given by:

$$d_j = \frac{\Delta_j}{\Delta_j + \left\| \delta_j(t) - \sigma_j(t) \right\|} \tag{5.33}$$

where, $\Delta_j > 0$ is another temporal factor that is selected according to the minimum attack weight that can lower the confidence factor of the converters. The mismatch of norms for converter j is given by:

$$M_j(t) = \left\| \delta_j(t) - \sigma_j(t) \right\| \tag{5.34}$$

Let $s_m^j = \left\{ m_i^j | i \in N \right\}$ be the set of finite sequences of $M_j(t)$ such that $|m_i^j| = T$, where T is a hyper-parameter indicating sequence window length. LSTM auto-encoder networks are trained for each converter j to minimize:

$$L_j(k) = \sum_{s_m^j \in M_j} \sum \left\| m_i^j - m_i^{j}(k) \right\| \tag{5.35}$$

At each iteration k of training by optimizing the linear transformation matrices and biases given in (5.20), (5.21), (5.22), (5.23), and (5.24) for each LSTM layer in the encoder-decoder network and for the output layer. $m_i^j(k)$ is the reconstruction of the sequence m_i^j at the kth step of training. Let $s_m^j = \left\{ m_i^j | i \in N \right\}$ be the final reconstruction of s_m^j after the fine-tuning of the encoder-decoder network employed for converter j. The mean absolute error can be defined for all reconstructed sequences as:

$$E_j = \left\{ e_i^j | i \in N \right\} \tag{5.36}$$

And $e_i^j = \left\langle \left\| m_i^j - m_i^{j} \right\| \right\rangle$. For each E_j, we estimate the parameters μ_{E_j} and σ_{E_j} of Normal distribution $N\left(\mu_{E_j}, \sigma_{E_j} \right)$ using Maximum Likelihood Estimation. The penalty threshold for converter j is given by:

$$T_j = \max\left(\left(e_i^j - \mu_{E_j} \right)^2 / \sigma_{E_j} \right) \tag{5.37}$$

During the online phase, the mismatch of norms for each converter is reconstructed. Note that $|m_j| = T$ for each input sequence m_j. If Δt is the simulation time step, then the penalty score for the converter between time window t and $t + T\Delta t$ is given by:

$$P_j = \frac{\left(\langle \| m_j - \overline{m}_j \| \rangle - \mu_{E_j} \right)^2}{\sigma_{E_j}} \tag{5.38}$$

Remark 1: When the system is under normal operation,

$$\lim_{n \to \infty} P_j(n) < T_j \forall j \tag{5.39}$$

Here $P_j(n)$ represent the penalty score for the jth converter between the time window $t + n\Delta t$ and $t + (n + T)\Delta t$.

Remark 2: When the system is under FDI,

$$\lim_{n\to\infty} P_j(n) \nless T_j \forall j \tag{5.40}$$

5.6   Implementation of technique and verification

The suggested control method is validated using MATLAB with the microgrid configuration depicted in Figure 5.7. Specification details for which are provided in Table 5.1. The DC microgrid's validation is accomplished by utilizing the DG model, and its DC/DC converters establish a crucial node within the microgrid system. Each distinct DG with DC–DC converters is integrated into the DC bus as illustrated in Figure 5.7. To assess the effectiveness of the control approach, various scenarios are investigated.

5.6.1   Case 1: Load disturbance

Here, the control scheme's efficacy is assessed through load variation tests, involving sudden connections of DC loads to the DC bus. This assessment aims to

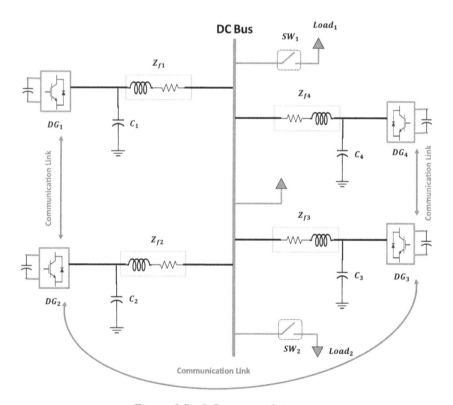

Figure 5.7   DC microgrid structure

Table 5.1   System properties and specifications

Droop gain	K_i	$K_1 = 3$, $K_2 = 6$, $K_3 = 1$, $K_4 = 4$
LC filter	Z_{f1-4}	$0.015 + j0.01$
Load	$R_L 1,2$	5Ω, 5Ω
V–I loop	k_{vp}, k_{vI}	5,150
	k_{cp}, k_{cI}	4,120

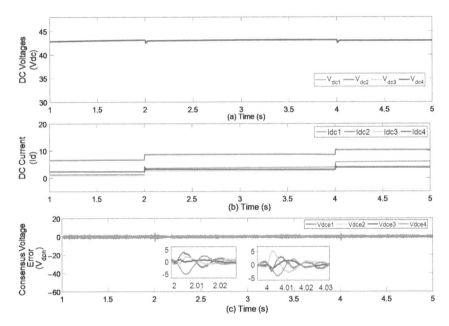

Figure 5.8   Effect of load perturbation. (a) DC Bus voltage. (b) Current in the converters. (d) Consensus voltage error.

evaluate the scheme's ability to manage transient effects. Load perturbations are conducted to examine the impact of uncertain loads on current distribution as well as DC bus voltage. The switches connecting the loads, labeled as sw_1 and sw_2, are activated at times $t = 2$ s and $t = 4$ s, respectively. During these perturbations, the scheme demonstrates robust performance with rapid suppression of errors, as demonstrated in Figure 5.8(a) and (b) during the secondary control operation of the DC microgrid. Figure 5.8(a) illustrates the swift stabilization of the bus voltage at $t = 2$ s and $t = 4$ s. The decrease in bus voltage is minimal, around 0.25 volts. The scheme exhibits a smooth transient response, and the current sharing capability operates effectively, as shown in Figure 5.8(b). Figure 5.8(c) demonstrates the suppression of consensus voltage errors. The triggering as per the triggering condition is illustrated in Figure 5.9, here the control input is renewed by satisfying the condition (5.16).

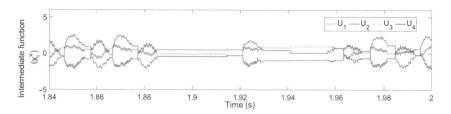

Figure 5.9   Triggering action in χ_i^v

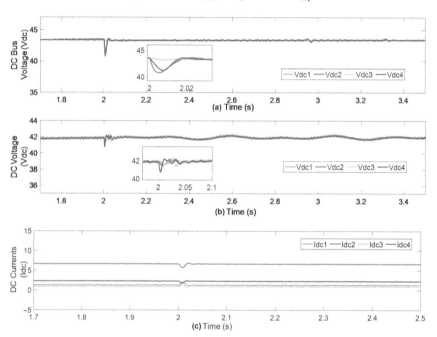

Figure 5.10   Actuation disturbance (fault) at t = 2 s. (a) Bus voltages under fault-tolerant scheme. (b) Bus voltages without fault-tolerant scheme. (c) Current in the converters.

5.6.2   Case 2: Impact of secondary control fault uncertainty

Here, the fault-tolerant scheme is evaluated for its efficacy in stabilizing DC voltage against faults occurring in control actuation. The fault in actuation is introduced at $t = 2$ s by incorporating the term $\varphi_i(t)$ into the control actuating signal. This introduced fault is assumed to be sudden and rapid, simulating an abrupt change. Consequently, the microgrid responds to this perturbation as instances of fault scenarios, resulting in partial effectiveness loss, as depicted in Figure 5.10(a) and (b). A comparative analysis is conducted without fault-tolerant control strategies. The investigation reveals that the actuation will experience reduced effectiveness during the occurrence of fault uncertainty, characterized by a moderate loss of functionality. The fault affecting the actuation is like "attack", wherein the fault-tolerant method manages anomalies in data reception and transmission. To simulate the actuation fault, it is assumed that initially, all agent actuations are functioning normally with parameter values $\varphi_i(t) = 0$ and $\omega_i(t) = 0$. However, the actuator malfunction is introduced at $t = 2$ s, featuring $\varphi_i(t) = 0.6 \& \omega_i(t) = 3\sin(20t)$ to replicate the fault in the actuation.

5.6.3   Case 3: Performance under FDI attack

The effectiveness of the scheme for identification of FDI is examined through four DG-based Simulink models with the consensus-based cooperative control topology as mentioned in the previous section. Data is collected at a specific time from the running test system for the training phase. The time during which the data was collected was 5 s, sampled every 0.2 ms. The training phase data contains load change and increased communication delay perturbation. Thus, 25,000 samples of M_j for each DER unit ($j = 1,2,3$ and 4) are collected. Twenty percent of the data was used as a validation set for fine-tuning the encoder-decoder network. Early stopping was used to prevent the network from over-fitting. The tuned hyper-parameters for encoder-decoder networks (one for reconstructing voltage norm mismatch and one for reconstructing current norm mismatch) are given in Tables 5.2 and 5.3. The reconstruction errors are used to estimate the parameters of the normal distribution and penalty thresholds for each DER unit for the reconstruction of voltage and current mismatch norm, as in (5.37), are given in Table 5.4.

Three separate scenarios are taken to examine the performance of the attack detection method. The analysis of the effectiveness of the explained strategy in detecting FDI with minimal weights to bypass the detection system is also presented.

Table 5.2   Hyper-parameters of LSTM autoencoder used for each converter for reconstructing voltage norm mismatch

Hyper-parameter	Value
Layers of LSTM in encoder	4
Layers of LSTM in decoder	4
LSTM units in each layer	128
T	30
Activation function	sigmoid
Recurrent activation function	Hyperbolic tangent

Table 5.3   Hyper-parameters of LSTM autoencoder used for each converter for reconstructing current norm mismatch

Hyper-parameter	Value
Layers LSTM in encoder	3
Layers LSTM in decoder	3
LSTM units in each layer	64
T	15
Activation function	sigmoid
Recurrent activation function	Hyperbolic tangent

*Table 5.4   Estimated parameters of normal distribution and reconstruction
          penalty thresholds for each DER unit*

DER unit	$\mu_{E_j}^v$	$\sigma_{E_j}^v$	T_j^v	$\mu_{E_j}^i$	$\sigma_{E_j}^i$	T_j^i
1	0.82849	0.7630	47.51	0.9325	1.0986	57.326
2	0.83703	0.54680	28.544	0.7156	0.6989	83.114
3	0.86231	0.52676	29.861	0.7331	0.6788	33.456
4	0.70250	0.71236	55.189	0.8070	0.5925	47.189

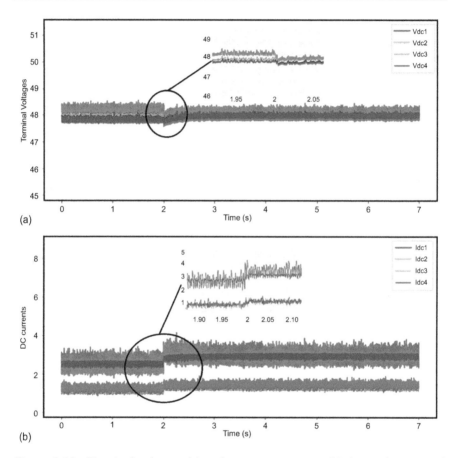

Figure 5.11   Terminal voltages (a) and converter currents (b) (normal operation)

5.6.3.1   Scenario 1: Operation in usual conditions

In this scenario, the microgrid is operating and is in a steady state before $t = 0$. At
$t = 2$ s, the local load at DER unit 1 is increased to120 Ω. The penalty scores for
the weighted norm of voltage and current mismatch are calculated for each DER
unit in every time window of T according to (5.37) and based on (5.39), the
reconstruction penalties should lie within the respective thresholds. Figure 5.11

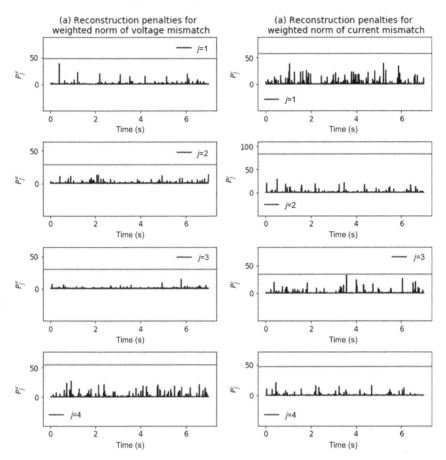

Figure 5.12 *Reconstruction penalties of the weighted norm of voltage mismatch (a) and current mismatch (b) (normal operation)*

shows the (a) terminal voltages real value and (b) DC currents. Figure 5.12 shows the reconstruction penalties for the weighted norm of voltage mismatch (a) and weighted norm of current mismatch (b). Values of P_j^v and P_j^i, both satisfy (5.39).

5.6.3.2 Scenario 2: FDI 1

In this scenario, the DER unit 1 is under FDI and the attack model is based on (5.28) with $\chi = 3.2 \sin 60t$. The voltage sensor of the mentioned DER unit is targeted immediately after $t = 2$ s. Figure 5.13 shows (a) terminal voltages and (b) DC currents. Figure 5.14 shows (a) the reconstruction penalties value of the weighted norm of voltage mismatch and (b) the weighted norm of current mismatch. The values P_j^v cross the threshold immediately after the injection of FDI as expected as per (5.40). The reconstruction penalty of the DER unit 4 and its neighbors cross their respective thresholds to very high magnitudes immediately after FDI indicating that the neighbors of the attacked DER unit are also affected.

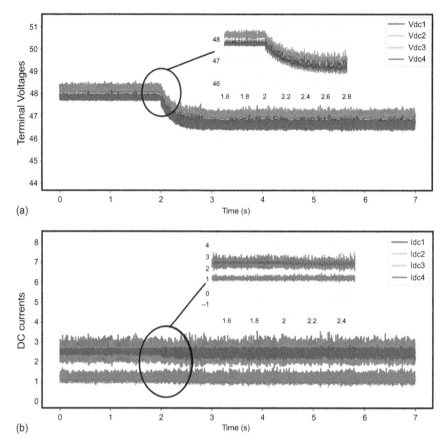

Figure 5.13   Terminal voltages (a) and DC currents (b) (FDI 1)

5.6.3.3   Scenario 3: FDI 2

This scenario targets the current sensor of the DER unit 3 and the attack model is based on (5.28) with $\chi = 1.5 \sin 60t$. The false data is injected immediately after $t = 2$ s. Figure 5.15 shows (a) values of terminal voltages and (b) DC currents. Figure 5.16 shows (a) values of reconstruction penalties of the weighted norm of voltage mismatch and (b) current mismatch. The values P_j^i follow (5.40) as expected and the immediate rise P_j^i of for DER unit 3 and its neighbors shows that the current sensor of that DER unit has been targeted.

5.6.4   Minimal weight FDI detection

An attack sequence, that may avoid the FDI identification construction, can be fabricated by the intruder by using extra small weights for the attack vector in the FDI models. The attack model is based on (5.28) with χ being the minimal. The intruder targets the voltage sensor of the DER unit 4 immediately after $t = 2$ s. The terminal

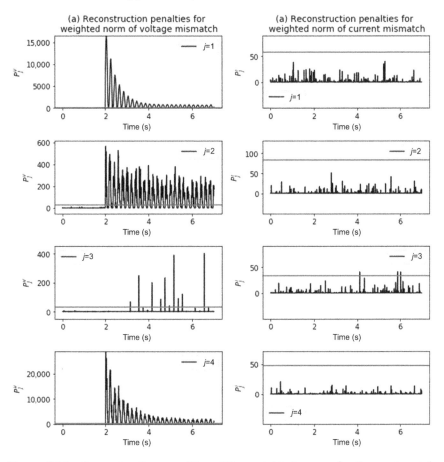

*Figure 5.14   Reconstruction penalties of the weighted norm of voltage mismatch
(a) and current mismatch (b) (FDI 1)*

voltages and DC currents are respectively shown in Figure 5.17(a) and (b). The reconstruction penalties for weighted voltage norm and current norm mismatch are respectively illustrated in Figure 5.18(a) and (b). The values P_j^v still cross the threshold according to (5.40). Thus, it is verified that the FDI with least or zero destabilizing validities can also be identified through the described framework.

5.6.5   FDI mitigation

Various mitigation strategies with consensus control can be used to suppress the impacts of the attack. Using (5.32), the confidence values for each converter can be calculated during the DC microgrid operation [29]. For a converter j under normal scenario:

$$\lim_{t\to\infty} \left| M_j(t) - \overline{M}_j(t) \right| < \sqrt{T_j \sigma_{E_j}} + \mu_{E_j} \tag{5.41}$$

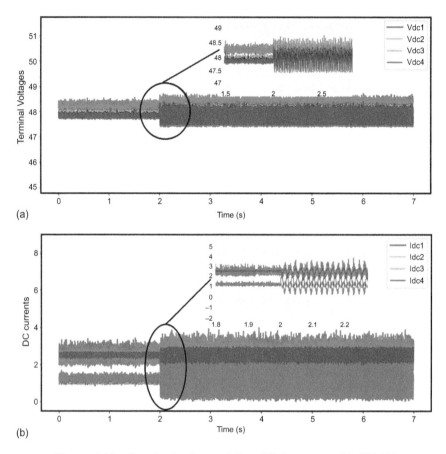

Figure 5.15   Terminal voltages (a) and DC currents (b) (FDI 2)

This can be deduced using (5.37). The weighted norm of state mismatch for a given converter j always lies in finite bounds during normal operation. Hence, the local confidence values of those converters reach a steady state value. When a converter j is under FDI, $M_j(t)$ increases above the bound and the local confidence value of the affected converter falls below the steady state value. We define the converter's trust value k, allocated by jth converter as:

$$T_{jk}(t) = \max\left(C_j(t), b_{jk}(t)\right) \tag{5.42}$$

where

$$\dot{b}_{jk}(t) = \gamma_j\left(s_{jk}(t) - b_{jk}(t)\right) \tag{5.43}$$

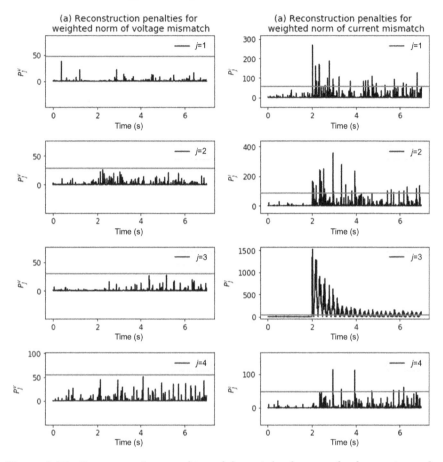

Figure 5.16 Reconstruction penalties of the weighted norm of voltage mismatch (a) and current mismatch (b) (FDI 2)

where $\gamma_i > 0$ is a temporal factor.

$$s_{jk}(t) = \frac{\theta_j}{\theta_j + \left\| X_k - 1/N_j \sum_{l \in N_j, l \neq k} X_l \right\|} \qquad (5.44)$$

Hence, the state update protocol of the DC microgrid can be given as:

$$\bar{v}_i(t) = v_i(t) + \int_0^t \sum_{j \in N_i} T_{ij} a_{ij} \big(\bar{v}_j(\tau) - \bar{v}_i(\tau)\big) d\tau \qquad (5.45)$$

$$\delta_i = \sum_{j \in N_i} T_{ij} c a_{ij} \big(i_j^{pu} - i_i^{pu}\big) \qquad (5.46)$$

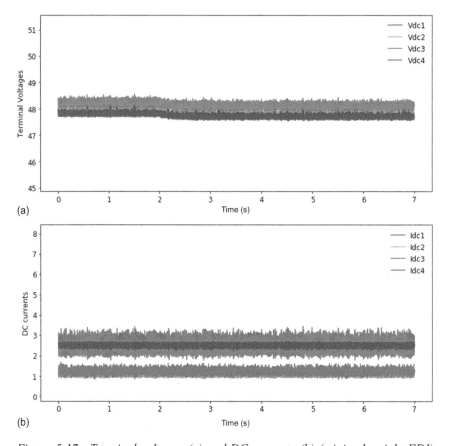

Figure 5.17   Terminal voltages (a) and DC currents (b) (minimal weight FDI)

The DC microgrid prototyped in the case study is employed with the confidence mechanism. The DER unit 3 is targeted with a node FDI at $t = 2$ s using the attack model given by (5.28) with $\chi = 3.2 \sin 60t$. The mitigation parameters are given as:

$$\theta_1 = 0.375, \ \theta_2 = 0.415, \ \theta_3 = 0.375, \ \theta_4 = 0.35$$

$$\Delta_1 = \Delta_2 = \Delta_3 = \Delta_4 = 0.65$$

$$\gamma_1 = \gamma_2 = \gamma_3 = \gamma_4 = 1.0$$

The terminal voltages and DC currents are shown in Figure 5.19. The modified trust-based mechanism is put into action after at least 150 continuous values of reconstruction penalties cross the threshold for a given unit. The weighted trust values $(T_{jk}a_{jk})$ assigned by each converter j to its neighbor are shown in Figure 5.20. As we can see, the trust factor assigned to a neighbor goes down when it is close to the attacked unit or is an attacked unit. Also, the trust assigned by the attacked unit to its neighbors increases.

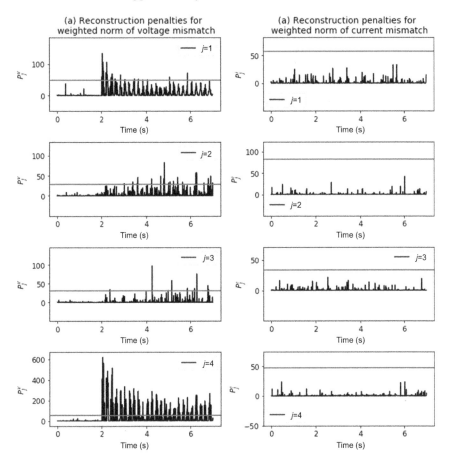

Figure 5.18 *Reconstruction penalties of the weighted norm of voltage mismatch (a) and current mismatch (b) (minimal weight FDI)*

The converter used in each DER unit is a step-down (buck) converter. Each of them has $L = 2.64$ mH and $C = 2.2$ mF and $F_s = 10$ kHz. The local loads are $R_1 = 30\ \Omega$ and $R_2 = R_3 = R_4 = 20\ \Omega$. The transmission line impedances are $Z_{12} = Z_{34} = 0.5 + (50\ \mu\text{H})s$. Other control specifications are as follows:

$$I_{rated} = diag\{6, 3, 3, 6\}$$

$$A_g = \begin{bmatrix} 0 & 90 & 0 & 110 \\ 90 & 0 & 100 & 0 \\ 0 & 100 & 0 & 120 \\ 110 & 0 & 120 & 0 \end{bmatrix}$$

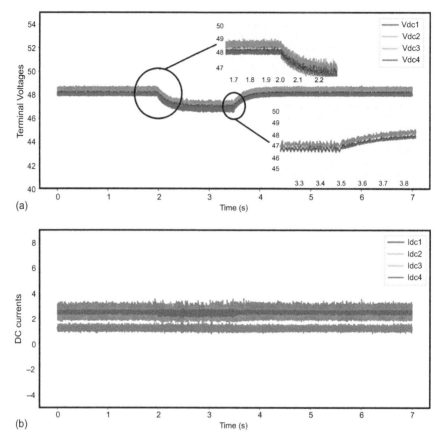

Figure 5.19 Terminal voltages (a) and DC currents (b) (attack resilient operation)

$$H_p = \begin{bmatrix} 0.1 & 0 & 0 & 0 \\ 0 & 0.09 & 0 & 0 \\ 0 & 0 & 0.08 & 0 \\ 0 & 0 & 0 & 0.11 \end{bmatrix}$$

$$r = \begin{bmatrix} 6 & 3 & 3 & 6 \end{bmatrix}, c = 0.005$$

$$H_I = \begin{bmatrix} 6 & 0 & 0 & 0 \\ 0 & 5 & 0 & 0 \\ 0 & 0 & 5.4 & 0 \\ 0 & 0 & 0 & 5.6 \end{bmatrix}$$

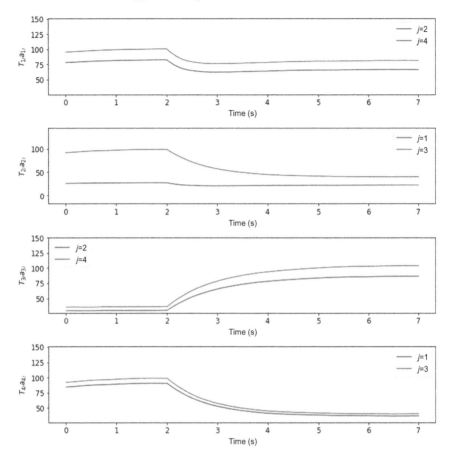

Figure 5.20   Weighted trust factor assigned by each converter to its neighbors (attack resilient operation)

$$G_P = \begin{bmatrix} 1.1 & 0 & 0 & 0 \\ 0 & 1.0 & 0 & 0 \\ 0 & 0 & 1.2 & 0 \\ 0 & 0 & 0 & 1.1 \end{bmatrix}, G_I = \begin{bmatrix} 7 & 0 & 0 & 0 \\ 0 & 7.4 & 0 & 0 \\ 0 & 0 & 6.6 & 0 \\ 0 & 0 & 0 & 7 \end{bmatrix}$$

5.7   Conclusion

In this chapter, the ANN scheme for FDI detection in microgrids is discussed. The method explains the realizing and reconstructing capability of the LSTM auto-encoder in reconstructing the weighted norm of state mismatches for all the DER units. When the system was in normal operating condition, the reconstruction penalties of the weighted norm of state mismatches lay within the calculated thresholds for all the DER units. When the system was under FDI, the reconstruction penalties crossed the thresholds for the targeted DER unit and its neighbors. This method successfully detects FDIs with minimal weights which were supposed to bypass the detection framework. The obtained results indicate that the method can distinguish between the normal grid transients and FDIs effectively. The effectiveness of the method is assessed across a range of disturbance scenarios. The simulation outcomes demonstrate the effectiveness of the approach when dealing with both system faults and disruptions. Furthermore, the application scope of the technique is expanded to offer robustness against cyberattacks, particularly those involving false data injection.

References

[1]   L. Xing, Y. Mishra, Y-C. Tian, *et al.*, "Dual-consensus-based distributed frequency control for multiple energy storage systems," *IEEE Trans. Smart Grid*, vol. 10, no. 6, pp. 6396–6403, 2019.

[2]   P. Lin, T. Zhao, B. Wang, Y. Wang, and P. Wang, "A semi-consensus strategy toward multi-functional hybrid energy storage system in DC microgrids," *IEEE Trans. Energy Conversion*, vol. 35, no. 1, pp. 336–346, 2020.

[3]   H. Cai and G. Hu, "Distributed nonlinear hierarchical control of ac microgrid via unreliable communication," *IEEE Trans. Smart Grid*, vol. 9, no. 4, pp. 2429–2441, 2018.

[4]   C. Peng, J. Li, and M. Fei, "Resilient event-triggering h1 load frequency control for multi-area power systems with energy-limited dos attacks," *IEEE Trans. Power Syst.*, vol. 32, no. 5, pp. 4110–4118, 2017.

[5]   Q. Xu, X. Hu, P. Wang, J. Xiao, P. Tu, C. Wen, and M. Y. Lee, "A decentralized dynamic power sharing strategy for hybrid energy storage system in autonomous dc microgrid," *IEEE Trans. Ind. Electron.*, vol. 64, no. 7, pp. 5930–5941, 2017.

[6]   R. Han, L. Meng, J. M. Guerrero, and J. C. Vasquez, "Distributed nonlinear control with event-triggered communication to achieve current sharing and voltage regulation in dc microgrids," *IEEE Trans. Power Electron.*, vol. 33, no. 7, pp. 6416–6433, 2018.

[7]   Q. Xu, C. Zhang, C. Wen, and P. Wang, "A novel composite nonlinear controller for stabilization of constant power load in dc microgrid," *IEEE Trans. Smart Grid*, vol. 10, no. 1, pp. 752–761, 2019.

[8]   M. A. Setiawan, A. Abu-Siada, and F. Shahnia, "A new technique for simultaneous load current sharing and voltage regulation in dc microgrids," *IEEE Trans. Ind. Inform.*, vol. 14, no. 4, pp. 1403–1414, 2018.

[9]   S. Augustine, M. K. Mishra, and N. Lakshmi narasamma, "Adaptive droop control strategy for load sharing and circulating current minimization in low-voltage standalone dc microgrid," *IEEE Trans. Sustain. Energy*, vol. 6, no. 1, pp. 132–141, 2015.

[10] V. Kumar and I. Ali, "Fractional order sliding mode approach for chattering free direct power control of dc/ac converter," *IET Power Electron.*, vol. 12, no. 13, pp. 3600–3610, 2019.

[11] X. Lu, K. Sun, J. M. Guerrero, J. C. Vasquez, and L. Huang, "State of-charge balance using adaptive droop control for distributed energy storage systems in dc microgrid applications," *IEEE Trans. Ind. Electron.*, vol. 61, no. 6, pp. 2804–2815, 2014.

[12] J. M. Guerrero, J. C. Vasquez, J. Matas, L. G. De Vicuna, and M. Castilla, "Hierarchical control of droop-controlled ac and dc microgrids-a general approach toward standardization," *IEEE Trans. Ind. Electron.*, vol. 58, no. 1, pp. 158–172, 2011.

[13] A. A. Hamad, M. A. Azzouz, and E. F. El-Saadany, "Multiagent supervisory control for power management in dc microgrids," *IEEE Trans. Smart Grid*, vol. 7, no. 2, pp. 1057–1068, 2016.

[14] C. Wang, J. Duan, B. Fan, Q. Yang, and W. Liu, "Decentralized high-performance control of dc microgrids," *IEEE Trans. Smart Grid*, vol. 10, no. 3, pp. 3355–3363, 2018.

[15] J. Peng, B. Fan, J. Duan, Q. Yang, and W. Liu, "Adaptive decentralized output-constrained control of single-bus dc microgrids," *IEEE/CAA J. Autom. Sin.*, vol. 6, no. 2, pp. 424–432, 2019.

[16] S. Anand, B. G. Fernandes, and J. Guerrero, "Distributed control to ensure proportional load sharing and improve voltage regulation in low voltage dc microgrids," *IEEE Trans. Power Electron.*, vol. 28, no. 4, pp. 1900–1913, 2013.

[17] V. Nasirian, S. Moayedi, A. Davoudi, and F. L. Lewis, "Distributed cooperative control of dc microgrids," *IEEE Trans. Power Electron.*, vol. 30, no. 4, pp. 2288–2303, 2014.

[18] M. Dong, L. Li, Y. Nie, D. Song and J. Yang, "Stability analysis of a novel distributed secondary control considering communication delay in DC microgrids," *IEEE Trans. Smart Grid*, vol. 10, no. 6, pp. 6690–6700, 2019.

[19] X.-K. Liu, H. He, Y.-W. Wang, Q. Xu, and F. Guo, "Distributed hybrid secondary control for a dc microgrid via discrete-time interaction," *IEEE Trans. Energy Convers.*, vol. 33, no. 4, pp. 1865–1875, 2018.

[20] F. Guo, Q. Xu, C. Wen, L. Wang, and P. Wang, "Distributed secondary control for power allocation and voltage restoration in islanded dc micro-grids," *IEEE Trans. Sustain. Energy*, vol. 9, no. 4, pp. 1857–1869, 2018.

[21] L. Xing, Y. Mishra, F. Guo, *et al.*, "Distributed secondary control for current sharing and voltage restoration in DC microgrid," *IEEE Trans. Smart Grid*, vol. 11, no. 3, pp. 2487–2497, 2020.

[22] Z. Liu, M. Su, Y. Sun, W. Yuan, H. Han, and J. Feng, "Existence and stability of equilibrium of dc microgrid with constant power loads," *IEEE Trans. Power Syst.*, vol. 33, pp. 6999–7010, 2018.

[23] T. Dragičević, X. Lu, J. C. Vasquez, and J. M. Guerrero, "Dc microgrids—part ii: A review of power architectures, applications, and standardization issues," *IEEE Trans. Power Elect.*, vol. 31, pp. 3528–3549, 2016.

[24] Y. Zeng, Q. Zhang, Y. Liu, H. Guo, and F. Zhang, "Distributed cooperative control strategy for stable voltage restoration and optimal power sharing in islanded DC microgrids," *IEEE Trans. Power Syst.*, vol. 39, no. 2, pp. 3431–3443, 2024, doi: 10.1109/TPWRS.2023.3274932.

[25] X. Lu, J. M. Guerrero, K. Sun, and J. C. Vasquez, "An improved droop control method for DC microgrids based on low bandwidth communication with dc bus voltage restoration and enhanced current sharing accuracy," *IEEE Trans. Power Electron.*, vol. 29, no. 4, pp. 1800–1812, 2014.

[26] S. Anand and B. G. Fernandes, "Steady state performance analysis for load sharing in dc distributed generation system," in *Proc. 10th Int. Conf. Environ. Elect. Eng.*, 2011, pp. 1–4.

[27] Y. W. Li and C. N. Kao, "An accurate power control strategy for power electronics-interfaced distributed generation unit's operation in a low voltage multi-bus microgrid," *IEEE Trans. Power Electron.*, vol. 24, no. 12, pp. 2977–2988, 2009.

[28] J. He and Y. W. Li, "Analysis, design and implementation of virtual impedance for power electronics interfaced distributed generation," *IEEE Trans. Ind. Appl.*, vol. 47, no. 6, pp. 2525–2538, 2011.

[29] S. Abhinav, H. Modares, F. L. Lewis, and A. Davoudi, "Resilient cooperative control of dc microgrids," *IEEE Trans. Smart Grid*, vol. 10, pp. 1083–1085, 2019.

[30] S. Saha, T. Roy, M. Mahmud, M. Haque, and S. Islam, "Sensor fault and cyber-attack resilient operation of dc microgrids," *Int. J. Electr. Power Energy Syst.*, vol. 99, pp. 540–554, 2018.

[31] S. Hu, P. Yuan, D. Yue, C. Dou, Z. Cheng, and Y. Zhang, "Attack-resilient event-triggered controller design of DC microgrids under DoS attacks," *IEEE Trans. Circuits Syst. I: Regular Papers*, vol. 67, pp. 699–710, 2020.

[32] V. Kumar and S. R. Mohanty, "Resilient optimal gain control and continuous twisting observer for enhanced power system performance under uncertainties," *IEEE Syst. J.*, vol. 17, no. 2, pp. 2733–2744, 2023.

[33] Y. Liu, P. Ning, and M. K. Reiter, "False data injection attacks against state estimation in electric power grids," *ACM Trans. Inf. Syst. Secur.*, vol. 14, pp. 13:1–13:33, 2011.

[34] S. Liu, Z. Hu, X. Wang, and L. Wu, "Stochastic stability analysis and control of secondary frequency regulation for islanded microgrids under random denial of service attacks," *IEEE Trans. Ind. Inform.*, vol. 15, pp. 4066–4075, 2019.

[35] B. Chatfield, R. J. Haddad, and L. Chen, "Low-computational complexity intrusion detection system for jamming attacks in smart grids," *2018 International Conference on Computing, Networking and Communications* (ICNC), March 2018.

[36] Y. Mo and B. Sinopoli, "Secure control against replay attacks," in *2009 47th Annual Allerton Conference on Communication, Control, and Computing (Allerton)*, pp. 911–918, September 2009.

[37] K. Manandhar, X. Cao, F. Hu, and Y. Liu, "Detection of faults and attacks including false data injection attack in smart grid using kalman filter," *IEEE Trans. Control of Netw. Syst.*, vol. 1, no. 4, pp. 370–379, 2014.

[38] D. B. Rawat and C. Bajracharya, "Detection of false data injection attacks in smart grid communication systems," *IEEE Signal Process. Lett.*, vol. 22, pp. 1652–1656, 2015.

[39] G. Chaojun, P. Jirutitijaroen, and M. Motani, "Detecting false data injection attacks in ac state estimation," *IEEE Trans. Smart Grid*, vol. 6, pp. 2476–2483, 2015.

[40] S. K. S. Tyagi, R. Yadav, D. K. Jain, Y. Tu, and W. Zhang, "Paired swarm optimized relational vector learning for FDI attacks detection in IoT-aided smart grid," *IEEE Internet Things J.*, vol. 10, no. 21, pp. 18708–18717, 2023, doi: 10.1109/JIOT.2022.3224671.

[41] J. Zhao, G. Zhang, M. La Scala, Z. Dong, C. Chen, and J. Wang, "Short term state forecasting-aided method for detection of smart grid general false data injection attacks," *IEEE Trans. Smart Grid*, vol. 8, no. 4, pp. 1580–1590, 2017, doi: 10.1109/TSG.2015.2492827.

[42] M. Esmalifalak, L. Liu, N. Nguyen, R. Zheng, and Z. Han, "Detecting stealthy false data injection using machine learning in smart grid," *IEEE Syst. J.*, vol. 11, no. 3, pp. 1644–1652, 2017, doi: 10.1109/JSYST.2014.2341597.

[43] S. Li, Y. Yilmaz, and X.Wang, "Quickest detection of false data injection attack in wide-area smart grids," *IEEE Trans. Smart Grid*, vol. 6, no. 6, pp. 2725–2735, 2015.

[44] S. Sahoo, S. Mishra, J. C. Peng, and T. Dragicevic, "A stealth cyber-attack detection strategy for dc microgrids," *IEEE Trans. Power Elect.*, vol. 34, pp. 8162–8174, 2019.

[45] S. Sahoo, J. C. Peng, D. Annavaram, S. Mishra, and T. Dragicevic, "On detection of false data in cooperative dc microgrids-a discordant element approach," *IEEE Trans. Ind. Electr.*, vol. 67, no. 8, pp. 6562–6571, 2020, doi: 10.1109/TIE.2019.2938497.

[46] Y. He, G. J. Mendis, and J. Wei, "Real-time detection of false data injection attacks in smart grid: A deep learning-based intelligent mechanism," *IEEE Trans. Smart Grid*, vol. 8, pp. 2505–2516, 2017.

[47] M. Ozay, I. Esnaola, F. T. Yarman Vural, S. R. Kulkarni, and H. V. Poor, "Machine learning methods for attack detection in the smart grid," *IEEE Trans. Neural Netw. Learn. Syst*, vol. 27, pp. 1773–1786, 2016.

Chapter 6

Bidirectional virtual inertia through decentralized virtual synchronous generator control in hybrid AC–DC microgrids

Bandla Krishna Chaithanya[1] and Prasad Padhy Narayana[1]

The rapid penetration of renewable sources has laid the foundation for providing synthetic inertia through power electronic converters interfacing renewables and energy storage systems with the grid. However, within network-connected hybrid AC–DC microgrids, the stability of the DC bus is contingent upon local inertial support from rapid-response energy storage systems. This dependence has been identified as a source of power imbalance among the DC buses during transient states in network-connected microgrids. This chapter introduces an advanced control strategy for virtual synchronous generators tailored for hybrid AC–DC microgrids, specifically designed to mitigate the aforementioned challenge. The proposed control scheme addresses the need for precise power-sharing among multiple microgrids in a decentralized manner, countering the destabilizing effects of transient states. The strategy not only ensures coordinated power-sharing among hybrid energy storage systems across diverse microgrids but also facilitates plug-and-play capability. The proposed control strategy is validated through theoretical analysis and Power Hardware In Loop (PHIL) simulations. This chapter presents a comprehensive exploration of the proposed control strategy, offering valuable insights into its effectiveness and adaptability within the context of network-connected hybrid microgrids.

6.1 Introduction

6.1.1 Motivation

Traditional power generation has long been the sole source of energy for daily needs and industrial operations. However, rising energy demands, concerns over global warming, and the dwindling availability of fossil fuels have accelerated the integration of renewable energy sources (RES) into the power grid. Hybrid AC–DC microgrids have emerged as a promising solution to seamlessly incorporate RES,

[1]Department of Electrical Engineering, Indian Institute of Technology Roorkee, India

offering benefits such as reduced power conversions, enhanced system reliability, and flexible power control [1].

Central to the effective management of hybrid AC–DC microgrids is the flexible control of power electronic converters that link the AC and DC buses. Power management within these systems remains a persistent challenge, with ongoing efforts to achieve precise power sharing between AC and DC buses. Various power management techniques have been developed, broadly classified as centralized [2], distributed [3], and decentralized [4–6], depending on the required control and communication infrastructure.

While existing control strategies primarily focus on power equalization during steady-state conditions, the increasing penetration of renewable energy in these microgrids raises concerns about system stability. The rapid response of power converters without traditional rotational inertia presents a potential threat [7,8]. To address this, a novel approach utilizing power electronic converters that mimic the behavior of synchronous generators has been proposed, known as Virtual Synchronous Generators (VSGs) or Synchronverters. These VSG-controlled converters can operate in parallel, ensuring proportional active and reactive power sharing based on frequency and voltage droop. They can also serve as voltage sources during islanded modes, aligning with grid codes that mandate virtual inertia provision as a critical ancillary service in renewable integration [9].

6.1.2   State-of-the-art technology

The current body of research has predominantly delved into various methodologies for interlinking converters aimed at generating transient responses, often termed virtual inertia, to ensure the stable operation of AC power systems. In parallel, akin to AC microgrids, the transient response also plays a pivotal role in bolstering the stability of DC microgrids. This is achieved by curbing the rate of change in DC bus voltage, thereby mitigating power perturbations in voltage-sensitive loads [10,11].

A power management strategy tailored for multiple energy storage systems within DC microgrids, based on virtual inertia, has been proposed. Its objective is to guarantee proportional sharing of transient responses injected into the DC bus [12]. Additionally, researchers in [13] introduced a control strategy aimed at infusing transient responses from the grid to enhance the inertia of grid-connected DC microgrids. Moreover, a distributed virtual inertia control strategy for DC microgrids was suggested in [14], leveraging mechanical inertial power from permanent magnet synchronous generator (PMSG) wind turbines to diminish the rate of variation in DC bus voltage.

The inertial response observed in hybrid AC–DC microgrids differs markedly from standalone AC or DC microgrids. In a network-connected hybrid AC–DC microgrid, energy sources integrated into the DC bus and those in neighboring microgrids can promptly bolster load transients in the AC bus, delivering the necessary transient and steady-state response via traditional VSG control. However, the support for a load transient in the DC bus of one microgrid from energy sources in other microgrids is limited, unless the interlinking converter is specifically directed to transfer power from the AC bus to the DC bus. This constraint hinders power response from the AC bus to the DC bus during transient and steady-state conditions.

To address these challenges, researchers in [15] proposed a unified power flow control strategy for the interlinking converter in hybrid AC–DC microgrids, aimed at managing transient and steady-state responses during load transients on the AC side. Nevertheless, this approach does not extend the same level of response to load disturbances in the DC bus. To overcome this limitation, an improved power flow control strategy based on the virtual synchronous machine concept was introduced in [16], providing power responses from AC to DC buses and vice versa, including transient responses. However, deriving both transient and steady-state responses from distributed sources can potentially reduce the lifespan of storage devices.

Further studies, such as those by the authors in [17–19], have allocated transient responses to super-capacitors and steady-state responses to batteries through comprehensive control strategies tailored for hybrid AC–DC microgrids. In a similar vein, an enhanced control strategy for the interlinking converter was proposed in [20], aiming to synthesize transient and steady-state responses for both AC and DC buses, while considering the dynamics of AC bus frequency and DC bus voltage. Nonetheless, these strategies necessitate the communication of AC-side parameters to the DC side and vice versa for coordinated power management. This communication requirement impedes the plug-and-play operation of batteries and super-capacitors and compromises system stability due to the heightened risk of single points of failure. Additionally, in these approaches, the interlinking converter primarily functions as a power source, relinquishing its capacity to maintain the AC bus during islanded operations.

While transmitting AC bus frequency to the DC bus proves unnecessary due to the natural injection of active power from the DC bus as a transient response through VSG control, deviations in the DC bus voltage must be communicated to the AC side. This communication ensures that other hybrid AC–DC microgrids can provide the necessary inertial response for the DC load transient. Typically, transmitting DC bus voltage information from each hybrid AC–DC microgrid to the others would necessitate a substantial communication infrastructure. To circumvent this requirement, deviations in DC bus voltage are injected as power transients into the AC bus connected to other hybrid microgrids. Subsequently, these hybrid AC–DC microgrids interpret these transients as load changes in the AC bus and respond with an inertial response akin to a conventional VSG.

6.1.3   Contributions

To address the previously mentioned drawbacks, this chapter proposes an enhanced decentralized virtual synchronous generator control for the interlinking converter in a hybrid AC–DC microgrid [21]. The major contributions are outlined as follows:

- The enhanced decentralized control scheme incorporates an enhanced unified power flow control strategy for the interlinking converter. This ensures the transmission of inertial and steady-state responses from the DC bus to the AC bus during AC load transients, and from the AC bus to the DC bus during DC load transients.
- The control strategy ensures accurate power sharing between steady-state energy sources and fast-acting energy sources such as super-capacitors.

- The method is feasible for implementation, as it is derived from and similar in nature to the conventional VSG control scheme.
- The control scheme ensures coordinated transient and steady-state power management among multiple hybrid AC–DC microgrids, effectively managing load transients in any network-connected hybrid AC–DC microgrid.
- The scheme is decentralized and requires no communication of DC bus quantities to the AC bus and vice versa.

6.2 Bidirectional virtual inertia for hybrid AC–DC microgrids

The configuration of a standard network-connected hybrid AC–DC microgrid is depicted in Figure 6.1, incorporating renewable energy sources, energy storage

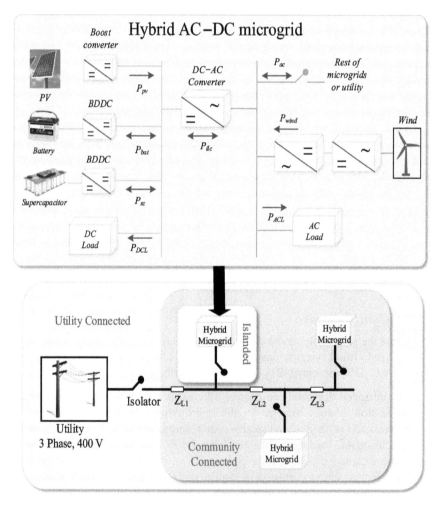

Figure 6.1   Architecture of network-connected hybrid AC–DC microgrids and the configurations during various modes of operation

systems, and loads. Renewable sources like PV panels and wind turbines are designed to operate at maximum capacity, channeling available power into their respective buses. Each microgrid's DC bus integrates a Hybrid Energy Storage System (HESS), comprising high-power density and high-energy density storage devices such as super-capacitors and batteries. The HESS supports load demands and upholds power equilibrium within the DC bus. The interlinking converter enables bidirectional power flow between the AC and DC buses. This network-connected hybrid AC–DC microgrid seamlessly transitions between utility-connected, community-connected, and island modes, adapting effortlessly from one mode to another and vice versa.

6.2.1 Power flow analysis in the AC side of hybrid AC–DC microgrid

The power flow equation governing the injection of active power into the grid at the connection point of the hybrid AC–DC microgrids is represented by Equation (6.1). This equation delineates the power balance within the system, considering contributions from the AC buses, interlinking converters, renewable sources, and the grid.

$$\sum_{i=1}^{n} P_{ac}^i + P_{grid} = 0; \sum_{i=1}^{n} (P_{ilc}^i + P_w^i - P_{acl}^i) + P_{grid} = 0 \tag{6.1}$$

The net power exchange by the interlinking converter for AC bus power transitions is captured in Equation (6.2). It illustrates how the interlinking converter manages the power flow between the AC bus, renewable sources, and the grid to maintain stability and balance.

$$\sum_{i=1}^{n} P_{ilc}^i = \sum_{i=1}^{n} (P_{acl}^i - P_w^i) - P_{grid} \tag{6.2}$$

Equation (6.3) details the active power injected by the interlinking converter in response to deviations in AC bus frequency. This power injection is crucial for maintaining stability and proper functioning during different operational modes.

$$P_{ilc}^i = k_{ac}^i (\omega_r - \omega); k_{ac}^i = \frac{P_r^i}{(\omega_{max}^i - \omega_{min}^i)} \tag{6.3}$$

However, as renewable sources and energy storage systems become more integrated into these microgrids, there is a growing need for interlinking converters to provide transient responses akin to synchronous generators. Equation (6.4) outlines the transient power response expected from the interlinking converter, simulating the behavior of a synchronous generator to ensure system stability during sudden changes.

$$P_{ilc}^i = \frac{2H^i S_r^i}{\omega_r} \frac{d}{dt} (\omega_r - \omega) + D_p^i (\omega_r - \omega) \tag{6.4}$$

The unified power response required by the interlinking converter for AC bus power deviations is expressed in (6.5). This equation incorporates the necessary power response for different operational modes, accounting for the integrator constant and the power reference command (P^{ac}_{set}) that dictates secondary power responses.

$$P^i_{ilc} = \frac{2H^i S^i_r}{\omega_r} \frac{d}{dt}(\Delta\omega) + (D^i_p + k^i_{ac})(\Delta\omega) + P^{ac}_{set}$$

$$P^{ac}_{set} = \frac{M^{ac}}{S}(\Delta\omega) \Rightarrow \text{islanded}; 0 \Rightarrow \text{community and grid connected}$$

(6.5)

where $\Delta\omega = \omega_r - \omega$. The power reference command for various operational modes is detailed in (6.5), where M^{ac} represents the integrator constant.

6.2.2 *Power flow analysis in the DC side of hybrid AC–DC microgrid*

Moving to the DC side analysis, Equation (6.6) describes the active power flow at the DC link capacitor, considering power generation from renewable sources and power consumption by DC loads. It also factors in the damping coefficient and voltage variations in the DC bus.

$$P_{pv} - P_{dl} = C_{dc}V^r_{DC}\frac{d(V_{DC} - V^r_{DC})}{dt} + D_{dc}(V_{DC} - V^r_{DC})$$

(6.6)

The DC link capacitor's damping is ensured by a large resistor connected alongside it. However, the equilibrium of the DC bus voltage depicted by this setup often experiences significant deviations in its steady state, demanding active power assistance from controllable elements within the DC bus, such as the energy storage system (ESS) and the interlinking converter (ILC). While the capacitor swiftly addresses transient fluctuations for brief periods, the consistent power response necessary for stable DC bus operation is orchestrated in a decentralized fashion by the ESS (P_{ESS}) and the ILC (P_{ILC}). This dynamic is captured through voltage deviations and can be mathematically expressed as:

$$P_{ESS} + P_{ILC} = k_{dc}(V^r_{DC} - V_{DC})$$

(6.7)

The steady-state power relation at DC link capacitor is represented as

$$P_{pv} - P_{dl} = -P_{ESS} - P_{ILC}$$

(6.8)

Combining (6.6) and (6.8), (6.9) presents a unified expression for the transient and steady-state power responses required at the DC bus. This equation encapsulates the dynamic interactions between the physical DC link capacitance, virtual capacitance, and power references across different operational modes.

$$P_{pv} - P_{dl} = \underbrace{C_{dc}V_{DC}^r \frac{d(\Delta V_{DC})}{dt}}_{\text{transient response}} + \underbrace{k_{dc}(\Delta V_{DC}) + P_{set}^{dc}}_{\text{steadystate response}}$$

(6.9)

$$P_{set}^{dc} = \begin{cases} \dfrac{M^{dc}}{S}(\Delta V_{DC}) \Rightarrow \text{islanded and grid connected} \\ 0 \Rightarrow \text{community connected} \end{cases}$$

The damping coefficient D_{dc} is neglected in the above relationship as it is significantly smaller compared to the droop coefficient k_{dc}. Similar to the AC bus, the power reference command is used to provide secondary power reference during various operational modes, as detailed in (6.9), where M_{dc} represents the integrator constant. It is important to note that the responsibility of providing a transient response for load demand in both the DC and AC buses falls on the DC link capacitor. The transient power demand of the DC load can be minimized by quickly initiating a steady-state response from the energy storage systems and AC bus through the interlinking converter. However, the transient power required by the AC bus depends on the inertia constant of the virtual synchronous generator control managing the ILC. Therefore, to determine the DC link capacitance needed to fulfill the transient power demand, the instantaneous power requirements on the DC side and the AC side are equated as

$$C_{dc}V_{DC}^r \frac{d(V_{DC}^r - V_{DC})}{dt} = \frac{2HS_r}{\omega_r}\frac{d}{dt}(\omega_r - \omega)$$

(6.10)

Upon simplifying the above relation and linearizing, the DC link capacitance can be represented as

$$C_{dc} = \frac{2HS_r(\omega_r - \omega_{min})}{w_r V_{DC}^r(V_{DC}^r - V_{DC}^{min})}$$

(6.11)

To overcome the utilization of high DC-link capacitance, the power relation expressed in (6.9) is represented as

$$P_{pv} - P_{dl} = \underbrace{C_{dc}V_{DC}^r \frac{d(\Delta V_{DC})}{dt}}_{\text{transient response from DC link Capacitor}}$$

$$+ \underbrace{C_{dc}^v V_{DC}^r \frac{d(\Delta V_{DC})}{dt}}_{\text{transient response from super-capacitor}} + \underbrace{k_{dc}(\Delta V_{DC}) + P_{set}}_{\text{steadystate response}}$$

(6.12)

The virtual capacitance C_{dc}^v is significantly larger than the physical DC link capacitance C_{dc}. The transient response from the virtual capacitance is also shared between the combination of ESS and ILC. The ESS comprises a combination of batteries and super-capacitors, with the transient response allocated to the super-capacitor due to its high-power density. Additionally, the ILC is responsible for providing a transient response extracted from the super-capacitors integrated into the rest of the hybrid AC–DC microgrid through the AC bus. It is important to note that power-sharing among the battery, super-capacitor, and ILC depends on the operational mode, and the allocation of power reference among these entities is outlined as follows:

6.2.2.1 Islanded mode

The Energy Storage System (ESS) integrated into the DC bus is tasked with providing both transient and steady-state responses during a load transient in the DC bus, while the ILC remains inactive since the AC bus is disconnected from the network. The power references for the battery (P_b), supercapacitor (P_{sc}), and interlinking converter (P_{ILC}) are denoted as:

$$P_b = k_{dc}(V_{DC}^r - V_{DC}) - P_{set}^{dc}$$
$$P_{sc} = C_{dc}^v V_{DC}^r \frac{d(V_{DC}^r - V_{DC})}{dt}; P_{ILC} = 0 \tag{6.13}$$

6.2.2.2 Community connected mode

The energy storage systems and the interlinking converter together supply the necessary transient and steady-state responses for a DC load transient. The corresponding power references are denoted as:

$$P_b + P_{ILC} = k_{dc}(V_{DC}^r - V_{DC}) - P_{set}^{dc}$$
$$P_{sc} = C_{dc}^v V_{DC}^r \frac{d(V_{DC}^r - V_{DC})}{dt} \tag{6.14}$$

6.2.2.3 Grid-connected mode

During grid-connected mode, the task of maintaining the power balance in the DC bus is assigned to the grid via the interlinking converter. The battery remains inactive, while the super-capacitor only offers a transient response to power deviations in the DC bus. The power references for the battery, super-capacitor, and interlinking converter during grid-connected mode are indicated as:

$$P_{ILC} = k_{dc}(V_{DC}^r - V_{DC}) - P_{set}^{dc}$$
$$P_{sc} = C_{dc}^v V_{DC}^r \frac{d(V_{DC}^r - V_{DC})}{dt}; P_b = 0 \tag{6.15}$$

6.3 Decentralized virtual synchronous generator control scheme for ILC

The decentralized virtual synchronous generator control scheme for the ILC is pivotal in maintaining power equilibrium between the AC and DC buses in hybrid

AC–DC microgrids. Equations (6.5), (6.12), (6.13), and (6.14) outline the power response needed from the ILC during various operational modes. The unified power response generated by the ILC across these modes is expressed as:

$$P_{ilc}^{ac} = \frac{2HS_r}{\omega_r}\frac{d(\omega_r - \omega)}{dt} + (D_\omega + k_{ac})(\omega_r - \omega) + P_{set}^{ac}$$

$$P_{ilc}^{dc} = C_{dc}^v V_{dc}^r \frac{d(V_{dc}^r - V_{dc})}{dt} + (k_{dc})(V_{dc}^r - V_{dc}) - P_{set}^{dc}$$

(6.16)

Here, P_{ilc}^{ac} represents the power injected into the AC bus from the DC bus during an AC load transient, while P_{ilc}^{dc} represents the power injected from the AC bus to the DC bus during a DC load transient. Combining these responses yields the unified power response of the ILC, as shown in (6.16).

$$P_{ilc} = \frac{2HS_r}{\omega_r}\frac{d\Delta\omega}{dt} + (D_\omega + k_{ac})\Delta\omega + P_{set}^{ac}$$

$$+ C_{dc}^v V_{dc}^r \frac{d\Delta V_{dc}}{dt} + k_{dc}\Delta V_{dc} + P_{set}^{dc}$$

(6.17)

The above equation resembles a conventional swing equation of a virtual synchronous generator, determining the reference frequency for seamless power transfer between AC and DC buses during transient and steady-state conditions. Equation (6.18) is derived from (6.17) and represents the balanced power response required for AC and DC load transients in the microgrid.

$$P_{in} - P_{ilc} = \frac{2HS_r}{\omega_r}\frac{d(\omega - \omega_r)}{dt} + (D_\omega + k_{ac})(\omega - \omega_r)$$

$$P_{in} = P_{set}^{ac} + P_{set}^{dc} + C_{dc}^v V_{dc}^r \frac{d(\Delta V_{dc})}{dt} + k_{dc}(\Delta V_{dc})$$

(6.18)

This equation determines the reference frequency for the ILC, considering the transient response necessary to compensate for DC bus load demand. Equation (6.20) modifies (6.18) for practical application, accounting for the differential component of DC bus voltage deviation.

$$\omega = \left(\frac{1}{1 + st_f}\right)\left(\frac{P_{in} - P_{ilc}}{D_\omega + k_{ac}}\right); t_f = \frac{2HS_r}{\omega_r(D_\omega + k_{ac})}$$

$$\omega = \left(\frac{P_{set}^{ac} + P_{set}^{dc} + k_{dc}(\Delta V_{dc})}{(D_\omega + k_{ac})(1 + st_f)}\right) + \left(\frac{C_{dc}^v V_{dc}^r s\Delta V_{dc}}{(D_\omega + k_{ac})(1 + st_f)}\right)$$

(6.19)

Equation (6.19) indicates that the reference frequency for the interlinking converter includes a component to generate the transient response needed to compensate for the load demand in the DC bus. However, the equation mentioned in (6.18) necessitates calculating the differential component of the DC bus voltage deviation $\frac{d(\Delta V_{dc})}{dt}$, which is inherently complex. Therefore, the equation in (6.18) can be adjusted as:

$$\frac{2HS_r}{\omega_r}\frac{d\Delta\omega}{dt} - C_{dc}^v V_{dc}^r \frac{d\Delta V_{dc}}{dt} =$$

$$P_{set}^{ac} + P_{set}^{dc} - P_{ilc} + k_{dc}\Delta V_{dc} - k_{ac}\Delta\omega$$

$$\frac{2HS_r}{\omega_r}s\left(\Delta\omega - \frac{C_{dc}^v V_{dc}^r \omega_r}{2HS_r}\Delta V_{dc}\right) =$$

$$P_{set}^{ac} + P_{set}^{dc} - P_{ilc} + k_{dc}\Delta V_{dc} - k_{ac}\Delta\omega \qquad (6.20)$$

$$\Delta\omega = \left(\frac{(P_{set}^{ac} + P_{set}^{dc} - P_{ilc} + k_{dc}\Delta V_{dc} - k_{ac}\Delta\omega)}{\left(\frac{2HS_r}{\omega_r}\right)s}\right)$$

$$+ \omega_r + M\Delta V_{dc}; \; M = \frac{C_{dc}^v V_{dc}^r \omega_r}{2HS_r}$$

The relationship described above bears a resemblance to the conventional virtual synchronous generator but with adjusted power reference commands. The unified power reference command applicable to islanded, community-connected, and grid-connected modes of operation is derived from (6.5) and (6.9), and it is represented as:

$$P_{set} = \frac{M^{ac}}{S}\left((\omega_r - \omega) - \frac{M^{dc}}{M^{ac}}(V_{dc}^r - V_{dc})\right) \qquad (6.21)$$

The decentralized Virtual Synchronous Generator (VSG) control strategy for interlinking converters within a hybrid AC–DC microgrid is outlined in Figure 6.2. This approach continuously monitors terminal voltages across the AC and DC buses, alongside the output current of the interlinking converter, to compute the active and reactive power injected into the AC bus. These metrics, along with the monitored terminal voltage at the DC side of the ILC, are inputted into the VSG control scheme, designed per (6.20). Subsequently, the VSG control scheme generates the reference frequency and, consequently, the reference phase angle for the ILC.

The AC bus's required voltage magnitude is determined using the Q- droop method, emulating a synchronous generator's excitation system, as shown in Figure 6.2. Here, Q, Q_{set}, and D_q denote the reactive power output, reactive power set-point, and reactive power droop coefficient, respectively. These, along with the reference phase angle and voltage (v_{ref}), are fed into dual voltage and current loop controllers. These controllers ascertain the reference three-phase signal necessary to generate the control signal for the interlinking converter.

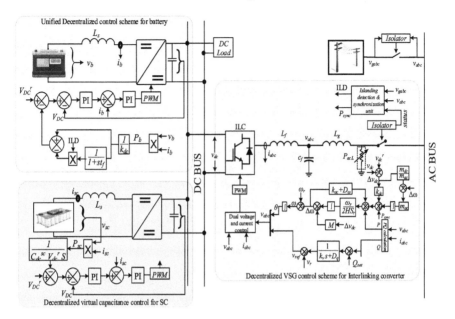

Figure 6.2   Proposed enhanced decentralized virtual synchronous generator control scheme in a hybrid AC–DC microgrid

6.4   Decentralized control of energy storage systems

In grid-connected and community-connected modes, Energy Storage Systems (ESS) play a vital role in maintaining stability by responding to both steady-state and transient load fluctuations across the DC and AC buses. Collaborating closely with the interlinking converter, ensure efficient management of these responses. Conversely, in the islanded mode of operation, the ESS takes on the sole responsibility of maintaining the power balance.

6.4.1   *Decentralized virtual capacitance control scheme for super-capacitor*

The transient response requirements are specifically allocated to the super-capacitor, as outlined in (6.13), (6.14), and (6.15).

Real-time computation of the differential component of the DC bus voltage, however, can be complex. Hence, the power response is adjusted to derive the reference DC voltage that the converter integrated into the super-capacitor should uphold. This adjustment can be expressed linearly as the slope of the DC bus voltage at each time step, akin to the concept presented in (6.13).

$$V_{dc}^* = V_{dc}^r - \frac{P_{sc}}{C_{dc}^{sc} V_{dc}^r S} \tag{6.22}$$

Here, V_{dc}^* and C_{dc}^{sc} denote the reference DC link voltage and the virtual capacitance emulated by the super-capacitor, respectively. In this equation, P_{sc} signifies the super-capacitor's output power and the resulting reference voltage is fed into a cascaded voltage and current control loop, as depicted in Figure 6.2. This loop generates the control signal for the bidirectional converter, ensuring the necessary transient response is achieved.

6.4.2   *Unified decentralized control scheme for battery*

The battery's role in managing the DC bus's power and the hybrid AC–DC microgrid, alongside the interlinking converter, is pivotal. It shares the responsibility of stabilizing the DC bus voltage at its rated value with the ILC during various operational modes. However, unlike the interlinking converter, the battery's power response varies across modes, as detailed in (6.13), (6.14), and (6.15). It plays a crucial role in facilitating smooth transitions between operational modes, although excessive mode switching can occasionally strain its integrator capacity. To address this challenge, a unified reference signal is established alongside an Islanding Detection Signal (ILD) that indicates the operating mode of the hybrid AC–DC microgrid, ensuring seamless transitions between modes. In this configuration, the battery acts as a potential voltage source, injecting power as needed rather than simply supplying power. The reference voltage (V_{DC}^*) for the bidirectional DC–DC converter (BDDC) is derived from the power response described in (6.13), (6.14), and (6.15), and it is formulated as follows:

$$V_{DC}^* = \begin{cases} V_{DC}^r - P_b \dfrac{1}{k_{dc}} \Rightarrow \text{community} * \text{\&grid connected} \\ V_{DC}^r - P_b \left(\dfrac{S}{k_{dc}S + M^{dc}} \right) \Rightarrow \text{islanded} \end{cases} \tag{6.23}$$

The unified voltage reference applicable to all modes of operation, incorporating the islanding detection signal, can be expressed as:

$$V_{DC}^* = V_{DC}^r - \frac{P_b}{k_{dc}} \left(1 - ILD \left(\frac{1}{1 + st_f} \right) \right); t_f = \frac{k_{dc}}{M^{dc}} \tag{6.24}$$

The islanding detection signal toggles the power response between islanded mode and either community or grid-connected mode, and vice versa. The resultant unified reference voltage is then fed into a cascaded voltage and current controller to generate the necessary steady-state power response, as illustrated in Figure 6.2.

6.5   Small signal modeling and stability analysis

The small signal modeling of multiple hybrid AC–DC microgrids connected to a network and operating in community-connected mode is explained in this section. The modeling is segregated into two sections comprising of DC bus and AC bus along with an interlinking converter.

6.5.1 Modeling of DC microgrid

The DC bus of a hybrid AC–DC microgrid consists of PV, battery, supercapacitors, and DC loads. The PV is treated as a constant current source injecting maximum power extracted from the renewables. To reduce the complexity of the small signal model, the DC–DC converters integrating the battery and supercapacitor to the DC bus are tuned as per the method given in [22,23]. Hence the battery and supercapacitors are treated as power sources injecting real power with respect to the DC bus voltage deviations. The voltage of the DC bus measured at the DC link capacitor with reference to the current injected by PV (i_{PV}), battery (i_{bat}), supercapacitor (i_{sc}), and power drawn by DC load (i_{DCL}) and interlinking converter (i_{ilc}) is represented as:

$$C_{dc}^i \Delta V_{DC}^{\bullet i} = \Delta i_{pv}^i + \Delta i_{bat}^i + \Delta i_{sc}^i - \Delta i_{dload}^i - \Delta i_{ilc}^{DC^i} \tag{6.25}$$

$$\Delta V_{DC}^{\bullet i} = \frac{\Delta P_{pv}^i + \Delta P_{bat}^i + \Delta P_{sc}^i - \Delta P_{dload}^i - \Delta P_{ilc}^{DC^i}}{C_{dc}^i V_{DC}^r} \tag{6.26}$$

$$\Delta P_{ilc}^{DC^i} = \frac{\Delta P_{ilc}^i}{\eta}; \Delta P_{bat}^i = -k_{dc}^i \Delta V_{DC}^i \tag{6.27}$$

$$\Delta P_{sc}^i = -C_{dc}^{sc^i} V_{dc}^r \Delta V_{DC}^{\bullet i} \tag{6.28}$$

6.5.2 Modeling of AC microgrid and interlinking converter

The AC sub-grid comprises multiple renewable sources injecting available active power to the AC bus and loads operating on AC power. The interlinking converter intertwines the AC and DC bus and regulates the flow of active and reactive power in the system. The generalized model of multiple hybrid AC DC microgrids operating in network-connected mode is modeled as follows:

6.5.2.1 Modeling of VSG control of interlinking converter

The small signal model of the modified VSG control scheme proposed in this paper is modeled similarly to the active and reactive power control scheme of conventional VSG control. The reference operating frequency of the interlinking converter with respect to output power, DC link voltage can be represented as

$$\frac{2H^i S_r}{\omega_r} \Delta \dot{M}_i = k_{dc}^i \Delta V_{dc}^i - \Delta P_{ilc}^i - k_{ac}^i \Delta \omega^i \tag{6.29}$$

$$\Delta M_i = \Delta \omega^i - \frac{C_{dc}^{vi} V_{dc}^r \omega_r}{2H^i S_r} \Delta V_{dc}^i \tag{6.30}$$

$$\Delta \dot{\delta}_i = \Delta \omega_i \tag{6.31}$$

Similarly, the relation between the output reactive power and the magnitude of the terminal voltage is represented as

$$k_v \Delta \dot{E}_i = \Delta Q_{set}^i - \Delta Q_{ilc}^i - d_q^i \Delta E_i \tag{6.32}$$

6.5.2.2 Modelling of voltage and current control schemes

The derived AC side reference frequency and magnitude of terminal voltage are used to generate the equivalent d-q component of reference voltage and represented as

$$\Delta v_d^{i*} = \Delta E_i; \; \Delta v_q^{i*} = 0 \tag{6.33}$$

due to the local reference frame, however, all the interlinking converters are to be modeled with respect to a common reference frame. The common reference frame considered here is too well known to all the interlinking converters and the d-q reference quantities when transformed into a common reference frame are represented as

$$\Delta V_d^{i*} = v_d^{i*} \sin \, \delta_i(0)\Delta\delta_i + \Delta v_d^{i*} \cos \, \delta_i(0) \tag{6.34}$$

$$\Delta V_q^{i*} = v_d^{i*} \cos \, \delta_i(0)\Delta\delta_i + \Delta v_d^{i*} \sin \, \delta_i(0) \tag{6.35}$$

The voltage controller inputs the reference voltage and measured voltage signals to determine reference current quantities through the PI controller and represented as

$$\begin{aligned}
\Delta I_d^{i*} &= -\omega_r C_f^i \Delta V_{oq}^i + k_{pv}(\Delta V_d^{i*} - \Delta V_{od}^i) + k_{iv}\Delta\alpha_d^i \\
\Delta I_q^{i*} &= \omega_r C_f^i \Delta V_{od}^i + k_{pv}(\Delta V_q^{i*} - \Delta V_{oq}^i) + k_{iv}\Delta\alpha_q^i \\
\Delta\alpha_d^{\bullet i} &= \Delta V_d^{i*} - \Delta V_{od}^i; \; \Delta\alpha_q^{\bullet i} = \Delta V_q^{i*} - \Delta V_{op}^i
\end{aligned} \tag{6.36}$$

The derived reference current quantities are used to generate the reference voltage utilized in pulse-width modulation (PWM) to generate the terminal voltage of the ILC. The relation is represented as

$$\begin{aligned}
\Delta V_d^{iref} &= -\omega_r L_f^i \Delta I_{lq}^i + k_{pi}(\Delta I_d^{i*} - \Delta I_{ld}^i) + k_{ii}\Delta\beta_d^i \\
\Delta V_q^{iref} &= \omega_r L_f^i \Delta I_{ld}^i + k_{pi}(\Delta I_q^{i*} - \Delta I_{lq}^i) + k_{ii}\Delta\beta_q^i \\
\Delta\dot{\beta}_d^i &= \Delta I_d^{i*} - \Delta I_{ld}^i; \; \Delta\dot{\beta}_q^i = \Delta I_q^{i*} - \Delta I_{lq}^i \\
\Delta V_{id}^i &= \Delta V_d^{iref}; \; \Delta V_{iq}^i = \Delta V_q^{iref}
\end{aligned} \tag{6.37}$$

The generated reference voltage can be assumed equivalent to the terminal voltage of the interlinking converter as the PWM block is a unity function.

6.5.2.3 Modelling of output LC filter and networked connection

The output current of the interlinking converters with reference to the LC filter is represented as

$$\begin{aligned}
L_f^i \Delta \dot{I}_{ld}^i &= \omega_r L_f^i \Delta I_{lq}^i - r_f^i \Delta I_{ld}^i + \Delta V_{id}^i - \Delta V_{od}^i \\
L_f^i \Delta \dot{I}_{lq}^i &= -\omega_r L_f^i \Delta I_{ld}^i - r_f^i \Delta I_{lq}^i + \Delta V_{iq}^i - \Delta V_{oq}^i
\end{aligned} \tag{6.38}$$

The output voltage observed at the capacitance of the LC filter with respect to the connected load of the power injected into the network is represented as

$$
\begin{aligned}
\Delta \dot{V}^i_{od} &= \omega_r \frac{\Delta V^i_{oq}}{C^i_f} + \frac{\Delta I^i_{ld}}{C^i_f} + \frac{\Delta I^i_{wd}}{C^i_f} - \frac{\Delta I^i_{aload\_d}}{C^i_f} - \frac{\Delta I^i_{od}}{C^i_f} \\
\Delta \dot{V}^i_{oq} &= -\omega_r \frac{\Delta V^i_{od}}{C^i_f} + \frac{\Delta I^i_{lq}}{C^i_f} + \frac{\Delta I^i_{wq}}{C^i_f} - \frac{\Delta I^i_{aload\_q}}{C^i_f} - \frac{\Delta I^i_{oq}}{C^i_f}
\end{aligned}
\tag{6.39}
$$

The load connected to the AC bus and the active and reactive injected by the wind into the AC bus are represented as equivalent current quantities in the d-q frame as follows

$$
I^i_{Zd} = \frac{V^i_{od}P^i_Z + V^i_{oq}Q^i_Z}{\left(V^i_{od}\right)^2 + \left(V^i_{oq}\right)^2}; \; I^i_{Zq} = \frac{V^i_{op}P^i_Z - V^i_{od}Q^i_Z}{\left(V^i_{od}\right)^2 + \left(V^i_{oq}\right)^2}
\tag{6.40}
$$

where Z represents both AC load and wind quantities respectively. The currents injected into the rest of the network with respect to the changes in loads of rest of microgrids is represented as

$$
\begin{aligned}
L^i_o \Delta \dot{I}^i_{od} &= \omega_r L^i_o \Delta I^i_{oq} - r^i_o \Delta I^i_{od} + \Delta V^i_{od} - \Delta V^i_{bd} \\
L^i_o \Delta \dot{I}^i_{oq} &= -\omega_r L^i_o \Delta I^i_{od} - r^i_o \Delta I^i_{oq} + \Delta V^i_{oq} - \Delta V^i_{bq}
\end{aligned}
\tag{6.41}
$$

The junction voltage is represented as ΔV^i_{bdq} derived with respect to the network configuration. However, to simplify the analysis the hybrid AC–DC microgrids are assumed to be connected at the point of coupling and the voltage at the PCC is represented as

$$
\Delta V^i_{pccd} = R_{pcc} \sum_{i=1}^{n} \Delta I^i_{od}; \; \Delta V^i_{pccq} = R_{pcc} \sum_{i=1}^{n} \Delta I^i_{oq}
\tag{6.42}
$$

The output active and reactive power of the interlinking converter $\Delta P^i_{ilc}, \Delta Q^i_{ilc}$ required for the VSG control scheme is represented as

$$
\begin{aligned}
\Delta P^i_{ilc} &= \Delta P^i_{aload} - \Delta P^i_w + \Delta P^i_{out} \\
\Delta Q^i_{ilc} &= \Delta Q^i_{aload} - \Delta Q^i_w + \Delta Q^i_{out} \\
\Delta P^i_{out} &= V^i_{od}\Delta I^i_{od} + I^i_{od}\Delta V^i_{od} + V^i_{oq}\Delta I^i_{oq} + I^i_{oq}\Delta V^i_{oq} \\
\Delta Q^i_{out} &= V^i_{od}\Delta I^i_{oq} + I^i_{oq}\Delta V^i_{od} - V^i_{oq}\Delta I^i_{od} - I^i_{od}\Delta V^i_{oq}
\end{aligned}
\tag{6.43}
$$

6.5.3   Small signal model of network-connected hybrid microgrids

The small signal model of multiple networked connected hybrid AC–DC microgrids can be represented as

$$\Delta \dot{X} = A\Delta X + B\Delta U$$

$$\Delta X = [\Delta X_1, \Delta X_2, \Delta X_3]^T; \Delta U = [\Delta U_1, \Delta U_2, \Delta U_3]^T$$

$$\Delta X_i = \begin{bmatrix} \Delta V_{DC}^i, \Delta M_i, \Delta \delta_i, \Delta E_i, \Delta \alpha_d^i, \Delta \alpha_q^i, \Delta \beta_d^i, \\ \Delta \beta_q^i, \Delta I_{ld}^i, \Delta I_{lq}^i, \Delta V_{od}^i, \Delta V_{oq}^i, \Delta I_{od}^i, \Delta I_{oq}^i \end{bmatrix} \tag{6.44}$$

$$\Delta U_i = \begin{bmatrix} \Delta P_{pv}^i, \Delta P_{Dload}^i, \Delta P_{Aload}^i, \Delta Q_{Aload}^i, \\ \Delta P_w^i, \Delta Q_w^i, \Delta Q_{set}^i \end{bmatrix}$$

$$a_1 = \frac{-k_{dc}^i}{W}; a_2 = \frac{-I_{od}^i}{W}; a_3 = \frac{-I_{oq}^i}{W}; a_4 = \frac{-V_{od}^i}{W}$$

$$W = (C_{dc}^i + C_{dc}^{sc^i})V_{dc}^r; a_5 = \frac{-V_{oq}^i}{W}$$

$$a_6 = k_{dc}^i - \frac{k_{ac}^i C_{dc}^{v^i} V_{dc}^r \omega_r}{2H^i S_r}; a_7 = \frac{-I_{od}^i \omega_r}{2H^i S_r}$$

$$a_8 = \frac{-I_{oq}^i \omega_r}{2H^i S_r}; a_9 = \frac{-V_{od}^i \omega_r}{2H^i S_r}; b_1 = \frac{-V_{oq}^i \omega_r}{2H^i S_r}$$

$$b_2 = \frac{C_{dc}^{v^i} V_{dc}^r \omega_r}{2H^i S_r}; b_3 = E_i \text{Sin}\delta_i(0); b_4 = \cos \delta_i(0);$$

$$b_5 = E_i \cos \delta_i(0); b_6 = \text{Sin}\delta_i(0); \tag{6.45}$$

$$b_7 = \frac{-(1 + k_{pi}k_{pv})}{L_f^i}; b_8 = \frac{-\omega_r C_f^i k_{pi}}{L_f^i};$$

$$b_9 = \frac{-(r_o^i + R_{pcc})}{L_f^i}; c_1 = 1/W; c_2 = \frac{-\omega_r}{2H^i S_r}$$

$$c_3 = \frac{-V_{od}^i}{N}; c_4 = \frac{-V_{oq}^i}{N}; N = C_f^i((V_{od}^i)^2 + (V_{oq}^i)^2)$$

$$B_i = \begin{bmatrix} c_1 & -c_1 & -c_1 & 0 & c_1 & 0 & 0 \\ 0 & 0 & c_2 & 0 & -c_2 & 0 & 0 \\ 0 & 0 & 0 & \frac{-1}{k_v^i} & 0 & \frac{1}{k_v^i} & \frac{1}{k_v^i} \\ 0 & 0 & 0 & 0 & 0 & 0 & 0 \\ 0 & 0 & 0 & 0 & 0 & 0 & 0 \\ 0 & 0 & 0 & 0 & 0 & 0 & 0 \\ 0 & 0 & 0 & 0 & 0 & 0 & 0 \\ 0 & 0 & 0 & 0 & 0 & 0 & 0 \\ 0 & 0 & 0 & 0 & 0 & 0 & 0 \\ 0 & 0 & 0 & 0 & 0 & 0 & 0 \\ 0 & 0 & 0 & 0 & 0 & 0 & 0 \\ 0 & 0 & c_3 & c_4 & -c_3 & -c_4 & 0 \\ 0 & 0 & c_4 & c_3 & -c_4 & -c_3 & 0 \\ 0 & 0 & 0 & 0 & 0 & 0 & 0 \end{bmatrix} \tag{6.46}$$

The coefficients of the A matrix are represented as follows

$$
A = \begin{bmatrix}
a_1 & 0 & 0 & 0 & 0 & 0 & 0 & 0 & 0 & 0 & a_2 & a_3 & a_4 & a_5 \\
a_6 & \dfrac{k^i_{ac}\omega_r}{2H^iS_r} & 0 & 0 & 0 & 0 & 0 & 0 & 0 & 0 & a_7 & a_8 & a_9 & b_1 \\
b_2 & 1 & 0 & 0 & 0 & 0 & 0 & 0 & 0 & 0 & 0 & 0 & 0 & 0 \\
0 & 0 & 0 & \dfrac{-d^i_q}{k^i_v} & 0 & 0 & 0 & 0 & 0 & 0 & \dfrac{-I^i_{oq}}{k^i_v} & \dfrac{I^i_{od}}{k^i_v} & \dfrac{V^i_{oq}}{k^i_v} & \dfrac{-V^i_{od}}{k^i_v} \\
0 & 0 & b_3 & b_4 & 0 & 0 & 0 & 0 & 0 & -1 & 0 & 0 & 0 & 0 \\
0 & 0 & b_5 & b_6 & 0 & 0 & 0 & 0 & 0 & 0 & -1 & 0 & 0 & 0 \\
0 & 0 & k_{pv}b_3 & k_{pv}b_4 & k_{iv} & 0 & 0 & 0 & -1 & 0 & -k_{pv} & -\omega_r C^i_f & 0 & 0 \\
0 & 0 & k_{pv}b_5 & k_{pv}b_6 & 0 & k_{iv} & 0 & 0 & 0 & -1 & \omega_r C^i_f & -k_{pv} & 0 & 0 \\
0 & 0 & \dfrac{k_{pv}k_{iv}b_3}{L^i_l} & \dfrac{k_{pv}k_{iv}b_4}{L^i_l} & \dfrac{k_{pv}k_{iv}}{L^i_l} & 0 & \dfrac{k_{ii}}{L^i_l} & 0 & \dfrac{-r^i_f}{L^i_l} & 0 & b_7 & b_8 & 0 & 0 \\
0 & 0 & \dfrac{k_{pv}k_{iv}b_5}{L^i_l} & \dfrac{k_{pv}k_{iv}b_6}{L^i_l} & 0 & \dfrac{k_{pv}k_{iv}}{L^i_l} & 0 & \dfrac{k_{ii}}{L^i_l} & \dfrac{-r^i_f}{L^i_l} & b_8 & b_7 & 0 & 0 \\
0 & 0 & 0 & 0 & 0 & 0 & 0 & 0 & \dfrac{1}{C^i_f} & 0 & 0 & \dfrac{\omega_r}{C^i_f} & \dfrac{-1}{C^i_f} & 0 \\
0 & 0 & 0 & 0 & 0 & 0 & 0 & 0 & 0 & \dfrac{1}{C^i_f} & \dfrac{-\omega_r}{C^i_f} & 0 & 0 & \dfrac{1}{C^i_f} \\
0 & 0 & 0 & 0 & 0 & 0 & 0 & 0 & 0 & 0 & \dfrac{1}{L^i_o} & 0 & b_9 & \omega_r \\
0 & 0 & 0 & 0 & 0 & 0 & 0 & 0 & 0 & 0 & 0 & \dfrac{1}{L^i_o} & -\omega_r & b_9
\end{bmatrix}
$$

$$(6.47)$$

6.6 Power hardware in loop simulation

The proposed method for achieving bidirectional virtual inertia in a hybrid AC–DC microgrid was put to the test using Power Hardware In Loop (PHIL) simulation. This simulation involved a real hardware prototype of the microgrid, shown in Figure 6.3, interacting with a network of simulated hybrid AC–DC microgrids running on a Real-Time Digital Simulator (RTDS). This interaction was made possible through a power amplifier, as seen in Figure 6.3(a). The details of the prototype's parameters and the control strategy can be found in Table 6.1.

The proposed control scheme is validated under a wide variety of scenarios, such as grid-connected mode, community-connected mode, and islanded mode with versatile load changes. A comparison of the proposed control scheme with the unified control scheme for interlinking converters proposed in [23] is presented for better illustration.

6.6.1 Grid-connected mode

During grid-connected mode, the responsibility for maintaining the stability of both AC and DC buses in terms of frequency and DC voltage falls on the utility. As a result, the hybrid AC–DC microgrid operates in a power dispatch mode. In this setup, the grid acts as both a source and a sink for surplus and deficit power within the system. Consequently, the battery, which is responsible for providing steady-state response, remains inactive while the super-capacitor steps in to provide the necessary transient response during load transients.

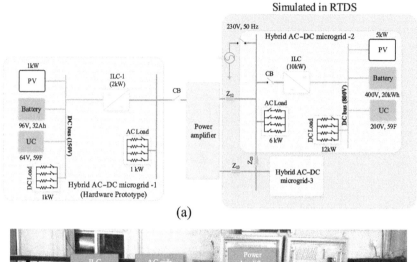

(a)

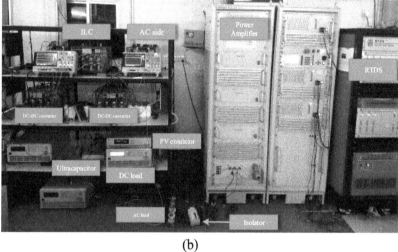

(b)

Figure 6.3   (a) PHIL simulation setup featuring a hybrid AC–DC microgrid. (b) A hybrid AC–DC microgrid prototype utilized in the simulation.

Table 6.1   Parameters of the prototype Hybrid AC–DC microgrid

Subsystem	Description	Range
DC bus	PV	8 kW
	Battery	400 V, 65 Ah
	Supercapacitor	100 V, 109 F
	Rated DC bus voltage (vdcr)	800 V
	DC load	5 kW
	DC droop coefficient (kdc)	400
	Integrator Constant (Mdc)	2

(Continues)

Table 6.1   (Continued)

Subsystem	Description	Range
	DC link capacitance (Cdc)	1 mF
	Virtual capacitance emulated by super-capacitor (Cdcsc)	0.1 F
	Bandwidth of cascaded	100 Hz (outer loop)
	Control loops of battery	1 kHz (inner loop)
	Bandwidth of cascaded control	500 Hz (outer loop)
	Loops of supercapacitor	2 kHz (inner loop)
AC bus	Capacity of wind Pw	2 kW
	AC load	7 kW
	rated frequency (wr)	314.1592 rad/s
	AC bus voltage	230 V RMS
Interlinking	Inertia constant (H)	2
Converter	Rated capacity (Sr)	10 kVA
	Damping coefficient (Dw)	100
	Droop coefficient (kac)	2,000
	Integrator constant (Mac)	10
	Virtual capacitance emulated interlinking converter (cdci)	0.1 F

The performance of the network-connected microgrid during grid-connected mode, utilizing the proposed control scheme, is depicted in Figure 6.5(a). Initially, all hybrid microgrids (HMGs) are feeding surplus power into the grid. However, as soon as a DC load of 2.5 kW is activated at $t = 2$ s and subsequently deactivated at $t = 12$ s, the interlinking converter swiftly supplies the required transient response from the rest of the HMGs in the network, in coordination with the super-capacitor in the specific HMG where the load transient occurs. Nonetheless, the steady-state response is managed by the grid, as it maintains control over AC frequency and DC voltage. The transient response needed for the load demand is proportionally shared among all hybrid AC–DC microgrids in the system.

6.6.2   Community connected mode

The network-connected hybrid AC–DC microgrids transit to a community-connected mode in the absence of the grid. During this scenario, the responsibility of power management is shared among the HMGs integrated into the network. The performance of the network-connected HMGs during community-connected mode is shown in Figure 6.5(b). The network seamlessly transitioned from grid-connected to utility-connected mode at $t = 3$ s, upon which the HMGs in the network supported the load demand and maintained the power balance in the system. The performance and collaborative power management of the network-connected HMGs are validated through a DC load transient of 2.5 kW in two different HMGs at $t = 13$ s and $t = 23$ s in HMG-1 and HMG-2, respectively. It is evident that the proposed control strategy enabled the interlinking converter to provide the required transient response initially and shift to a steady-state response over time. The performance of the proposed strategy is compared with the control scheme proposed in [23], which is shown in Figure 6.4. The

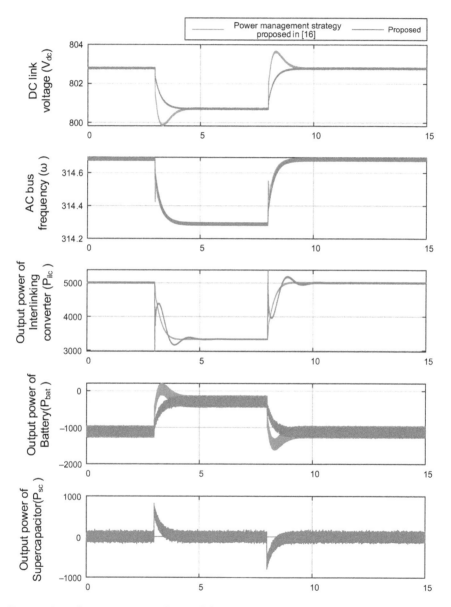

Figure 6.4   Comparative analysis of the performance of the proposed control scheme during community-connected mode

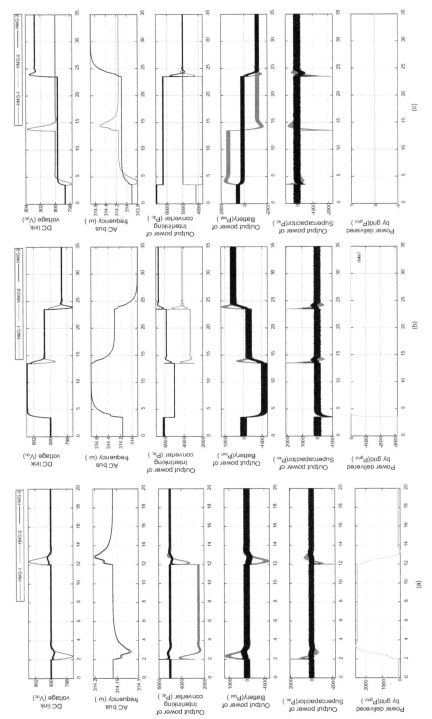

Figure 6.5 PHIL simulation of network-connected hybrid AC–DC microgrid during (a) grid-connected mode, (b) community-connected mode, and (c) islanded mode

comparative analysis validates that the proposed strategy ensures low deviations in DC voltages and AC frequency and enhanced transient response. The enhanced transient response translates to the improved life span of the steady-state energy storage devices.

6.6.3   Islanded mode

Hybrid AC–DC microgrid islands from the network during unavoidable circumstances, during which the energy storage systems hold the responsibility to maintain power balance through constant DC voltage and AC frequency in the microgrid. The interlinking convert operates in grid-forming mode to regulate the AC bus during this scenario. The battery regulates the DC link voltage at the rated value, with transient power response support from a super-capacitor, while the renewable sources operate at their rated power. The HMG-1 is islanded from the network at $t = 3$ s, which seamlessly transitioned from community mode to islanded, as shown in Figure 6.5(c). To demonstrate the transient and steady-state response during islanded mode, a load demand of 2.5 kW is tuned off at $t = 23$ s, and similarly, a DC load shedding of 2.5 kW in HMG-2 in the network at $t = 23$ s shows the inability of HMG-1 to participate in power management due to islanded mode.

6.7   Conclusion and future scope

This chapter has introduced an advanced virtual synchronous generator control scheme designed to generate bidirectional virtual inertia in hybrid AC–DC microgrids. The described control strategy effectively synthesizes both transient and steady-state power responses required for both AC and DC buses. Moreover, it ensures a harmonized management of transient and steady-state responses across all energy storage systems within the interconnected hybrid AC–DC microgrids.

The proposed control scheme is validated through power hardware-in-the-loop experimentation on a prototype. The experimental outcomes unequivocally confirmed the efficacy of the control scheme, showcasing its ability to provide bidirectional power support during transient as well as steady-state conditions within the hybrid AC–DC microgrid.

Looking ahead, the future scope involves enhancing the proposed control strategy by factoring in the state of charge (SoC) of batteries and super-capacitors. This enhancement aims to improve the determination of local power references and facilitate global power sharing through a decentralized approach. These developments are poised to further refine the performance and efficiency of hybrid AC–DC microgrids.

References

[1]   Q. Xu, J. Xiao, P. Wang, and C. Wen, "A decentralized control strategy for economic operation of autonomous ac, dc, and hybrid ac/dc microgrids," *IEEE Transactions on Energy Conversion*, vol. 32, no. 4, pp. 1345–1355, 2017.

[2]   J. Wang, C. Jin, and P. Wang, "A uniform control strategy for the inter-linking converter in hierarchical controlled hybrid ac/dc microgrids," *IEEE Transactions on Industrial Electronics*, vol. 65, no. 8, pp. 6188–6197, 2018.

[3]   H. Yoo, T. Nguyen, and H. Kim, "Consensus-based distributed coordination control of hybrid ac/dc microgrids," *IEEE Transactions on Sustainable Energy*, vol. 11, no. 2, pp. 629–639, 2020.

[4]   Y. Xia, W. Wei, M. Yu, X. Wang, and Y. Peng, "Power management for a hybrid ac/dc microgrid with multiple subgrids," *IEEE Transactions on Power Electronics*, vol. 33, no. 4, pp. 3520–3533, 2018.

[5]   Y. Xia, W. Wei, M. Yu, Y. Peng, and J. Tang, "Decentralized multi-time scale power control for a hybrid ac/dc microgrid with multiple subgrids," *IEEE Transactions on Power Electronics*, vol. 33, no. 5, pp. 4061–4072, 2018.

[6]   S. Peyghami, H. Mokhtari, and F. Blaabjerg, "Autonomous operation of a hybrid ac/dc microgrid with multiple interlinking converters," *IEEE Transactions on Smart Grid*, vol. 9, no. 6, pp. 6480–6488, 2018.

[7]   U. Tamrakar, D. Shrestha, M. Maharjan, B. Bhattarai, T. Hansen, and R. Tonkoski, "Virtual inertia: Current trends and future directions," *Applied Sciences*, vol. 7, no. 654, 2017.

[8]   Q. Zhong, "Virtual synchronous machines: A unified interface for grid integration," *IEEE Power Electronics Magazine*, vol. 3, no. 4, pp. 18–27, 2016.

[9]   J. Xiao, P. Wang, and L. Setyawan, "Hierarchical control of hybrid energy storage system in dc microgrids," *IEEE Transactions on Industrial Electronics*, vol. 62, no. 8, pp. 4915–4924, 2015.

[10]   Y. Huang, X. Yuan, J. Hu, P. Zhou, and D. Wang, "Dc-bus voltage control stability affected by ac-bus voltage control in vscs connected to weak ac grids," *IEEE Journal of Emerging and Selected Topics in Power Electronics*, vol. 4, no. 2, pp. 445–458, 2016.

[11]   E. Unamuno, J. Paniagua, and J. A. Barrena, "Unified virtual inertia for ac and dc microgrids: And the role of interlinking converters," *IEEE Electrification Magazine*, vol. 7, no. 4, pp. 56–68, 2019.

[12]   P. J. d. S. Neto, T. A. d. S. Barros, J. P. C. Silveira, E. R. Filho, J. C. Vasquez, and J. M. Guerrero, "Power management strategy based on virtual inertia for dc microgrids," *IEEE Transactions on Power Electronics*, vol. 35, no. 11, pp. 12 472–12 485, 2020.

[13]   W. Wu, Y. Chen, A. Luo, *et al.*, "A virtual inertia control strategy for dc microgrids analogized with virtual synchronous machines," *IEEE Transactions on Industrial Electronics*, vol. 64, no. 7, pp. 6005–6016, 2017.

[14]   X. Zhu, Z. Xie, S. Jing, and H. Ren, "Distributed virtual inertia control and stability analysis of dc microgrid," *IET Generation, Transmission Distribution*, vol. 12, no. 14, pp. 3477–3486, 2018.

[15]   X. Li, L. Guo, Y. Li, *et al.*, "A unified control for the dc–ac interlinking converters in hybrid ac/dc microgrids," *IEEE Transactions on Smart Grid*, vol. 9, no. 6, pp. 6540–6553, 2018.

[16]   Z. Liu, S. Miao, Z. Fan, J. Liu, and Q. Tu, "Improved power flow control strategy of the hybrid ac/dc microgrid based on vsm," *IET Generation, Transmission Distribution*, vol. 13, no. 1, pp. 81–91, 2019.

[17]   L. He, Y. Li, J. M. Guerrero, and Y. Cao, "A comprehensive inertial control strategy for hybrid ac/dc microgrid with distributed generations," *IEEE Transactions on Smart Grid*, vol. 11, no. 2, pp. 1737–1747, 2020.

[18]   L. He, Y. Li, Z. Shuai, *et al.*, "A flexible power control strategy for hybrid ac/dc zones of shipboard power system with distributed energy storages," *IEEE Transactions on Industrial Informatics*, vol. 14, no. 12, pp. 5496–5508, 2018.

[19]   G. Melath, S. Rangarajan, and V. Agarwal, "A novel control scheme for enhancing the transient performance of an islanded hybrid ac–dc microgrid," *IEEE Transactions on Power Electronics*, vol. 34, no. 10, pp. 9644–9654, 2019.

[20]   X. Li, Z. Li, L. Guo, J. Zhu, Y. Wang, and C. Wangand, "Enhanced dynamic stability control for low-inertia hybrid AC/DC microgrid with distributed energy storage systems," *IEEE Access*, vol. 7, pp. 91234–91242, 2019.

[21]   K. Bandla, and N. Prasad Padhy. An improved virtual synchronous generator control for decentralized and coordinated sharing of transient response in hybrid AC–DC microgrids. *2022 IEEE IAS Global Conference On Emerging Technologies (GlobConET)*. pp. 769–774 (2022)

[22]   Q. Xu, J. Xiao, P. Wang, X. Pan, and C. Wen, "A decentralized control strategy for autonomous transient power sharing and state-of-charge recovery in hybrid energy storage system," *IEEE Transactions on sustainable energy*, vol. 8, no. 4, pp. 1443–1452, 2017.

[23]   Q. Xu, J. Xiao, P. Wang, X. Hu, and M.Y. Lee, "A decentralized power management strategy for hybrid energy storage system with autonomous bus voltage restoration and state-of-charge recovery," *IEEE Transactions on Industrial Electronics*, vol. 64, no. 9, pp. 7098–7108, 2017.

Chapter 7

Power quality issues in microgrids

Azizulrahman Shafiqurrahman[1], Preetha Sreekumar[2]
and Vinod Khadkikar[1]

This chapter addresses the pivotal challenge of maintaining power quality within microgrids, a critical component for their effective and sustainable operation. It presents a comprehensive review of the various types of microgrids and the primary obstacles they encounter. Additionally, it explores various strategies to maintain power quality, including droop control, centralized and decentralized load sharing, negative virtual harmonic impedance (NVH-Z) technique for harmonic current sharing, and dynamic voltage restorer (DVR) which are aimed at preserving power quality. Furthermore, the chapter incorporates case studies to illustrate the efficacy of these strategies in improving the performance and power quality of microgrids across multiple scenarios, including linear, nonlinear, and unbalanced load sharing in islanded configurations. These case studies demonstrate the presented solutions' success and highlight the significance of the discussed control mechanisms in ensuring the operational effectiveness and sustainability of the microgrids.

7.1 Introduction

Microgrid is a low voltage system that incorporates distributed generators (DGs) including traditional generators powered by fossil fuels such as diesel or natural gas generators, or renewable energy sources (RES) based DGs like solar photovoltaics (PVs), wind turbines, biomass generators, energy storage elements mainly batteries, and load. It is an integrated system that operates on a local scale, comprising various components consisting of loads, distributed energy resources (DERs), intelligent switching mechanisms, protective devices, and sophisticated communication, control, and automation systems. DERs are compact energy sources installable at utility sites, near where the energy is consumed, or directly on customer properties, offering localized electricity supply. DER is not limited to just DG as it also encompasses other small energy resources, including energy storage systems (ESSs) and electric vehicles (EVs) [1].

[1]Electrical Engineering Department, Khalifa University, UAE
[2]Faculty of Engineering Technology and Science, Higher Colleges of Technology, UAE

Microgrid is a small-scale system compared to a main power grid and offers several services used in several applications such as integration of RES to reduce the dependency on fossil fuels, grid stability, grid support or grid independence, and reduce cost in the long-term. Figure 7.1 presents a brief schematic of a microgrid architecture, featuring essential components comprising various loads, an ESS, RES, electric vehicle supply equipment (EVSE) connected to EVs, and a central control unit. This control unit plays a pivotal role in regulating energy distribution among these elements and overseeing the microgrid's integration with the broader utility grid. The diagram highlights the interconnected nature of the microgrid, illustrating how it harmonizes traditional and renewable energy sources, storage capabilities, and the emerging demand for electric vehicles, ensuring a sustainable and efficient energy ecosystem.

Microgrid functions as a single controllable entity that is both capable of operating in tandem with the central electricity grid and autonomously, thereby supporting connectivity in grid-tied or independent islanded modes. In grid-connected mode, the microgrid exchanges power with the main utility grid, however, in the islanded mode, they function independently from the utility grid. Furthermore, due to their dependency on specific environmental conditions and constrained energy production, DGs require integration with the current power infrastructure, including the utility grid and ESSs to achieve optimal performance [2].

Renewable DG systems, though advantageous for their sustainability and eco-friendliness, encounter significant challenges due to their reliance on inherently variable weather conditions. This variability results in an unpredictable and often fluctuating power output, making it difficult to guarantee a steady, reliable, and consistent supply of energy. Consequently, the integration of such systems into the broader energy grid necessitates ESSs and advanced planning to mitigate these fluctuations and enhance the reliability and efficiency of renewable DGs [3].

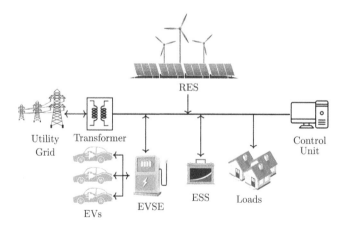

Figure 7.1   Structure diagram of a microgrid

7.2   Microgrid types

Figure 7.2 presents the classification of microgrid systems, segmented into five different categories: connectivity, DG dispatchability, control, power source, and load type. Under connectivity, microgrids are distinguished by their operational mode, either as grid-connected or islanded, indicating if they are attached to a power grid or functioning independently.

Microgrid DGs are classified into dispatchable and non-dispatchable, reflecting the ability to control the generation output. Dispatchable DGs can be regulated by the microgrid's master controller, adhering to specific technical constraints. Conversely, non-dispatchable DGs are beyond the control of the microgrid master controller due to their reliance on unpredictable sources. Predominantly, these non-dispatchable units are RESs such as solar and wind, characterized by their fluctuating and intermittent power output [4].

Microgrids can be categorized based on their control structures and the type of current they utilize. Specifically, they are distinguished into three main types: alternating current (AC), direct current (DC), and hybrid systems that combine AC and DC. Additionally, microgrids are classified according to their control mechanisms, which can either be centralized, where control is managed from a single point, or decentralized, allowing for distributed management across various points within the system. In [5], an in-depth analysis of both AC and DC microgrid systems is presented, emphasizing their integration with DG units that harness RESs, with the discussion of ESSs' role in managing power fluctuations and ensuring a steady power supply. The power source category differentiates between renewable and non-RESs that supply the microgrid. Lastly, the load type category breaks down the consumers of the microgrid energy into residential, commercial, or industrial, denoting the end-use application of the supplied power.

Microgrids can be classified based on their connection with the electrical grid into two modes; islanded mode which operates independently, and grid-connected mode which is integrated into the main power grid. Such connection depends on the purpose of the microgrid. The classification of microgrids based on their connection to the electrical grid into islanded or grid-connected modes is fundamental to understanding their operation, purpose, and benefits [6]. This chapter focuses on connectivity-based classification as the connection to the main grid remains a defining aspect that shapes the control techniques and design considerations employed in microgrid systems.

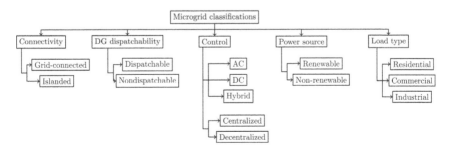

Figure 7.2   Classification of microgrids

7.3   Grid-connected microgrids

When operating in grid-connected mode, the microgrid is linked to the primary electrical grid, allowing it to either supply power to the grid or draw power from it. The grid-connected mode allows for the trade of electricity. For DGs to work their best, they need to be connected to the main power grid and use ESS since DGs can only produce a limited amount of power and highly depend on weather and other environmental factors. By linking them with the grid and storage systems, their efficiency can be ensured [7].

Microgrids are characterized by their ability for bidirectional power flow, the integration of DG sources, demand-side management strategies, and a significant incorporation of power electronics. These characteristics introduce complexities in controlling the microgrid, both in terms of managing its day-to-day operations in each mode and in switching from being connected to the grid to operating independently [6]. It is essential to tackle these control issues to optimize the performance of microgrids and to fully harness their potential benefits.

Grid-connected microgrids provide numerous benefits, as they integrate smoothly with local distribution networks and greatly enhance their reliability. By reducing the strain on transmission and distribution infrastructure, they not only cut costs associated with these systems but also enhance the power quality to meet the reactive and nonlinear demands of local loads. Moreover, the diversification of resources within these microgrids paves the way for a reduction in carbon emissions, aligning with global efforts to mitigate environmental impact. This holistic approach to energy management underscores the pivotal role of grid-connected microgrids in fostering a sustainable and efficient power supply framework [8].

7.4   Islanded microgrids

When operating in islanded mode, a microgrid functions autonomously, supplying electricity to its immediate surroundings. Originally, the concept of islanding, or off-grid operation, was proposed as a contingency plan for significant disturbances within the main electrical grid. Extensive research has been dedicated to identifying instances of islanding and ensuring a smooth transition between being connected to the main grid to operate independently, and vice versa. Typically, disconnection from the primary grid is reserved for moments of critical grid instability and is not a frequent occurrence. Consequently, initial investigations into islanded mode operation are concentrated on ensuring stable functionality over brief periods.

However, advancements in research have broadened the scope of applications for microgrids operating in islanded mode significantly. These extended applications include not only rural electrification, which promises to bring reliable power to remote areas without access to the main grid, but also specialized sectors such as aviation, automotive, and maritime industries. In these contexts, islanded microgrids offer a versatile and reliable power solution, capable of enhancing operational efficiency, safety, and sustainability. This evolution in the understanding and application of islanded microgrid technology underscores its potential to revolutionize energy distribution and consumption across a diverse range of environments and industries. Key difficulties encountered by an isolated microgrid include maintaining voltage stability, distributing the load evenly, and ensuring the quality of power.

7.5 Control techniques

Power quality refers to the process of measuring, analyzing, and improving the bus voltage to ensure it maintains a sinusoidal waveform at a standard voltage and frequency. Issues related to power quality in distributed power systems have become more pronounced with the growing use of power electronic-based switches and non-linear loads. Additionally, the integration of DGs, such as solar and wind energy sources, further impacts the stability and quality of power within a microgrid [7]. Therefore, it is crucial to adopt sophisticated control strategies aimed at countering the negative repercussions associated with DG-integrated microgrids [9]. Implementing these advanced techniques ensures the efficient and reliable operation of the power grid, maintaining power quality and supporting the sustainable incorporation of renewable energy sources [10].

7.5.1 Grid-connected microgrid control techniques

Grid-connected DG inverters are essential for integrating renewable energy into the electrical grid, distinguished by three main types: grid-forming (GFM), grid-feeding, and grid-supporting converters, each defined by its interaction with the grid and operational approach [8].

7.5.1.1 Grid-forming (GFM) control

Grid-forming (GFM) power converters essentially serve as controlled voltage sources, exemplified by systems like standby Uninterruptible Power Supplies (UPS). These converters, operating in a voltage-controlled mode, enable DG inverters to function in islanded and grid-connected scenarios, providing flexibility and reliability in power delivery [11].

7.5.1.2 Grid-feeding control

The most prevalent operational mode for grid-connected converters is the grid-feeding mode, where the DG inverter operates as a current source. The primary aim here is maximizing the injection of active power into the grid, typically maintaining a unity power factor, though this mode does not support standalone or off-grid operations [12].

7.5.1.3 Grid-supporting control

Grid-supporting converters play a crucial role in enhancing the quality of power in the network. Capable of operating either as a current source with shunt impedance or a voltage source with series impedance, these converters supply power to local loads and regulate the voltage amplitude and frequency for both the main grid and microgrids, highlighting their importance in maintaining grid stability and power quality [13].

7.5.2   *Islanded microgrid control techniques*

Control techniques play a crucial role in ensuring stability and reliability when addressing islanded mode operations. These techniques are broadly categorized into centralized and decentralized methods, distinguished primarily by the deployment of communication links to execute control actions. Centralized control involves a singular control center that receives and processes information from various parts of the grid to make comprehensive decisions, thereby necessitating a robust communication infrastructure. On the other hand, decentralized control is characterized by the distribution of control tasks among multiple controllers within the system, significantly reducing the reliance on communication networks [9].

Several techniques are designed to address critical issues such as the balancing of load variations across different phases to avoid the scenarios of overloading or underloading, which could potentially lead to system inefficiencies or even failures.

The load sharing among DGs is primarily determined by the impedance characteristics of the lines, which can lead to unequal distribution of load if impedances vary. To address this, two primary strategies have been developed which are centralized and decentralized control strategies.

7.5.2.1   **Centralized control**

In centralized control systems, each DG unit receives its reference current either directly from a central controller or through communication links with other DG units, enabling a cohesive operational framework [14]. This approach includes various active load-sharing strategies such as concentrated control, master-slave configuration, circular chain control (3C), and average load sharing (ALS) each with distinct mechanisms for managing the distribution of power [15].

Concentrated control: It divides the total load current equally among DG units, relying on a central controller to measure load current in real time and calculate the necessary reference currents. In this control system, the reference current for each DG is determined by equally distributing the total load current among all DG units. This method necessitates the real-time monitoring of the load current and relies on a central controller to execute the calculation of the reference currents. Enhancements to this approach could involve implementing robust communication networks to ensure rapid and reliable data exchange between the central controller and each DG unit [14].

Master-slave: In a Master-Slave control setup, one designated DG unit operates as the master, essentially serving as a voltage source for the system. This master unit is responsible for dispatching current references to other DG units, which are designated as slaves, ensuring a harmonized operation. In case of a failure in the master unit, an alternative unit swiftly assumes the master's role, thereby preventing a complete system breakdown. Additionally, the master-slave control strategy can be executed by employing both active and reactive power references, enhancing the flexibility and reliability of the system's overall operation. This method ensures a stable and efficient distribution of power, adapting to changes in demand or unit functionality seamlessly [16,17].

Circular chain control (3C): This strategy links the DG units in a control ring, where each unit sets its reference current to match the output of the preceding unit, promoting a chain of consistent power distribution. Active load sharing stands out for its democratic approach, averaging the output current of all active DG units to determine each unit's reference current, offering flexibility and scalability. It typically focuses on the output current's fundamental component to avoid the pitfalls of averaging harmonic components, which could lead to circulating currents. This technique, while innovative, underscores the need for precise control in managing both average active and reactive powers instead of merely current references [18].

Average load sharing (ALS): ALS stands out as a proper method for managing power distribution, allowing each DG unit to adjust its reference current to the collective average of all participating DG units. This approach underscores ALS's flexibility and scalability within centralized control methods. Notably, ALS is versatile enough to accommodate average active and reactive powers instead of relying solely on current references. However, its primary focus on the fundamental component of the output current means that ALS does not address harmonic components. This oversight can lead to inaccuracies, as averaging harmonic components may generate misleading signals and potentially cause circulating currents, undermining system efficiency and stability [19].

Mitigating the voltage unbalance in loads has been introduced in [20,21]. This method involves the collection and transmission of load voltage data via a low-bandwidth communication channel directly to each DG unit. The technique employs the measurement of the voltage unbalance factor (VUF) to gauge the degree of imbalance present within the load voltages. Following this assessment, it focuses on adjusting the VUF to an optimal level. To achieve this, a compensation signal for voltage unbalance, coupled with a virtual impedance loop, is implemented to refine the voltage control mechanism of the DG units. The control strategy proposed in [22] focuses on distributing the negative sequence current across DGs through the implementation of a virtual negative sequence impedance controller. This approach involves sensing the load currents and incorporating them into the DG control loop, necessitating communication technology. By utilizing both local and non-local load currents, the strategy adjusts the DGs' output negative sequence impedance, effectively diminishing the negative sequence currents in the lines.

7.5.2.2 Decentralized control

Decentralized compensation methods, as discussed in the literature offer alternatives to relying on communication infrastructure. The decentralized strategy enables the independent operation of each DG [23], adopting a droop control approach based on direct local measurements for effective load sharing [24]. A modified model predictive control (MPC) strategy has been presented in [25], to correct voltage imbalances at the DG terminals. Additionally, it addresses the issue of active power overloading in DG units caused by unequal power distribution. Implementation of the MPC techniques in microgrids is further discussed in detail in [26–28]. In [29], another decentralized approach for mitigating voltage unbalances at distributed generation DG terminals is introduced. This method involves computing the negative sequence reactive power using the positive sequence line voltage along with the negative sequence current, thereby deriving a reference value for conductance. The process incorporates the utilization of a newly formulated Q-G droop equation to determine this conductance value. Nonetheless, it is important to acknowledge that adopting this strategy entails a compromise between achieving voltage unbalance correction and maintaining effective voltage regulation, a point also explored in [22]. This inherent trade-off emphasizes the complexity of optimizing DG system performance, emphasizing the need for a balanced approach to system design and operation. In [30], a method is presented that combines virtual resistance with P/Q droop control mechanisms. This integration is achieved through the application of positive sequence currents, offering a sophisticated solution to address issues of load imbalance. This approach not only enhances the efficiency of power distribution systems but also ensures a more balanced load management by mitigating discrepancies in electrical current distribution.

A decentralized method for distributing load imbalances based on the line impedance of DGs has been introduced in [31,32]. This strategy enhances the efficiency and reliability of power distribution by optimizing the way electrical loads are balanced across the network. By taking into account the specific impedance characteristics of DG lines where line impedances are typically unknown, the presented method ensures a more effective and equitable distribution of power, leading to improved system stability. This approach, detailed in [31,32], represents a significant advancement in the management of distributed energy resources, offering a more adaptive and responsive solution to the challenges of modern power systems.

Overall, the extensive adoption of decentralized schemes, as opposed to relying on communication technology for coordinated power sharing, can be attributed to several factors. First, the complexity involved in implementing communication-based coordination systems poses significant challenges. Additionally, the high costs associated with these technologies deter their widespread use. Moreover, the limited reliability of supervisory systems further discourages reliance on communication technology for power distribution. Instead, decentralized control schemes offer a simpler, more cost-effective, and reliable alternative for managing power sharing, making them a preferred choice in various settings [7]. As a result, the control strategies based on communication links, as discussed, may fall short of being the optimal solution for islanded microgrids. This is due to the potential geographical separation of DG units and loads within a microgrid, which can span several kilometers. Consequently, decentralized control systems, which operate independently of communication links, emerge as a more suitable alternative. These systems offer enhanced reliability and efficiency in managing the distributed energy resources and loads across the expansive layout of microgrids, ensuring a more stable and self-sufficient operation [33].

7.6 Power quality challenges

7.6.1 *Inertia and voltage stability*

A notable limitation of microgrids is their low rotational inertia compared to conventional synchronous generators, which naturally offer significant inertia, especially in the islanded mode which lacks physical inertia in inverter-based DG units. This renders the microgrid vulnerable to fluctuations caused by disturbances in the system. Microgrids lack this inherent trait, and this issue becomes more pronounced with the extensive incorporation of RESs. The reliance on power converters over mechanical rotating devices in RES integration contributes to further diminishing the system's inherent inertia. Such reduction in inertia can lead to several operational challenges, including frequency instability, increased sensitivity to disruptions, and compromised effectiveness in grid management [10].

The absence of physical inertia within an islanded microgrid can lead to issues of voltage instability. To tackle this challenge, droop control methods have been employed. Droop-based control strategies involve managing inverter-based DG units to replicate the dynamics of synchronous generators. This replication is achieved by deliberately modifying the frequency and voltage amplitude in direct correlation with the active and reactive power outputs. The controlled, gradual adjustments in frequency and voltage amplitude contribute significantly to enhancing the stability of weak microgrids characterized by high impedance, thereby addressing the inherent instability concerns [24]. To tackle the issues of inertia in droop control mechanisms, the concept of a virtual synchronous generator (VSG) is introduced as a promising solution. This approach, discussed in [34,35], aims to mimic the operation of conventional synchronous generators, thereby improving the reliability and stability of the power systems. By integrating the VSG model, the microgrids can dynamically adjust to changes, ensuring a more resilient and efficient energy distribution.

7.6.2 *Load sharing*

In decentralized control systems, DG units function in a mode that controls voltage, employing droop-based methods to distribute the load for managing DG inverters' reference voltages. These droop-based load-sharing strategies are elaborately categorized into two main types: linear and nonlinear load sharing. Linear load sharing operates on a direct proportionality basis, where power output adjusts linearly with changes in load demand. Conversely, nonlinear load sharing responds to load changes in a nonlinear manner, often employing complex algorithms to optimize performance under varying load conditions. This approach enables a more flexible and efficient management of power sharing among DG units [36].

7.6.2.1 Linear load sharing

The droop-based method for load sharing is considered a prominently adopted decentralized strategy, especially for handling linear loads, as discussed in [37–39]. Delving into the mechanics of droop control requires examining the DG inverter's equivalent circuit model, which is coupled to the point of common coupling (PCC),

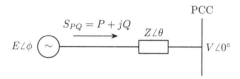

Figure 7.3   Circuit diagram representing a DG inverter linked to an AC bus at PCC

as depicted in Figure 7.3. This approach not only facilitates the effective distribution of loads across multiple sources but also supports the adaptability and efficiency of decentralized power systems in managing energy distribution, ensuring stability and reliability even in the absence of centralized control mechanisms.

Where E and V represent the magnitudes of the inverter output voltage and the voltage at the PCC respectively, and Z and θ denote the magnitude and phase angles of the line impedance. S_{PQ}, P, and Q represent the linear apparent power, the fundamental active power, and the fundamental reactive power, respectively. The basic expressions for the active and reactive power of the inverter can be represented as follows.

$$P = \frac{E}{R^2 + X^2}[R(E - V\cos\varnothing) + XV\sin\varnothing] \qquad (7.1)$$

$$Q = \frac{E}{R^2 + X^2}[-RV\sin\varnothing + X(E - V\cos\varnothing)] \qquad (7.2)$$

Therefore, by modifying the phase angle and voltage amplitude at the output of the inverter, the control over the flow of both active and reactive power is effectively achievable through a line. This principle forms the core of droop control, a method that leverages these relationships to maintain power system stability.

In islanded systems characterized by inductive lines, each DG inverter is meticulously configured to mimic the operational dynamics of synchronous generators. This emulation involves adjusting the phase angle and the voltage amplitude based on the active and reactive power outputs, respectively. Such adjustments adhere to the $P - f/Q - V$ droop principle, commonly referred to as conventional droop control. The most basic version of the droop equations, as outlined in the literature is presented as follows.

$$\omega = \omega^* - m(P - P^*) \qquad (7.3)$$

$$E = E^* - n(Q - Q^*) \qquad (7.4)$$

Where the output frequency and peak voltage amplitude of the inverter when there is no load are denoted by ω^* and E^* respectively. P^* and Q^* represent the references for active and reactive power. The droop coefficients are represented by m and n. By fine-tuning these parameters, droop control facilitates a responsive and adaptable microgrid, ensuring a balanced and consistent supply of power.

Figure 7.4 shows the implementation of droop control, while Figure 7.5 illustrates the variations in ω and E with traditional droop.

For effective sharing of load among DG units, it is crucial that their droop coefficients are meticulously calibrated in accordance with each unit's capacity. This means the product of the droop coefficient (m) and the unit's rating (S) should be consistent across all units, expressed as $m_1S_1 = m_2S_2 = \cdots = m_nS_n$. Similarly, this principle applies to another set of coefficients (n), ensuring $n_1S_1 = n_2S_2 = \cdots = n_nS_n$, where S_n represents the capacity of the nth DG unit. This alignment guarantees that load sharing is proportional and efficient, reflecting the capabilities and intended operation of each unit within the microgrid (Figure 7.5).

In a low-voltage microgrid with resistive lines, the relationship between $P - f/Q - V$ gets opposite. This is also described through the droop equations for lines that are mainly resistive, referred to as opposite droop.

$$\omega = \omega^* + mQ \tag{7.5}$$

$$E = E^* - nP \tag{7.6}$$

Enhancing the load-sharing precision in high-voltage lines can be significantly achieved by reducing the interdependence between $P - V/Q - f$. This enhancement involves modifying the line's impedance towards a more inductive nature. Utilizing virtual impedance allows for the effective increase of the line's inductance without physical alterations. This method involves the integration of a virtual impedance by

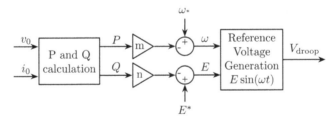

Figure 7.4   Block diagram of conventional droop control

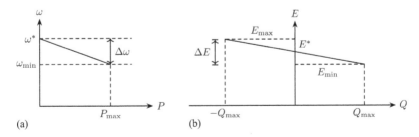

Figure 7.5   (a) Frequency and (b) voltage amplitude variations based on basic droop formulas

deducting a specific voltage component, derived from the product of the inverter's output current fundamental component and a predetermined constant impedance, from the voltage droop (v_{droop}) control. To effectively incorporate this virtual inductance, L_v, adjustments are made to the v_{droop} control, resulting in a revised voltage reference (v_{ref}), as detailed in [40]. This approach not only optimizes the distribution of loads across high-voltage lines but also enhances the overall system stability and efficiency. The sinusoidal reference voltage, known as droop voltage, is produced based on the frequency and voltage amplitude derived from droop equations.

$$v_{droop} = E\sin \omega t \qquad (7.7)$$

$$v_{ref} = v_{droop} - j\omega L_v i_{01} \qquad (7.8)$$

where i_{01} signifies the fundamental component of the inverter's output current. The term $j\omega L_v i_{01}$ refers to the voltage drop attributed to the virtual inductance. By deducting this from v_{droop}, it effectively integrates a virtual inductance L_v into the line. Likewise, to reduce the interaction between P and f, and Q and V in low-voltage circuits, virtual resistance is employed.

7.6.2.2   Nonlinear load sharing

The linear load sharing methods are designed to manage the distribution of fundamental active and reactive power. Yet, the impact of nonlinear loads and the resulting power quality problems in low voltage distribution systems needs to be considered as well.

The literature describes various techniques for nonlinear load sharing, which include the use of decentralized droop control for linear load distribution. Nonlinear loads, on the other hand, are managed either through communication networks or via decentralized control methods. Techniques that rely on communication for managing nonlinear loads are known as hierarchical control strategies. Meanwhile, methods that are independent of communication links for such purposes are referred to as decentralized harmonic control strategies.

Hierarchical control: A secondary level of control utilizes communication infrastructure to enable nonlinear sharing of loads effectively in hierarchical control systems. The strategies implementing hierarchical control prioritize linear load sharing as an essential service, managing it independently of communication links. In contrast, nonlinear load sharing is treated as a secondary service. This approach imposes significant restrictions on hierarchical control schemes because there are instances when the harmonic content in the load current surpasses the fundamental component. Maintaining high-quality voltage for the load is crucial, highlighting the limitations of this assumption in such control systems. Additionally, all hierarchical control strategies necessitate the real-time transmission of complete harmonic data, which might demand more bandwidth and increase costs [13].

Decentralized harmonic control: To explore the concept of harmonic load flow within a distribution network, a scenario involving two DGs supplying power to a shared, nonlinear load is presented in Figure 7.6 with pathways shown for both

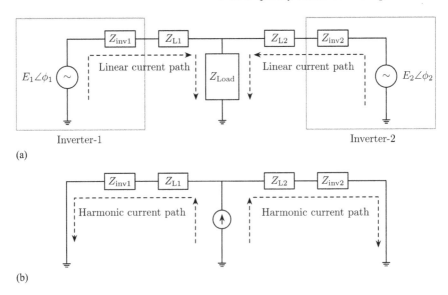

Figure 7.6   Two DG systems operating at (a) the fundamental frequency and (b) harmonic frequencies

linear and nonlinear currents in this configuration. At the fundamental frequency, the inverter is represented as an ideal voltage source combined with its inherent inverter impedance (Z_{inv}). Z_{L1} and Z_{L2} denote the impedances of the distribution lines, and the nonlinear load is depicted as an ideal current source. Given that the inverter's reference voltage is purely sinusoidal, the DGs function essentially as impedances to harmonic currents. The harmonic load current is distributed between the two inverters according to the combined impedances of their respective paths, denoted as $Z_{inv1} + Z_{L1}$ and $Z_{inv2} + Z_{L2}$.

   One of the fundamental factors contributing to voltage unbalance in electrical systems is the incorporation of unbalanced loads. Specifically, this phenomenon occurs when single-phase loads are connected in a manner that is not uniformly distributed across the three phases, either by attaching them to a single phase or spanning them between two phases. This unequal distribution of electrical load leads to variations in voltage levels across the phases, thereby causing a voltage unbalance. Such imbalances can significantly affect the efficiency and lifespan of electrical equipment, as they are designed to operate optimally under conditions of balanced voltage.

   In islanded microgrid configurations, it is essential for the overall power demand to be distributed proportionally among the various DGs taking into account their individual capacity ratings. Without the implementation of a strategic approach to load sharing, there exists a tendency for the DG unit located in closest proximity to the demand site to bear a disproportionate share of the power supply burden. Such a scenario can precipitate the risk of overburdening that specific DG unit, potentially leading to operational failures or reduced efficiency. The aforementioned droop control method plays a pivotal role in mitigating these risks by facilitating the equitable distribution of power, especially in handling linear loads. This technique ensures that the energy supply is balanced, preventing any single generator from being overstressed and thus maintaining the stability and reliability of the microgrid system [41].

7.6.3   Harmonics

7.6.3.1   Harmonic voltage sharing

In the context of integrating DG inverters with the main electrical grid, the grid-connected scenario presents a viable strategy for enhancing power quality. Specifically, these inverters can significantly contribute to mitigating harmonic voltages which is a common power quality issue within the utility grid. The ability of DG inverters to provide such support not only stabilizes the electrical grid but also ensures a more efficient and reliable power supply. This application is particularly valuable in modern electrical networks, where the demand for high-quality power is increasingly critical due to the sensitive nature of the loads being served. Through the implementation of grid-supporting DG inverters, utility providers can address power quality challenges more effectively, leading to improved system performance and customer satisfaction.

Several methods stand out as promising solutions for mitigating harmonic disturbances and enhancing the overall quality of power in electrical distribution networks. In [42], a method for controlling the voltage of DG systems is introduced to precisely align with the harmonic voltages found in the utility grid. By ensuring the DG voltage is in perfect harmony with grid harmonics, this approach effectively prevents the propagation of harmonic currents through the distribution feeder. As a direct consequence, the current flowing into the grid maintains a purely sinusoidal shape, leading to significant enhancements in the grid's voltage profile.

Another method in [43] involves utilizing DG units scattered across different locations in the electrical network, functioning as active power filters. The core mechanism behind this technique is the generation of a harmonic reference current designed to control the DG inverter. This process is governed by a control command, denoted as G_{vh}, where G symbolizes the virtual conductance, and vh represents the harmonic voltage observed at the PCC. In addition to this function, DG units equipped with this technology offer other significant benefits in terms of power quality solutions. Among these are the damping of harmonic resonance and the cancelation of harmonic voltage caused by grid-side inductance within a grid-connected microgrid setup [44,45]. Through these methods, DG units not only provide a reliable power supply to critical loads but also enhance the overall power quality.

7.6.3.2   Harmonic current sharing

The power quality within islanded microgrids is significantly impacted by the pervasive incorporation of electronic devices, including but not limited to computers, fluorescent lighting, and a myriad of household appliances [10]. These devices are known to draw substantial harmonic currents, which, when combined with the high impedance levels of feeders typical in microgrids, result in notable voltage distortions. This phenomenon links the issues of current and voltage harmonics, making them deeply interdependent. At the heart of power quality challenges in microgrids lies the unregulated flow of harmonic currents, which stands as the primary contributor to these disturbances [45].

To elaborate further, modern domestic and commercial environments are heavily reliant on electronic devices that operate with non-linear loads, generating harmonic currents as a byproduct of their functioning. These harmonics are electrical frequencies that can interfere with the grid's standard operating frequency, leading to inefficiencies and potential damage to both the infrastructure and the devices connected to it. The limited capacity and the specific electrical characteristics, such as high feeder impedance, amplify the effects of harmonic currents. Consequently, this results in voltage distortions that can compromise the stability and efficiency of the power supply, affecting everything from individual appliance performance to overall system reliability. The presence of uncontrolled harmonic currents thus emerges as a central factor undermining power quality in microgrid configurations.

To enhance the control of harmonic current flow within microgrids, several nonlinear load-sharing strategies have been developed. Despite these advancements, the complex challenges posed by voltage and current harmonics remain to be addressed further. This situation highlights the critical necessity to study and develop strategies for harmonic current sharing and control within islanded microgrids. This exploration represents a primary motivation for this chapter, highlighting the ongoing quest to mitigate these electrical disturbances and improve power quality in microgrid systems.

7.7 Negative virtual impedance technique for harmonic current sharing

In this section, a unique approach known as the virtual harmonic impedance (VHI) technique is introduced which is designed to minimize the effective line impedance through the application of negative VHI values. This method significantly enhances the voltage quality at the load by lowering overall line impedance. Additionally, an adaptive droop characteristic has been developed that dynamically determines the VHI value, ensuring optimal performance. To prevent overloading of DGs within the microgrid, the inverter rating is meticulously considered in the presented methodology, ensuring a balanced and efficient energy distribution system. The current techniques in series impedance control effectively facilitate harmonic sharing through the adjustment of the reference voltage via virtual harmonic resistance (VHR). This modification of the reference voltage, as detailed in [46,47] is shown as follows.

$$v_{ref,m} = v_{ref} - i_h R_h \tag{7.9}$$

The block diagram depicting the implementation of VHR within a dual-DG microgrid setup is illustrated in Figure 7.7. This diagram specifically represents a single-phase microgrid configuration with two DGs, which is the standard model

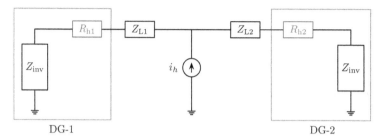

Figure 7.7   Schematic representation illustrating the implementation of VHR

used throughout the chapter. In this case, a constant resistance, denoted as R_h, is virtually integrated into the microgrid distribution line. When R_h is significantly higher than the line impedance, the impedance of both branches in Figure 7.7 tends to be nearly the same. As a result, the harmonic load is distributed evenly across the two DGs. Increasing R_h with higher values enhances the sharing process.

Introducing series resistance into the line leads to a range of power quality issues, with several key disadvantages associated with this approach. The application of VHR increases the total line resistance, which reduces the harmonic current in the lines, but negatively impacts current harmonics sharing. This line resistance elevation also deteriorates the load end's voltage quality. Additionally, the assumption that VHR holds a constant positive value fails to account for accurate harmonic sharing under dynamic scenarios, such as when load conditions fluctuate. Moreover, the absence of a defined methodology or procedure for determining the VHR value often results in a lack of sufficient rationale behind its selection, further complicating the application of this technique.

The proposed harmonic current sharing approach in [48], called the negative virtual harmonic impedance (NVH-Z) method, surpasses the constraints of the conventional VHR technique discussed earlier by employing the concept of negative impedance to adjust the line's effective impedance within an islanded microgrid. This adjustment, achieved by incorporating a voltage drop equivalent to the harmonic current into the inverter's reference voltage, aims to counteract the line's voltage drop, thereby diminishing voltage distortion. The theoretical analysis focuses on harmonic sharing across resistive lines, characterized by a uniform impedance response to all harmonics. In such scenarios, the adjustment leads to a modified reference voltage formula, tailored to maintain consistent harmonic distribution as follows.

$$v_{ref,m} = v_{ref} - i_h Z_h = v_{ref} + i_h R_h \qquad (7.10)$$

where Z_h denotes the NVH-Z and it exhibits negative values under normal operating conditions. If a line is purely resistive, then $Z_h = -R_h$. Figure 7.8 illustrates the schematic of the proposed NVH-Z method in [48].

Figure 7.9 illustrates the integration of NVH-Z within a single DG unit. Reference voltages are formulated through a droop-based, decentralized control

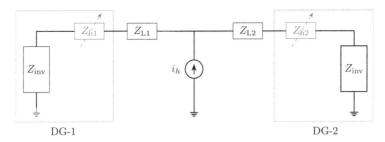

Figure 7.8   Block diagram illustrating the implementation of NVH-Z in [48]

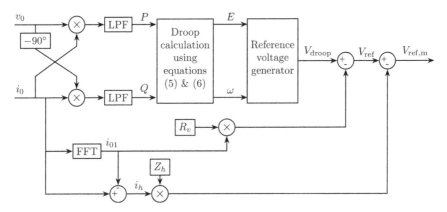

Figure 7.9   Control scheme using NVH-Z [48]

strategy to govern the inverters. Given the resistive nature of line impedances, modified droop equations, specifically Equations (7.5) and (7.6), are employed to achieve equitable load distribution. The reference voltage undergoes adjustments as outlined in (7.10). Subsequently, the adjusted reference voltage, which constitutes the output depicted in Figure 7.8, is accurately tracked utilizing a Proportional-Resonant (PR) controller. This controller selection is attributed to its exceptional efficacy in minimizing steady-state errors [42].

7.8   Voltage quality in weak grid-connected microgrids

In a strong grid, power withdrawals or injections have no impact on the grid's voltage stability. Additionally, the presence of nonlinear loads does not introduce any distortions into the grid's voltage, ensuring that it remains consistent, balanced, and sinusoidal. Conversely, a weak grid is susceptible to fluctuations caused by power being drawn from or supplied to it, leading to easily distorted grid voltages. As a result, unbalanced, distorted, and fluctuating voltages occur frequently in weak grids [49].

Voltage sags are defined as short-term reductions in the voltage level, ranging from 10% to 90% of the root mean square (RMS) value lasting for half a cycle to 1 min. Conversely, voltage swells are temporary increases above the standard voltage level [50]. Nonlinear loads are primarily responsible for distorting voltage and current waveforms through harmonics. The presence of harmonic distortion significantly impacts the grid-connected microgrids, particularly when load voltages become distorted. Such distortions manifest as voltage sags, transients, swells, and overall high distortion levels, compromising the power quality. Voltage sags and swells are notably critical disturbances within microgrids with industrial load. Disruptions such as the abrupt activation of large inductive loads or the charging of substantial capacitor banks can lead to voltage sags and swells. These fluctuations in voltage can severely impact sensitive machinery, leading to failures or forced shutdowns. Moreover, such conditions often result in the generation of unbalanced currents, posing a risk of blowing fuses or triggering circuit breakers in microgrids. These electrical disturbances not only compromise the reliability of the power supply but also endanger the integrity of connected equipment, necessitating robust preventive measures and system designs to mitigate their effects. The process of rectifying faults that lead to voltage sags is time-consuming, and these sags can propagate from the transmission and distribution system down to low voltage loads, further complicating matters. Total harmonic distortion (THD) quantifies the overall effect of harmonic distortion on voltage (THDv) or current (THDi) and provides a measure of harmonics in relation to the fundamental frequency. THD is a critical metric for assessing the quality of power in electrical systems, indicating the extent to which harmonics interfere with the system's operational efficiency.

Various strategies have been developed to address the challenges posed by voltage sags and swells in microgrids, with the most recent advancements focusing on the application of flexible alternative current transmission system (FACTS) devices. These technologies offer dynamic control over power systems, enhancing their stability and efficiency. By integrating FACTS devices into the grid, utilities can not only effectively manage fluctuations in voltage but also improve power quality and system reliability. The unified power quality conditioner (UPQC) [51], thyristor-controlled reactor (TCR), static VAR compensator (SVC), uninterruptible power supply (UPS), and dynamic voltage restorer (DVR) stand as the primary technologies employed for mitigating voltage fluctuations, including sags and swells, as well as for the elimination of harmonics. The DVR, as illustrated in Figure 7.10, is especially valued for its advanced ability to manage active power flow coupled with its advantages of high energy storage capacity, and cost-effectiveness when compared to its counterparts [52]. Operating as a specialized power device, the DVR is integrated into the microgrid line in a series configuration, where it produces an AC voltage that is precisely measured to compensate for voltage sags, thereby safeguarding sensitive electrical loads from potential disturbances.

Utilizing the existing PV system as a DC source for supplying DVR represents a promising area for future research and development. The inherent sustainability and renewability of PV systems, combined with the DVR's capability to protect

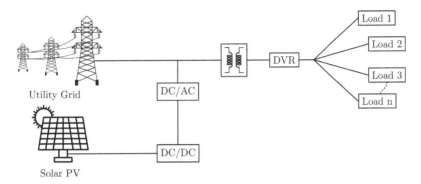

Figure 7.10   DVR integration with a weak grid-connected microgrid

electrical equipment from voltage sags, swells, and other disturbances, could lead to enhanced reliability and efficiency in microgrids. Such an integration not only supports the transition towards more sustainable energy sources but also offers the potential to improve the stability and quality of power in grids experiencing high penetration of renewable energy sources.

7.9   Case studies

7.9.1   Linear RL load sharing in islanded microgrids

The performance of islanded microgrids when employing linear resistive–inductive (RL) load sharing is presented in this section. This case study differentiates between the outcomes observed with and without the use of the linear load-sharing technique. To maintain both stability and efficiency in microgrids, linear load sharing guarantees that the DG units within the microgrid distribute the load fairly among themselves. This equitable distribution helps to prevent potential overloads, thereby boosting the entire system's reliability and performance.

7.9.1.1   System performance without linear load sharing

The absence of linear load-sharing mechanisms can result in uneven sharing of loads among DG units. Figure 7.11 demonstrates how uncoordinated DGs contribute to uneven current supply and power delivery, highlighting the critical need for efficient load-sharing strategies to ensure system balance. DG1 has a higher current compared to DG2, and therefore, delivers higher power as shown in Figure 7.11(c) and (d).

7.9.1.2   System performance with linear load sharing

Integrating linear load sharing enhances the operational stability of the microgrid by guaranteeing a fair distribution of load across DGs. This section showcases how linear load sharing corrects unbalances that are prevalent without this approach, and underscores the importance of droop control in refining the performance of microgrids. Figure 7.12(a) shows the equal sharing of current among DG1 and DG2, which results in sharing an equal amount of power as shown in Figure 7.12(c) and (d).

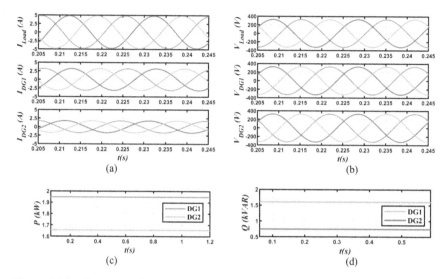

Figure 7.11   *System performance without linear load sharing: (a) current profiles,*
(b) voltage profiles, (c) active power profiles, and (d) reactive power
profiles

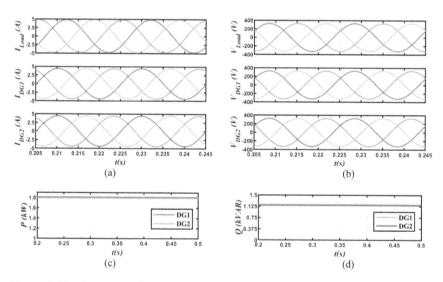

Figure 7.12   *System performance with linear load sharing: (a) current profiles,*
(b) voltage profiles, (c) active power profiles, and (d) reactive power
profiles

7.9.2 Nonlinear load sharing in islanded microgrids

This section explores the nonlinear load sharing method in islanded microgrids, highlighting the critical role of harmonics management in maintaining power quality and ensuring the stability of the system.

7.9.2.1 System performance without harmonic load sharing

Without effective harmonic load sharing, the system undergoes substantial distortions in voltage and current, due to nonlinear loads. The data in Table 7.1 highlights the negative effects of these distortions on the power quality. Figure 7.13 illustrates the operation of the two DG systems without employing a harmonic sharing technique. As shown in Figure 7.13(c), each DG exhibits a different harmonic power rating. Consequently, in the absence of a harmonic sharing scheme, DG2 being closer to the load, bears a greater share of the harmonic load. This includes a rise in THD levels, potentially affecting the durability and performance of sensitive devices within the microgrid.

Table 7.1   System parameters without harmonic load sharing

	V_{ab}	V_{bc}	V_{ca}	THD	I_a	I_b	I_c	THD
Load	317.8	317.8	317.8	1.4%	9.8	9.8	9.8	8.98%
DG_1	327	327	327	1.16%	4.12	4.12	4.12	3.88%
DG_2	323.5	323.5	323.5	1.33%	5.44	5.44	5.44	17.66%

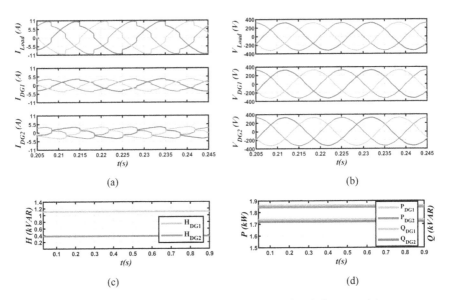

(a)

(b)

(c)

(d)

Figure 7.13   System performance without harmonic load sharing: (a) current profiles, (b) voltage profiles, (c) harmonic power profiles, and (d) DGs' active and reactive power profiles

7.9.2.2   System performance with NVH-Z harmonic load sharing

The implementation of the NVH-Z technique represents a notable progression in controlling harmonics within islanded microgrids. This approach, by modulating the effective line impedance, enables enhanced harmonic current distribution and improves power quality at the point of load. A detailed comparative analysis, reinforced with Table 7.2 and Figure 7.14, highlights the effectiveness of NVH-Z in overcoming the issues associated with nonlinear load sharing. To illustrate further, after applying the NVH-Z, both DGs supply harmonic power equally, at nearly 0.74 kVAR.

The discussions in subsections 7.9.1 and 7.9.2 emphasize the importance of implementing linear and nonlinear load balancing strategies in islanded microgrids

Table 7.2   System parameters with NVH-Z harmonic load sharing

	V_{ab}	V_{bc}	V_{ca}	THD	I_a	I_b	I_c	THD
Load	317.5	317.5	317.5	1.42%	9.98	9.98	9.98	8.35%
DG_1	325.4	325.4	325.4	1.5%	5.05	5.05	5.05	8.02%
DG_2	326	326	326	1.24%	4.96	4.96	4.96	8.21%

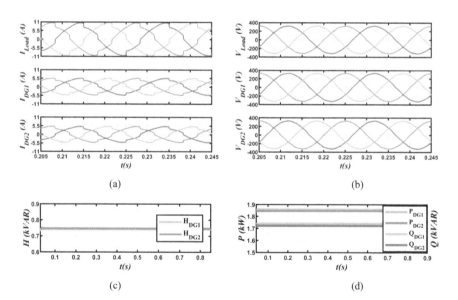

Figure 7.14   System performance with NVH-Z harmonic load sharing: (a) current profiles, (b) voltage profiles, (c) harmonic power profiles, and (d) DGs' active and reactive power profiles

to address the challenges of linear and nonlinear loads. By using advanced control techniques, such as droop control for linear load sharing and NVH-Z for harmonic management in non-linear scenarios, microgrids can achieve optimal performance and demonstrate stability, efficiency, and high power quality even without connection to the main power grid.

7.9.3 Unbalance load sharing in islanded microgrids

This section focuses on strategies for addressing unbalance load sharing, emphasizing both centralized and decentralized approaches. The centralized technique relies on communication links, while the decentralized approach distributes the unbalance load among DG units proportionally to their line impedances.

7.9.3.1 Uncompensated system performance

A two DG system feeding a common unbalance load has been implemented to compare both compensation techniques, as shown in Figure 7.15. The VUF and current unbalance factor (CUF) values of both the DG units and the load before any compensation are presented in Table 7.3. The load has VUF = 1.5% and CUF = 23.6%, as illustrated in Table 7.3 and Figure 7.16(c) and (d).

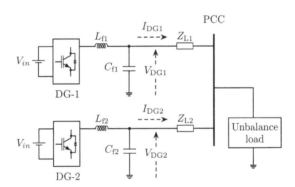

Figure 7.15   Schematic diagram of an islanded microgrid featuring two DGs and an unbalance load.

Table 7.3   Uncompensated system parameters

	V_{ab}	V_{bc}	V_{ca}	*VUF*	I_a	I_b	I_c	*CUF*
Load	366	370	376	1.5%	19.18	24.1	16.5	23.6%
DG_1	379	378	380	0.5%	3.2	5.4	3.7	38.1%
DG_2	377	377	384	1.2%	15.9	18.7	13	21.3%

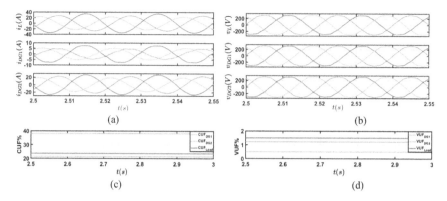

Figure 7.16   Uncompensated system performance: (a) current profile, (b) voltage profiles, (c) current unbalance factors, and (d) voltage unbalance factors

7.9.3.2   Compensated system performance (Centralized approach)

Various centralized methods that depend on communication links for unbalanced load compensation are reported in the literature for AC microgrids. To test the performance of a centralized compensation approach in an islanded microgrid, a two DG system feeding a common unbalanced load has been simulated in MATLAB®/Simulink® based on [20]. This method aims to compensate for the load voltage unbalanced. The load voltage is measured to determine the VUF, then a reference value is given to control the load VUF at any desired level. The unbalance voltage compensation signal is then used to modify the DG's voltage controller reference as shown in Figure 7.17.

Centralized compensation is applied in this case and a reference VUF (VUF_{ref}) of 0.9 is assigned to the secondary controller, to achieve a load VUF of 0.9%. The results after compensation are presented in Table 7.4, with graphical representations in Figure 7.18. These results indicate the successful regulation of the reference unbalance to the desired level. However, it is important to note that while the VUF at the load bus was effectively compensated, the VUF and CUF at the DG terminals increased significantly. Specifically, DG1's VUF increased from 0.5% to 1.8%, and DG2's VUF increased from 1.2% to 2.2%. Simultaneously, the DG1's CUF increased from 38.1% to 63.2%, and DG2's CUF increased from 21.3% to 36.5%.

Although the centralized compensation method successfully addressed voltage unbalance at the load bus, it had the unintended consequence of introducing unbalance at the DG terminals. Also, it could result in overloading one of the DG units or one of its phases. Additionally, it led to a substantial increase in current unbalance, which, in turn, caused the output filters to generate harmonics, as shown in Figure 7.17(a). In addition, deploying this control mechanism requires a low-bandwidth communication framework between the load and the DG units which

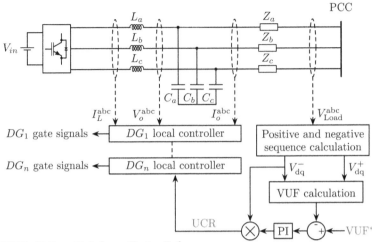

Figure 7.17   Centralized control scheme [20]

Table 7.4   System parameters with centralized approach [20]

	V_{ab}	V_{bc}	V_{ca}	VUF	I_a	I_b	I_c	CUF
Load	376.7	371.1	375.7	0.9%	19.62	24.67	16.44	25.1%
DG_1	386.2	387.3	376.6	1.8%	1.75	5.24	6.08	63.2%
DG_2	389.7	375.5	385	2.2%	18.32	20.40	10.44	36.5%

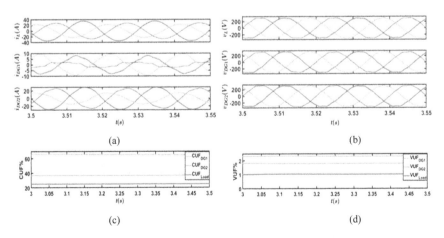

Figure 7.18   Compensated system performance with centralized approach:
(a) current profiles, (b) voltage profiles, (c) current unbalance
factors, and (d) voltage unbalance factors

raises worries about dependability and additional expenses associated with applying such a technique.

7.9.3.3   Compensated system performance (Decentralized approach)

Numerous decentralized control strategies for compensating unbalance in islanded
microgrids have been explored within the literature. The decentralized method
presented in [32], as outlined in Table 7.5 and depicted in Figure 7.19, actively
controls the output voltages of the DG units to ensure their balance by mitigating
any undesirable voltage unbalance caused by unbalanced load currents.

The presented method prioritizes load balance among DG units based on their
respective line impedances, thus preventing overload in any connected DG unit. By
ensuring proportional sharing of load imbalances, individual phase overloads
within the DG are effectively avoided. To achieve the above-mentioned goals, the
controller aims to eliminate the DG's contribution to the voltage unbalance at the
PCC bus by maintaining a balanced voltage at the DG terminal.

Consequently, in a microgrid with multiple DG units, shared load, and varying
line impedances, the proposed method ensures the alignment of current imbalance
factors across all lines and loads. Moreover, the actual current distribution among

Table 7.5   System parameters with decentralized approach [32]

	V_{ab}	V_{bc}	V_{ca}	VUF	I_a	I_b	I_c	CUF
Load	367	372	372	0.9%	19.1	24.3	16.5	24.4%
DG_1	379	379	379	0.15%	3.8	4.8	3.3	24.2%
DG_2	379	379	379	0.15%	15.3	19.4	13.2	24.4%

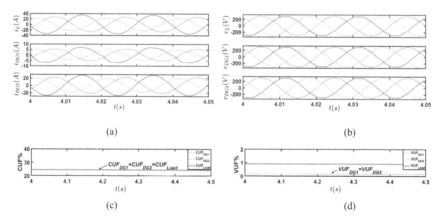

*Figure 7.19   Compensated system performance with decentralized approach: (a)
current profiles, (b) voltage profiles, (c) current unbalance factors,
and (d) voltage unbalance factors*

parallel DG units is proportionate to their respective line impedances. Importantly, the proposed technique operates seamlessly without necessitating any communication links between the load and DG units for compensation signal generation.

By addressing current imbalances directly, the presented decentralized method indirectly aids in reducing load bus voltage imbalances. This holistic approach not only enhances system stability but also simplifies implementation by eliminating the need for complex communication infrastructure.

The CUF serves as a key metric for assessing the effectiveness of the proposed methodology. It has been established that through the utilization of this approach, the load's CUF can be equitably shared, aligning the magnitudes of individual phase currents with the line impedance ratio. An added advantage of the proposed method is its capacity to enhance the balance of load voltages as well.

The control strategies that rely on communication, as previously mentioned, might not be the ideal choice for an islanded microgrid. This is because the DG units and loads within a microgrid could be spread out over several kilometers. As a result, decentralized control, which does not require communication links, is seen as a preferable alternative.

7.10 Conclusion

This chapter explores the power quality challenges inherent in microgrids, emphasizing their low inertia characteristics, voltage and current harmonics, and unbalanced load-sharing techniques. As integrated systems comprising loads, DERs, and advanced control mechanisms, microgrids offer substantial benefits for localized energy distribution and grid stability. However, their reliance on RESs and inverter-based DGs introduces complexity in maintaining power quality, especially in islanded microgrids. Through detailed analysis and case studies, this chapter discusses strategies for mitigating these challenges, focusing on droop control methods, centralized and decentralized control strategies for load sharing, and techniques to overcome voltage sags and swells in weak grid-connected microgrids.

The chapter categorizes microgrids based on connectivity, control mechanisms, power sources, and load types, providing insights into their operational dynamics. It delves into the nuances of grid-connected and islanded modes, highlighting the distinct control challenges and mitigation techniques required in each scenario. The discussion extends to innovative control strategies for ensuring equitable load distribution among DGs, critical for optimizing microgrid performance and reliability.

Case studies on linear, nonlinear, and unbalance load sharing provide practical insights into the effectiveness of the presented solutions, including NVH-Z, centralized, and decentralized control approaches. These examples underscore the significance of tailored control mechanisms and the potential of microgrids to revolutionize power distribution with improved efficiency, stability, and sustainability.

In conclusion, the chapter underscores the importance of addressing power quality issues through advanced control and mitigation techniques, ensuring microgrids can fulfill their promise as reliable, efficient, and sustainable solutions for modern energy challenges.

References

[1]   Q. Jiang, M. Xue, and G. Geng, "Energy Management of Microgrid in Grid-Connected and Stand-Alone Modes," *IEEE Transactions on Power Systems*, vol. 28, no. 3, pp. 3380–3389, 2013, doi: 10.1109/TPWRS.2013.2244104.

[2]   S. Parhizi, H. Lotfi, A. Khodaei, and S. Bahramirad, "State of the Art in Research on Microgrids: A Review," *IEEE Access*, vol. 3, pp. 890–925, 2015, doi:10.1109/ACCESS.2015.2443119.

[3]   J. T. Bialasiewicz, "Renewable Energy Systems With Photovoltaic Power Generators: Operation and Modeling," *IEEE Transactions on Industrial Electronics*, vol. 55, no. 7, pp. 2752–2758, 2008, doi: 10.1109/TIE.2008.920583.

[4]   A. T. D. Perera, V. M. Nik, D. Mauree, and J.-L. Scartezzini, "Electrical hubs: An Effective Way to Integrate Non-Dispatchable Renewable Energy Sources with Minimum Impact to the Grid," *Applied Energy*, vol. 190, pp. 232–248, 2017, doi: https://doi.org/10.1016/j.apenergy.2016.12.127.

[5]   J. J. Justo, F. Mwasilu, J. Lee, and J.-W. Jung, "AC-Microgrids Versus DC-Microgrids with Distributed Energy Resources: A Review," *Renewable and Sustainable Energy Reviews*, vol. 24, pp. 387–405, 2013, doi:https://doi.org/10.1016/j.rser.2013.03.067.

[6]   S. M. Ashabani and Y. A.-R. I. Mohamed, "A Flexible Control Strategy for Grid-Connected and Islanded Microgrids With Enhanced Stability Using Nonlinear Microgrid Stabilizer," *IEEE Trans Smart Grid*, vol. 3, no. 3, pp. 1291–1301, 2012, doi: 10.1109/TSG.2012.2202131.

[7]   M. H. Saeed, W. Fangzong, B. A. Kalwar, and S. Iqbal, "A Review on Microgrids' Challenges and Perspectives," *IEEE Access*, vol. 9, pp. 166502–166517, 2021, doi:10.1109/ACCESS.2021.3135083.

[8]   J. Rocabert, A. Luna, F. Blaabjerg, and P. Rodríguez, "Control of Power Converters in AC Microgrids," *IEEE Transactions on Power Electronics*, vol. 27, no. 11, pp. 4734–4749, 2012, doi: 10.1109/TPEL.2012.2199334.

[9]   S. Sen and V. Kumar, "Microgrid Control: A Comprehensive Survey," *Annual Reviews in Control*, vol. 45, pp. 118–151, 2018, doi: https://doi.org/10.1016/j.arcontrol.2018.04.012.

[10]   J. Hu, Y. Shan, K. W. Cheng, and S. Islam, "Overview of Power Converter Control in Microgrids—Challenges, Advances, and Future Trends," *IEEE Transaction on Power Electronics*, vol. 37, no. 8, pp. 9907–9922, 2022, doi:10.1109/TPEL.2022.3159828.

[11]   R. Rosso, X. Wang, M. Liserre, X. Lu, and S. Engelken, "Grid-Forming Converters: Control Approaches, Grid-Synchronization, and Future

Trends—A Review," *IEEE Open Journal of Industry Applications*, vol. 2, pp. 93–109, 2021, doi:10.1109/OJIA.2021.3074028.

[12] Á. Borrell, M. Velasco, J. Miret, A. Camacho, P. Martí, and M. Castilla, "Collaborative Voltage Unbalance Elimination in Grid-Connected AC Microgrids with Grid-Feeding Inverters," *IEEE Transaction on Power Electronics*, vol. 36, no. 6, pp. 7189–7201, 2021, doi:10.1109/TPEL.2020. 3039337.

[13] Z. Li, C. Zang, P. Zeng, H. Yu, and S. Li, "Fully Distributed Hierarchical Control of Parallel Grid-Supporting Inverters in Islanded AC Microgrids," *IEEE Transactions on Industrial Informatics*, vol. 14, no. 2, pp. 679–690, 2018, doi:10.1109/TII.2017.2749424.

[14] A. G. Tsikalakis and N. D. Hatziargyriou, "Centralized Control for Optimizing Microgrids Operation," *IEEE Transactions on Energy Conversion*, vol. 23, no. 1, pp. 241–248, 2008, doi: 10.1109/TEC.2007.914686.

[15] J. M. Guerrero, L. Hang, and J. Uceda, "Control of Distributed Uninterruptible Power Supply Systems," *IEEE Transactions on Industrial Electronics*, vol. 55, no. 8, pp. 2845–2859, 2008, doi: 10.1109/TIE.2008.924173.

[16] T. Caldognetto and P. Tenti, "Microgrids Operation Based on Master–Slave Cooperative Control," *IEEE Journal of Emerging and Selected Topics in Power Electronics*, vol. 2, no. 4, pp. 1081–1088, 2014, doi:10.1109/JESTPE. 2014.2345052.

[17] J. Lai, X. Lu, X. Yu, W. Yao, J. Wen, and S. Cheng, "Distributed Multi-DER Cooperative Control for Master-Slave-Organized Microgrid Networks with Limited Communication Bandwidth," *IEEE Transactions on Industrial Informatics*, vol. 15, no. 6, pp. 3443–3456, 2019, doi:10.1109/TII.2018. 2876358.

[18] T.-F. Wu, Y.-K. Chen, and Y.-H. Huang, "3C Strategy for Inverters in Parallel Operation Achieving an Equal Current Distribution," *IEEE Transactions on Industrial Electronics*, vol. 47, no. 2, pp. 273–281, 2000, doi:10. 1109/41.836342.

[19] X. Sun, L.-K. Wong, Y.-S. Lee, and D. Xu, "Design and Analysis of an Optimal Controller for Parallel Multi-Inverter Systems," *IEEE Transactions on Circuits and Systems II: Express Briefs*, vol. 53, no. 1, pp. 56–61, 2006, doi:10.1109/TCSII.2005.854136.

[20] M. Savaghebi, A. Jalilian, J. C. Vasquez, and J. M. Guerrero, "Secondary Control Scheme for Voltage Unbalance Compensation in an Islanded Droop-Controlled Microgrid," *IEEE Transactions on Smart Grid*, vol. 3, no. 2, pp. 797–807, 2012, doi: 10.1109/TSG.2011.2181432.

[21] M. Savaghebi, A. Jalilian, J. C. Vasquez, and J. M. Guerrero, "Autonomous Voltage Unbalance Compensation in an Islanded Droop-Controlled Microgrid," *IEEE Transactions on Industrial Electronics*, vol. 60, no. 4, pp. 1390–1402, 2013, doi:10.1109/TIE.2012.2185914.

[22] M. Hamzeh, H. Karimi, and H. Mokhtari, "A New Control Strategy for a Multi-Bus MV Microgrid Under Unbalanced Conditions," *IEEE*

Transactions on Power Systems, vol. 27, no. 4, pp. 2225–2232, 2012, doi: 10.1109/TPWRS.2012.2193906.

[23]   D. De and V. Ramanarayanan, "Decentralized Parallel Operation of Inverters Sharing Unbalanced and Nonlinear Loads," *IEEE Transactions on Power Electronics*, vol. 25, no. 12, pp. 3015–3025, 2010, doi:10.1109/TPEL.2010.2068313.

[24]   B. Hamad, A. Al-Durra, T. H. M. EL-Fouly, and H. H. Zeineldin, "Economically Optimal and Stability Preserving Hybrid Droop Control for Autonomous Microgrids," *IEEE Transactions on Power Systems*, vol. 38, no. 1, pp. 934–947, 2023, doi:10.1109/TPWRS.2022.3169801.

[25]   M. S. Golsorkhi and D. D.-C. Lu, "A Decentralized Control Method for Islanded Microgrids Under Unbalanced Conditions," *IEEE Transactions on Power Delivery*, vol. 31, no. 3, pp. 1112–1121, 2016, doi:10.1109/TPWRD.2015.2453251.

[26]   J. Hu, Y. Shan, J. M. Guerrero, A. Ioinovici, K. W. Chan, and J. Rodriguez, "Model Predictive Control of Microgrids – An Overview," *Renewable and Sustainable Energy Reviews*, vol. 136, p. 110422, 2021, doi:https://doi.org/10.1016/j.rser.2020.110422.

[27]   M. B. Abdelghany, A. Al-Durra, H. Zeineldin, and J. Hu, "Integration of Cascaded Coordinated Rolling Horizon Control for Output Power Smoothing in Islanded Wind–Solar Microgrid with Multiple Hydrogen Storage Tanks," *Energy*, vol. 291, p. 130442, 2024, doi:https://doi.org/10.1016/j.energy.2024.130442.

[28]   M. B. Abdelghany, A. Al-Durra, Z. Daming, and F. Gao, "Optimal Multi-Layer Economical Schedule for Coordinated Multiple Mode Operation of Wind–Solar Microgrids with Hybrid Energy Storage Systems," *Journal of Power Sources*, vol. 591, p. 233844, 2024, doi:https://doi.org/10.1016/j.jpowsour.2023.233844.

[29]   P.-T. Cheng, C.-A. Chen, T.-L. Lee, and S.-Y. Kuo, "A Cooperative Imbalance Compensation Method for Distributed-Generation Interface Converters," *IEEE Transactions on Industry Applications*, vol. 45, no. 2, pp. 805–815, 2009, doi:10.1109/TIA.2009.2013601.

[30]   S. Ghosh and S. Chattopadhyay, "Correction of Line-Voltage Unbalance by the Decentralized Inverters in an Islanded Microgrid," in *2020 IEEE Applied Power Electronics Conference and Exposition (APEC)*, 2020, pp. 622–628. doi:10.1109/APEC39645.2020.9124363.

[31]   S. A. Yahyaee, P. Sreekumar, and V. Khadkikar, "A Novel Decentralized Unbalance Load Sharing Approach For Islanded Microgrids," in *2022 IEEE Industry Applications Society Annual Meeting (IAS)*, 2022, pp. 1–7. doi:10.1109/IAS54023.2022.9940042.

[32]   A. Shafiqurrahman, S. A. Yahyaee, P. Sreekumar, and V. Khadkikar, "A Novel Decentralized Unbalance Load Sharing Approach For Islanded Microgrids," *IEEE Transactions on Industry Applications*, pp. 1–13, 2024, doi:10.1109/TIA.2024.3384462.

[33] E. Barklund, N. Pogaku, M. Prodanovic, C. Hernandez-Aramburo, and T. C. Green, "Energy Management in Autonomous Microgrid Using Stability-Constrained Droop Control of Inverters," *IEEE Transactions on Power Electronics*, vol. 23, no. 5, pp. 2346–2352, 2008, doi: 10.1109/TPEL.2008.2001910.

[34] Q.-C. Zhong and G. Weiss, "Synchronverters: Inverters That Mimic Synchronous Generators," *IEEE Transactions on Industrial Electronics*, vol. 58, no. 4, pp. 1259–1267, 2011, doi: 10.1109/TIE.2010.2048839.

[35] S. D'Arco and J. A. Suul, "Equivalence of Virtual Synchronous Machines and Frequency-Droops for Converter-Based MicroGrids," *IEEE Transactions on Smart Grid*, vol. 5, no. 1, pp. 394–395, 2014, doi: 10.1109/TSG.2013.2288000.

[36] F. Chen, R. Burgos, D. Boroyevich, J. C. Vasquez, and J. M. Guerrero, "Investigation of Nonlinear Droop Control in DC Power Distribution Systems: Load Sharing, Voltage Regulation, Efficiency, and Stability," *IEEE Transactions on Power Electronics*, vol. 34, no. 10, pp. 9404–9421, 2019, doi:10.1109/TPEL.2019.2893686.

[37] J. C. Vasquez, J. M. Guerrero, M. Savaghebi, J. Eloy-Garcia, and R. Teodorescu, "Modeling, Analysis, and Design of Stationary-Reference-Frame Droop-Controlled Parallel Three-Phase Voltage Source Inverters," *IEEE Transactions on Industrial Electronics*, vol. 60, no. 4, pp. 1271–1280, 2013, doi: 10.1109/TIE.2012.2194951.

[38] A. Kahrobaeian and Y. A.-R. I. Mohamed, "Analysis and Mitigation of Low-Frequency Instabilities in Autonomous Medium-Voltage Converter-Based Microgrids with Dynamic Loads," *IEEE Transactions on Industrial Electronics*, vol. 61, no. 4, pp. 1643–1658, 2014, doi: 10.1109/TIE.2013.2264790.

[39] H. Mahmood, D. Michaelson, and J. Jiang, "Accurate Reactive Power Sharing in an Islanded Microgrid Using Adaptive Virtual Impedances," *IEEE Transactions on Power Electronics*, vol. 30, no. 3, pp. 1605–1617, 2015, doi:10.1109/TPEL.2014.2314721.

[40] J. M. Guerrero, L. G. de Vicuna, J. Miret, J. Matas, and J. Cruz, "Output Impedance Performance for Parallel Operation of Ups Inverters Using Wireless and Average Current-Sharing Controllers," in *2004 IEEE 35th Annual Power Electronics Specialists Conference (IEEE Cat. No. 04CH37551)*, vol. 4, 2004, pp. 2482–2488. doi:10.1109/PESC.2004.1355219.

[41] F. Guo, C. Wen, J. Mao, J. Chen, and Y.-D. Song, "Distributed Cooperative Secondary Control for Voltage Unbalance Compensation in an Islanded Microgrid," *IEEE Transactions on Industrial Informatics*, vol. 11, no. 5, pp. 1078–1088, 2015, doi:10.1109/TII.2015.2462773.

[42] R. Teodorescu, F. Blaabjerg, M. Liserre, and P. Loh, "Proportional-Resonant Controllers and Filters for Grid-Connected Voltage-Source Converters," *IEE Proceedings-Electric Power Applications*, vol. 153, pp. 750–762, 2006, doi: 10.1049/ip-epa:20060008.

[43] P.-T. Cheng and T.-L. Lee, "Distributed Active Filter Systems (DAFSs): A New Approach to Power System Harmonics," *IEEE Transactions on Industrial Applications*, vol. 42, no. 5, pp. 1301–1309, 2006, doi:10.1109/TIA. 2006.880856.

[44] M. Hanif, V. Khadkikar, W. Xiao, and J. L. Kirtley, "Two Degrees of Freedom Active Damping Technique for LCL Filter-Based Grid Connected PV Systems," *IEEE Transactions on Industrial Electronics*, vol. 61, no. 6, pp. 2795–2803, 2014, doi: 10.1109/TIE.2013.2274416.

[45] X. Wang, F. Blaabjerg, and Z. Chen, "Autonomous Control of Inverter-Interfaced Distributed Generation Units for Harmonic Current Filtering and Resonance Damping in an Islanded Microgrid," *IEEE Transactions on Industrial Applications*, vol. 50, no. 1, pp. 452–461, 2014, doi:10.1109/TIA. 2013.2268734.

[46] T. Vandoorn, B. Meersman, J. De Kooning, and L. Vandevelde, "Controllable Harmonic Current Sharing in Islanded Microgrids: DG Units with Programmable Resistive Behavior toward Harmonics," *IEEE Transactions on Power Delivery*, vol. 27, no. 2, pp. 831–841, 2012, doi: 10.1109/TPWRD. 2011.2176756.

[47] S. Munir and Y. W. Li, "Residential Distribution System Harmonic Compensation Using PV Interfacing Inverter," *IEEE Transactions on Smart Grid*, vol. 4, no. 2, pp. 816–827, 2013, doi: 10.1109/TSG.2013.2238262.

[48] P. Sreekumar and V. Khadkikar, "A New Virtual Harmonic Impedance Scheme for Harmonic Power Sharing in an Islanded Microgrid," *IEEE Transactions on Power Delivery*, vol. 31, no. 3, pp. 936–945, 2016, doi:10. 1109/TPWRD.2015.2402434.

[49] S. B. Q. Naqvi, S. Kumar, and B. Singh, "Weak Grid Integration of a Single-Stage Solar Energy Conversion System With Power Quality Improvement Features Under Varied Operating Conditions," *IEEE Transactions on Industrial Applications*, vol. 57, no. 2, pp. 1303–1313, 2021, doi:10.1109/ TIA.2021.3051114.

[50] P. Zhongmei, H. Tonghui, and W. Yuqing, "Voltage Sags/Swells Subsequent to Islanding Transition of PV-Battery Microgrids," in *2016 IEEE 11th Conference on Industrial Electronics and Applications (ICIEA)*, 2016, pp. 2317–2321. doi:10.1109/ICIEA.2016.7603978.

[51] M. T. L. Gayatri, A. M. Parimi, and A. V. P. Kumar, "Utilization of Unified Power Quality Conditioner for voltage sag/swell mitigation in microgrid," in *2016 Biennial International Conference on Power and Energy Systems: Towards Sustainable Energy (PESTSE)*, 2016, pp. 1–6. doi:10.1109/ PESTSE.2016.7516475.

[52] M. T. L. Gayatri, A. M. Parimi, and A. V. P. Kumar, "Application of Dynamic Voltage Restorer in Microgrid for Voltage Sag/Swell Mitigation," in *2015 IEEE Power, Communication and Information Technology Conference (PCITC)*, 2015, pp. 750–755. doi:10.1109/PCITC.2015.7438096.

Chapter 8

Advanced fault direction identification strategy for AC microgrid protection

Rudranarayan Pradhan[1] and Premalata Jena[2]

A microgrid constitutes an integral component of the modern smart grid. Microgrid (MG) integrates several distributed energy sources and loads that behave with the grid as a single controllable entity and operate within predetermined electrical parameters. MG encourages the addition of varied loads and sources of renewable energy types. Numerous benefits of MG include decreased transmission losses, minimal carbon emissions, and increased system dependability. Due to the dynamic and inconsistent properties of renewable energy sources, the widespread use of Distributed Energy Resources (DERs), alters not only the power flow in the distribution system but also the level and direction of fault currents, which has a significant impact on the operation of the protection devices as the majority of the traditional distribution system's protection systems, including fuses, reclosers, relays, and circuit breakers, are developed using the detection of unidirectional fault current. Hence, these conventional protective devices cannot provide proper protection for the reliable and safe operation of MG. To enhance fault detection efficiency, it is crucial to ascertain the direction of fault currents precisely. Protection professionals find it challenging to determine the fault direction in an AC MG. Considering the problems associated with existing directional over current relays for the protection of MG, the primary objective of this research is to create a novel fault-directing element for AC MG. In this report, the various significant issues and challenges faced by microgrid protection are discussed, and to overcome these issues, an advanced fault direction estimation scheme based on the Dynamic value of Superimposed Positive Sequence Impedance Angle (SPSIA) is suggested. The recommended approach adeptly discerns the direction of symmetrical and asymmetrical faults with remarkable efficacy. The effectiveness of the proposed strategy has been thoroughly validated under diverse conditions, affirming its suitability for practical applications.

[1]School of Electrical Sciences, Odisha University of Technology and Research, India
[2]Department of Electrical Engineering, Indian Institute of Technology Roorkee, India

8.1 Introduction

Fossil fuels, including natural gas, oil, and coal, have propelled the world's economies for over 150 years. By 2020, these sources were still the world's leading energy source, supplying around 80% of it, according to the International Energy Agency (IEA). Data from the Central Electricity Authority (CEA), the Ministry of Power, and the Government of India as of October 31, 2023, show that 56.2% of India's power comes from fossil fuels, and 43.8% derives from non-fossil fuel sources [1]. Conventional power plants, nevertheless necessary for addressing energy demands, have several drawbacks. Conventional power plants, especially those that use non-renewable resources like coal, natural gas, oil, etc., release large volumes of greenhouse gases. This contributes to both air pollution and global warming. When fossil fuels burn, pollutants such as Sulfur Dioxide SO_2, Nitrogen Oxides (NO), and particulate matter are released into the air, which can cause respiratory disorders and poor air quality. As these non-renewable resources are limited, resource depletion and environmental degradation are caused by their extraction and usage. The regional concentration of these non-renewable resources raises questions regarding their long-term supply. Large volumes of water are needed for cooling in many conventional power plants. This may put a burden on nearby water supplies and endanger aquatic ecosystems.

Conventional power plant construction and its operation frequently necessitate large-scale land usage, which can disrupt habitats and perhaps result in biodiversity loss. The building of conventional power facilities, especially nuclear ones, can take many years, which causes delays in the addition of new grid capacity. Nuclear proliferation hazards are increased by the use of nuclear power, which sparks worries about the potential for the spread of nuclear technology and materials. Safety issues are associated with these nuclear facilities' susceptibility to natural calamities, including earthquakes, floods, and tsunamis. It may not be possible for many conventional power plants, particularly those that run on fossil fuels, to swiftly ramp up or down in response to changes in the electricity demand.

8.1.1 *The imperative for a superior standard of electrical quality*

Modern civilization depends heavily on an uninterrupted and high-quality supply of power. Maintaining a high electricity standard is essential to the proper operation and development of contemporary societies, not only as a matter of convenience. It supports economic activity, the dependability of vital services, and people's general well-being. A reliable and superior supply of electricity goes a long way towards satisfying customers overall [2]. A steady supply of electricity and no voltage variations improve people's quality of life, individually and collectively. A reliable and high-quality power source is essential to many critical infrastructure systems, such as hospitals, emergency services, and communication networks. Disruptions or variations in the electricity supply could jeopardize these vital services. High-quality power is essential for operating data centers, communication networks, and information technology infrastructure. Data loss and device damage are two consequences of power surges or sags. Energy waste can be minimized by having machines and appliances run at maximum efficiency thanks to properly controlled voltage and frequency levels. Inadequate power quality, such as high voltage swings, can cause overheating and raise the possibility of electrical fires in residences and commercial buildings.

8.1.2 The essential need for the seamless integration of renewable energy sources (RES)

To address the drawbacks of conventional power plants and the need for higher-quality, cleaner electricity, a move towards more sustainable and clean energy sources, greater energy efficiency, and the adoption of cutting-edge technologies for enhanced safety and environmental mitigation are necessary.

RES, for instance, solar, hydropower, biogas, wind, geothermal energy, fuel cells, and others, add to the renewable energy landscape by providing environmentally friendly substitutes for conventional fossil fuels and mitigating their adverse effects. The selection of a specific renewable energy source is frequently influenced by technological, economic, and geographic factors. When compared to conventional fossil fuels, renewable energy sources provide a means of producing power with substantially fewer carbon emissions, which is vital to the worldwide effort to slow down climate change. Another crucial factor is energy source diversification, which tries to improve energy security and lessen reliance on a small number of supplies. RES provides a decentralized approach to power generation by drawing on abundant and locally available resources, lessening the dependency on centralized, frequently distant power facilities. This enhances grid resilience and uses the local resources to the fullest.

Additionally, the usage of RES consistently reduces greenhouse gas emissions, improving air quality and its associated health benefits. When compared to more traditional energy sources, renewable energy sources are now significantly more competitive due to ongoing technology advancements that are typified by reduced costs, higher effectiveness, and adaptability. The geopolitical and economic advantages are notable since renewable energy sources contribute to long-term cost stability, job creation, and energy independence. Furthermore, the industry acts as a spark for technological advancement, directing research and development in the direction of more effective technologies and energy storage solutions. Incorporating renewable energy is also essential for adhering to legal frameworks and goals established by different countries to raise the proportion of clean energy in their total generation capacity.

Complying with the legislative frameworks and targets established by different countries to raise the proportion of clean energy in their total energy mix also requires the integration of renewables. All things considered, including renewable energy sources is essential to creating a contemporary power system that is resilient, sustainable, and forward-thinking. To supply society with dependable power, the future electricity grid must take energy efficiency, power quality, environmental protection, and supply cost into account. To meet the customer's need for an uninterrupted power supply, the power system today requires additional intelligence at the distribution level to integrate more renewable energy sources.

8.1.3   Initiation of a microgrid system

Research on the implementation of distributed generation on power systems is the common interest of power system engineers. Many different distributed generation (DG) such as fuel cells, wind turbines, and Photovoltaic (PV) are directly connected to the distribution grid to meet consumer demand, maintain voltage profile, and various technical and economic benefits.

The idea behind a microgrid is to build a localized power system on a smaller scale that can function both independently and in parallel with the main electrical grid. It is a useful way to incorporate distributed generation sources into the energy system as a whole.

1.   Most microgrids are located at the LV (Low Voltage) level, and their combined microgeneration capacity is usually less than a megawatt (MW) range. There may be some exceptions, though, since some MV (Medium Voltage) network segments might also be incorporated into a microgrid for connecting the needs of the power system.
2.   MG should be capable of operating in both grid feeding or grid forming mode (islanded mode).
3.   Long-term islanded operation demands an MG to meet strict necessitates for the size of storage and microgenerator capacity ratings. This is required to guarantee that all loads receive a constant power supply. On the other hand, the microgrid may rely on a high degree of demand flexibility. In the second scenario, the reliability benefits are measurable and result from the partial islanding of important loads.

A microgrid is a localized and decentralized group of electricity sources and loads that operates autonomously or is connected to the traditional centralized power grid. It is designed to provide reliable, resilient, and sustainable energy to a specific geographic area, such as a community, campus, or industrial complex. The concept of a microgrid incorporates various distributed energy resources (DERs) and technologies to enhance energy efficiency and grid reliability.

The structure of microgrids is based on the increased penetration of DERs, with microgenerators like photovoltaic (PV) arrays, microturbines, and fuel cells, along with storage devices like energy capacitors, batteries, flywheels, and adjustable (electric vehicles) loads, at the system's distribution level. Electricity produced from several smaller sources close to the point of consumption is referred to as distributed generation. Renewable energy systems like solar panels, wind turbines, and other on-site power sources can be among these sources.

From a customer's perspective, microgrids fulfill thermal and electricity requirements while bolstering local reliability, diminishing emissions, enhancing power quality through voltage support and dip reduction, and potentially reducing energy supply expenses. From the perspective of the grid operator, an MG can be seen as a managed entity within the power system. It has the capability to function as a unified, integrated load or generator. Additionally, with suitable compensation incentives, it can also serve as a minor power source or provider of ancillary services that support the stability and functionality of the network.

A microgrid fundamentally represents a concept of aggregating demand and supply-side resources within distribution grids. Utilizing the interaction between localized load and micro source generation, an MG can deliver various social, technical, environmental, and economic advantages to multiple individuals. When compared to other methods of aggregating micro sources, a microgrid stands out by providing the highest degree of ownership structure, enabling worldwide optimization of power system efficiency, and emerging as the best way to incentivize end users via a platform of common interest.

Integration of small *kW* scale combined heat and power and solar PV to the conventional distribution grid is required to deal with the world's climate change effort in reducing carbon emissions. They mainly utilize the waste heat for the generation of electricity and are connected near the load center. It also reduces the burden on the distribution and transmission grid by reducing losses and network facilities. MG enhances the quality of power supplied to the consumer as placed near the consumer and supports the network by restoring the network after a fault and reducing stress during congestion.

8.1.4 Classification of MG

Microgrids are classed based on their operational characteristics, energy sources, and applications. Figure 8.1 shows various types of microgrids. These classes emphasize the diversity of microgrids, emphasizing their adaptability to various contexts, applications, and energy goals. Based on the operating voltage they are broadly classified as

- AC Microgrid
- DC Microgrid
- Hybrid Microgrid

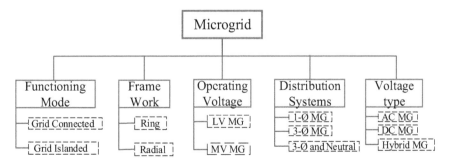

Figure 8.1   Classification of MG

8.1.4.1   AC microgrid

AC microgrids provide a flexible and adaptable power generation and distribution strategy, responding to the individual energy needs and goals of the areas or institutions they support. It is a localized and decentralized electrical system that runs on alternating current power as shown in Figure 8.2. It is intended to create, distribute, and manage electricity within a defined geographical area, giving autonomy and resilience. AC microgrids are made up of a variety of DERs, including power sources, energy storage devices, and loads, all of which are integrated into a smart and linked grid. It can operate in grid following or grid forming mode operation when desired. The overall cost of AC MG is less compared to DC MG.

8.1.4.2   DC microgrid

The demand for DC microgrids has arisen due to the prevalence of RES and the emergence of contemporary DC loads that necessitate a DC interface. Despite the predominant operation of distribution grids in AC, advancements in semiconductor technology and power electronics in recent decades have facilitated the adoption of DC grids as an attractive and competitive option for integrating DG systems. The developing landscape of energy technologies, the necessity for efficient power conversion, and compatibility with modern loads and renewable energy sources all contribute to the requirements for DC microgrids as shown in Figure 8.3. These microgrids offer a significant option in specific applications when their distinct benefits complement the peculiarities of the local energy ecology.

8.1.4.3   Hybrid microgrid (HMG)

A hybrid microgrid integrates different energy sources and can function in grid-dependent and grid-forming modes. The needs for a hybrid microgrid are varying,

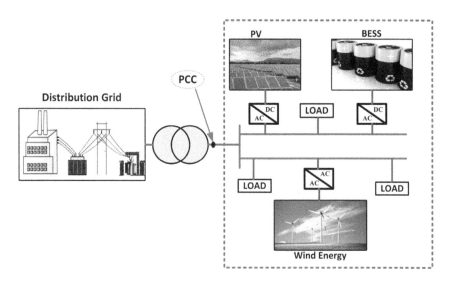

Figure 8.2   AC microgrid

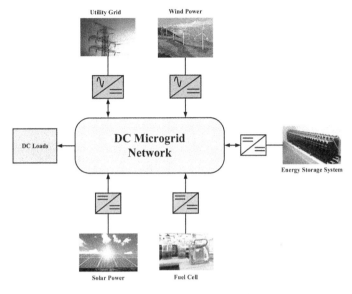

Figure 8.3   DC microgrid

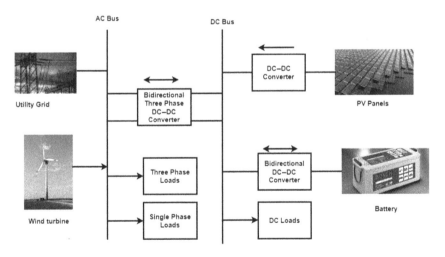

Figure 8.4   Hybrid microgrid

driven by the demand for enhanced energy resilience, sustainability, and efficiency. The general structure of hybrid MG is represented in Figure 8.4. Hybrid AC/DC microgrids present a compelling alternative by facilitating the incorporation of DC-based technologies with a dedicated DC subgrid while retaining the existing AC facilities. This approach allows for a synergistic blend of the benefits offered by both AC and DC grids, contributing to an effective and dependable integration of DG systems.

Beyond the advantages offered by standalone AC or DC MG, hybrid AC/DC infrastructure further enhances the installation of AC or DC-based devices. This improvement stems from their direct connection to the network with minimal interface components, reducing conversion steps and consequent energy losses. Such a configuration is particularly suitable for incorporating distributed storage, including emerging EV energy storage systems (EV-ES) designed for vehicle-to-grid (V-2-G) uses.

When it comes to managing voltage and frequency management, system stability, and maintaining both active and reactive power, EV storage has the potential to significantly enhance the overall efficiency of a hybrid microgrid. Nevertheless, optimizing the coordination of EV storages within microgrids poses a complex challenge due to their diverse control and configuration framework, coupled with the absence of standardized protocols for V-2-G applications.

8.1.5   Operational challenges of AC microgrid

The MG possesses numerous advantages; there are technical challenges to deal with for the reliable and efficient operation of MG.

(a) The power generated from renewable energy sources (solar, wind) varies and depends on day-to-day weather conditions
(b) MGs are associated with high installation cost
(c) Capacity of MGs is limited
(d) Space requirements are more
(e) Cost of batteries for storing electrical energy and its maintenance cost is very high.
(f) Vulnerable to cyber attack
(g) Maintaining stability when switching from grid-connected to grid-isolated mode of operation

8.1.6   Protection challenges in AC microgrid

Fault detection is crucial in a microgrid to maintain reliability, stability, and safety. It facilitates proactive problem-solving methods, such as protecting equipment, assuring uninterrupted power supply, and contributing to the overall efficiency of the microgrid. It helps maintain grid stability by isolating the faulty section and maintains the reliability of the system by providing power to the healthy section of the network. Identifying and clearing faults within a minimum time helps protect against cascading failure of the system, protects equipment as well as personnel, and reduces downtime, which in turn provides an uninterrupted power supply.

One of the most significant and challenging technological problems for integrating an MG into the electricity system is its protection. MG protection must respond efficiently to both internal and out-of-zone faults in both modes of operation. In the event of a fault on the primary grid during parallel operation, the ideal action is to separate the MG from the utility grid to safeguard the critical loads. When an internal fault arises in an MG, it separates the smallest section of the MG feeder to isolate the fault. The protection equipment like relays, fuses, circuit breakers (CB), and reclosers used in distribution system protection are based on identifying fault current in a particular direction, and most of them are based on over-current protection. The integration of DERs in the distribution system not only changes the magnitude but also the direction of fault current because of the dynamic properties and unpredictable characteristics of RES, which has an impact on the functioning and coordination of protection relays. As a result, typical protective devices are incapable of ensuring the reliability and safety of microgrid operation [3].

• The magnitude of fault currents exhibits notable variations between grid-feeding and islanded MG operations.
• Conventional protection strategies in distribution networks face issues due to the size and structure of DERs. While synchronous-based DG units produce fault currents ranging from (4–10) times the usual current, inverter-based DERs generate fault currents that are usually (1.2–2) times the standard current.

Most commonly used relays, i.e., directional overcurrent relays (D-O-C-R)s, cannot detect such a small amount of current or may take longer to respond to such fault currents. As a result, some basic requirements of the protection system such as speed of operation (delay in tripping), reliability, sensitivity (concealed), and selectivity (unnecessary tripping) are affected. This leads to several protection coordination issues in the microgrid among which the most significant issues are listed below.

1. Bidirectional fault current in MG
2. Blinding of protection in MG
3. Variation in fault current severity
4. Unnecessary/Sympathetic/Spurious tripping in MG

8.1.6.1 Bidirectional fault current in MG

A microgrid is an active distribution network where different types of DERs are located at different places of the MG. In a distribution network, the current normally flows from the source to load, which is typically unidirectional. The most popular D-O-C-R, which was earlier used to protect the distribution network, fails to protect the microgrid due to the bidirectional flow of fault current. Figure 8.5 shows a typical microgrid consisting of DER (which may be a renewable generation) and loads located at different places of the microgrid. When faults occur at location **F1**, as shown in Figure 8.5, the fault current is fed by the primary grid as well as the DER present. However, when the fault occurs at **F2**, the DER contributes to the fault in the opposite direction from the conventional direction. Therefore, the fault current turns bidirectional, and the relay present at CB2 senses less current when DERs are not present. When the value of the fault current is less than the threshold (D_{TH}), it results in delayed or no operation of the CB, and the protection systems fail to protect the microgrid in the event of a fault.

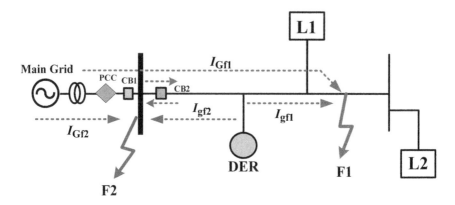

Figure 8.5   Bidirectional fault current in MG

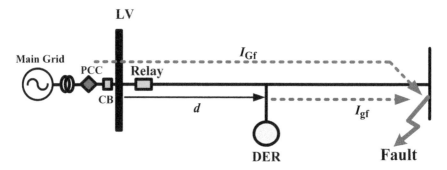

Figure 8.6   Blinding of the protection in MG

8.1.6.2   Blinding of protection in MG

When faults appear at the far end of the feeder within the microgrid, and DERs are in operation, these DERs contribute to the fault current, as illustrated in Figure 8.6. The fault current magnitude experiences a reduction, due to the participation of DERs. In the absence of DERs in the microgrid, the fault contribution from the main grid is more pronounced and easily detected by the relays. However, the presence of DERs diminishes the fault current, potentially leading to a failure in fault detection within the microgrid. This phenomenon, known as the "blinding of protection" in microgrids, underscores the importance of implementing robust protection schemes to mitigate such issues.

8.1.6.3   Variation in fault current severity

The fault current magnitude is different in an MG from grid-following to grid-isolated modes. The type of DERs, rating of DERs, and load rating also affect the fault current. Most of the DERs are connected to the MG through a converter, and the inherent topology of the converter restricts the fault current maximum (1.2–2)

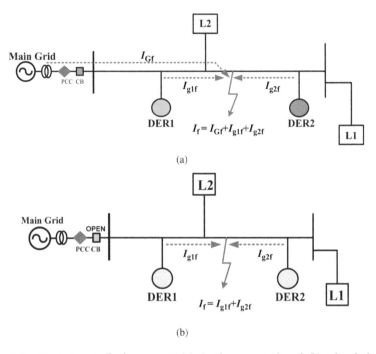

Figure 8.7 Variation in fault current. (a) Grid-connected and (b) islanded mode.

times the rated current. The radial and mesh topology also affect the fault current. Figure 8.7 represents the MG in a different mode of operation. In Figure 8.7(a), when a fault occurs in the MG in grid-connected mode, the **DER1** & **DER2** as well as the **main grid** contribute to the fault current. The fault current I_f can be expressed as $I_f = I_{Gf} + I_{g1f} + I_{g2f}$. In this situation, the main grid contribution to fault decreases, so the relay fails to detect this low-magnitude fault current, or sometimes the operation of the relay is delayed. Similarly, in grid forming mode shown in Figure 8.7(b), both the DER contribute to fault current, and the fault current I_f is represented as $I_f = I_{g1f} + I_{g2f}$. The value of the fault current at the relay location decreases significantly during both modes of operation, which adds challenges to the relay for fault detection.

8.1.6.4 Unnecessary/Sympathetic/Spurious/false tripping in MG

The sympathetic tripping occurs when a DER in an adjacent feeder supplies current to a faulty feeder and trips the healthy feeder in the MG. A microgrid with DER (solar/wind) present is shown in Figure 8.8. When a fault occurs in **Feeder 1** the DER present in **Feeder 2** is contributing to the fault current. The relay **R1** which is a directional relay, trips the healthy feeder. This phenomenon in the MG environment is known as Unnecessary/Sympathetic/Spurious/false tripping. This is highly undesirable in the power system. A number-switching incident may isolate the microgrid from the main grid, causing instability in the system. Changes in the

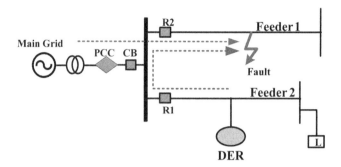

Figure 8.8   Unnecessary tripping in MG

network configuration of the microgrid, such as switching between different operational modes, may cause false tripping if the protective mechanisms are not adequately modified. Mitigating false tripping in microgrids necessitates using advanced protection methods, coordination tactics, and adaptive algorithms that consider the systems' dynamic and decentralized character. Furthermore, regular testing and calibration of protective devices are required to assure their dependability and reduce the frequency of false trips.

In the context of precise fault detection within a microgrid, the accurate determination of fault current direction becomes paramount. Particularly in microgrids where the fault current direction is dynamic, as opposed to fixed, discerning the fault current direction emerges as a primary challenge for protection engineers. This research introduces an innovative fault direction strategy tailored for AC microgrids. This novel scheme adeptly addresses the challenge, offering efficient fault direction determination specifically for the grid-connected mode of operation.

8.1.7   Review of microgrid fault direction estimation schemes

For accurate fault detection in a microgrid, the direction of the fault current must be correctly determined. In microgrids where the fault current direction is not specific, determining the direction of fault current is one of the most significant challenges for protection engineers. This research proposes a comprehensive fault direction approach for the AC microgrid, which can efficiently determine the fault direction in the grid-following mode of operation. The directional overcurrent relays (D-O-C-Rs) and distance relays require a minimum communication network, and low bandwidth to link relays plays an essential role in the protection of MG when power is only supplied by synchronous-based generation. However, detecting small fault currents when inverter-based DERs are present places a challenge to conventional relays. In addition to D-O-C-Rs, the directional element must be utilized in distance relays for the protection of MG. A dedicated study related to the directional element to enhance the performance of fault detection must be carried out, which is missing in many kinds of literature. Time domain analysis for fault direction estimation is presented in [4,5]. These techniques are simple and easy to implement, but their accuracy is low. Fault detection based on sequence components and phasor difference between voltage and current are represented in [6,7]. They usually use voltage and current data for MG fault direction estimation. These techniques are less affected by noise and CT saturation effects. These techniques are more reliable and accurate. However, these techniques use frequency analysis such as DFT and Fourier Transform, which reduce the speed of this technique and are practically slow. Zero sequence current (ZsC) is employed to determine the fault direction in [8]. However, the ZsC is not present in all types of faults. So, the proposed strategy can only detect the fault where the ground is involved, i.e., LG, L-L-G, and L-L-L-G.

A new fault direction estimate strategy for AC MG is proposed in [9]. The dynamic value of the superimposed negative sequence impedance angle (NSSIA) is used in this study to estimate the unbalanced fault direction in a grid-following mode. The sign of NSSIA decides if the fault is forward (F_D) or reverse (F_U) at the relay point. NSSIA is computed at different locations of the proposed MG by varying fault resistances, signals affected by noise, and when a phase is removed (pole tripping). Additionally, the recommended strategy was tested using various DG outputs and sample frequencies. NISSIA is positive for F_U faults but negative for F_D faults in all circumstances. The approach works effectively in all circumstances. In every instance, the proposed approach accurately predicts the direction of fault current. The recommended approach can be appropriately applied as the time required for the technical execution is less, and the implementation cost is less.

8.2   Suggested approach for fault direction estimation

8.2.1   *Microgrid system overview*

To analyze the suggested technology, a modified IEEE 34 test MG was designed in the RTDS (Real-Time Digital Simulator)/RSCAD (RTDS Simulator Software) platform. The RSCAD simulation environment's realistic operation provides an authenticated environment for physical relays. With RSCAD, protective devices can be tested in real-time while dynamic behaviors are accurately represented. High-fidelity modeling of generators, transformers, transmission lines, and distribution systems, among other power system components, is supported by RTDS, allowing for realistic and intricate simulations. It makes Hardware-in-the-Loop testing easier by enabling a real-time interface between physical devices like relays and controllers and the simulated power system for thorough testing. Because of its versatility, RTDS can be used for many different elements of power system analysis, such as advanced control strategy formulation, grid integration of renewable energy sources, and protection system testing. Engineers can verify and improve simulation models based on real-world performance data using RTDS for model validation.

The adapted IEEE 34-bus system stands out as an optimal choice for our endeavors due to its rich assortment of components, favorable topological properties, and abundance of extensive feeders. This unique configuration lends itself well to the seamless integration of various renewable sources. Additionally, the

inclusion of both $1 - \Phi$ and $3 - \Phi$ branches enhances its suitability for comprehensive analysis in rural distribution networks. This system's versatility makes it a
robust platform for our work, providing a diverse and comprehensive environment
for addressing the complexities associated with renewable energy integration and
rural distribution network assessments, as represented in Figure 8.9.

The suggested IEEE 34-bus system is added with solar-based generation at bus
no.: 860 and 890. The PV at the bus no. 860 and 890. In this research, each Solar
Photovoltaic (PV) unit at bus 860 and 890 is rated at 1.6 MW, using mono-
crystalline silicon semiconductor materials. The details regarding the PV used are
provided in Table 8.1.

The unbalanced nature of IEEE 34-bus system is transposed at different locations
to balance the inductance and capacitance along the line. It involves changing the
position of the conductor at regular intervals along the length of the line. It

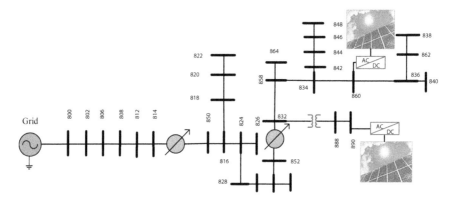

Figure 8.9   IEEE 34-bus test microgrid system

Table 8.1   Solar photovoltaic data

Details of PV parameter	Symbol used	Value
Rated Power	P_{PV}	1.6 MW
No. of series connected cells in each string and each module	N_C	36
Number of cells in a parallel string	N_{CP}	1
No. of module in series	N_{series}	230
No. of modules in parallel	$N_{parallel}$	132
O.C voltage	V_{OCref}	21.7 AV
S.C current	I_{SCref}	3.35 A
Voltage value at maximum power	V_{max}	17.4 V
Current value at maximum power	I_{max}	3.05 A
Semiconductor material of solar cell		Monocrystalline silicon
Series resistance	R_{se}	0.5 Ω
Shunt resistance	R_{sh}	100 Ω

accommodates a synchronous generator at the source end at bus no.: 800. Two voltage regulators are placed at 814–850, and 852–832 to maintain and stabilize the voltage within the prescribed limit to meet the requirements of utility and end users. Two shunt capacitor banks placed at nodes 844 and 848 are adaptable components essential for improving power systems' effectiveness and performance. Their uses include voltage support, loss reduction, power factor correction, and enhancing overall system stability. It consists of $1 - \Phi$ lines between nodes $816 - 818$, $862 - 838$, $814 - 850$, $808 - 810$, $852 - 832$, $854 - 856$, $820 - 822$, $818 - 820$, $824 - 826$, $858 - 864$ rest wires are $3 - \Phi$ with neutral wires.

8.2.2 Advanced technique for fault direction estimation

Most of the protection schemes used for fault direction estimation in MG applications fail to determine the direction of both symmetrical and unbalanced faults. It is of prime importance to design an algorithm that can be useful for the direction estimation of all types of faults for physical relays. The relay is positioned at bus no.: 800 and the instantaneous value of the 3-phase current (I_R, I_Y, I_B) and the voltage (V_R, V_Y, V_B) are measured at the relay. To understand the technique, a 3-Bus system is provided in Figure 8.10. The grid is connected at bus: U, and a PV is connected at bus: D. The relay is placed at position 'R'. The fault at (F_U) position is known as the reverse/upstream/backward fault, while at position (F_D) is known as the downstream/forward fault. This approach is instrumental in evaluating the efficacy of the proposed fault direction estimation scheme in real-world scenarios.

The innovative superimposed +ve sequence current and voltage phasors data can be estimated as below

$$v_{1,n} = \frac{1}{3}\left(V_{R,n} + \lambda V_{Y,n} + \lambda^2 V_{B,n}\right) \tag{8.1}$$

$$i_{1,n} = \frac{1}{3}\left(I_{R,n} + \lambda I_{Y,n} + \lambda^2 I_{B,n}\right) \tag{8.2}$$

where n symbolizes the bus number

and $\lambda = e^{\frac{j2\pi}{3}}$

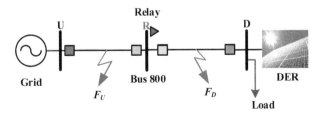

Figure 8.10   Test three bus system

For F_U, the fault at position U

$$\Delta \vec{i}_{1U} = \vec{i}_{1U \ postf} - \vec{i}_{1U \ pref} \tag{8.3}$$

$$\Delta \vec{v}_{1U} = \vec{v}_{1U \ postf} - \vec{v}_{1U \ pref} \tag{8.4}$$

$$\Delta \vec{\mathbb{Z}}_{1U} = \frac{\Delta \vec{v}_{1U}}{\Delta \vec{i}_{1U}} = \frac{|\Delta v_1| \angle \Phi_{1U}}{|\Delta i_1| \angle \Phi_{1UI}} = |\Delta \mathbb{Z}_1| \angle \Phi_U \tag{8.5}$$

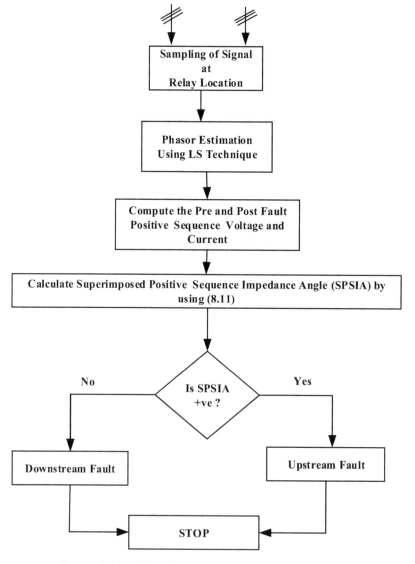

Figure 8.11   Flow diagram of the suggested algorithm

where

$$\angle\Phi_U = \angle\Phi_{1U} - \angle\Phi_{1UI} \tag{8.6}$$

For the fault at location D

$$\Delta\vec{i}_{1D} = \vec{i}_{1D\ postf} - \vec{i}_{1D\ pref} \tag{8.7}$$

$$\Delta\vec{v}_{1D} = \vec{v}_{1D\ postf} - \vec{v}_{1D\ pref} \tag{8.8}$$

$$\Delta\vec{\mathbb{Z}}_{1D} = \frac{\Delta\vec{v}_{1D}}{\Delta\vec{i}_{1D}} = \frac{|\Delta v_1|\angle\Phi_{1D}}{|\Delta i_1|\angle\Phi_{1DI}} = |\Delta\mathbb{Z}_1|\angle\Phi_D \tag{8.9}$$

where

$$\angle\Phi_D = \angle\Phi_{1D} - \angle\Phi_{1DI} \tag{8.10}$$

$$\angle\Phi = \angle\Phi_D \quad or \quad \angle\Phi_U \tag{8.11}$$

The $\angle\Phi$ is known as Superimposed Positive Sequence Impedance Angle (SPSIA)

The angle Φ ($\angle\Phi_D$ or $\angle\Phi_U$) determines the direction of fault in the proposed MG. The proposed scheme flow chart is shown in Figure 8.11.

8.3 Results assessment and interpretation

In the RTDS/RSCAD environment, we conducted simulations on an IEEE 34-bus system to validate our suggested technique under grid-following mode (GFM). Various symmetrical/unsymmetrical faults were intentionally induced at different points and operating environments of the MG, ensuring a comprehensive assessment of the proposed technique's efficacy under unbiased operation. The fault time was standardized to 0.15 seconds, with N_s (sampling frequency) of 12 kHz, while the MG operated at 60 Hz for analysis of this research.

To discern the sequence components within the MG, the least-square technique, employing a window of 200 samples, was applied. The placement of the relay, strategically positioned between bus no.: 800 and bus no.: 802, facilitated targeted observations. The voltage and current signal for our analysis were captured at the relay position, with pre-fault data simply transferred from the MG system.

Our suggested technique underwent rigorous verification for its performance in scenarios involving high-resistance faults, direction estimation amidst noise, and pole-tripping events. Notably, the technique demonstrated robustness across varying sample rates and diverse Distributed Generation (DG) output conditions. This comprehensive testing ensures the reliability and adaptability of our proposed technique in addressing real-world challenges within the grid-connected mode.

The typical calculation time needed to determine the direction is minimal, as only the sign $(+/-)$ of SPSIA is essential to estimate the fault direction, and the technique can be utilized in actual relays with significant accuracy and simplicity. In all the cases for F_U faults the SPSIA is positive $(+ve)$ while for F_D faults, SPSIA is negative $(-ve)$, confirming the proposed technique's accuracy.

Table 8.2   Results for SPSIA

S. No.	Fault type and fault resistance	Downstream Fault (F_U)	Upstream Fault (F_U)
1	LG fault (0.1 Ω)	−1.147	2.863
2	LL (0.1 Ω)	−0.385	1.229
3	LLG (0.1 Ω)	−3.031	0.242
4	LLL (0.1 Ω)	−1.497	2.095
5	LLLG (0.1 Ω)	−0.888	1.993
6	LG (10 Ω)	−1.144	2.375
7	LLL (10 Ω)	−2.356	1.57
8	LG (DG output variation)	−0.356	1.373
9	LL (10 Ω)	−1.283	2.273
10	LG (20 Ω)	−1.363	1.675
11	LLG (20 Ω)	−2.732	2.521
12	LL (20 Ω)	−3.556	2.47
13	LLL (30Ω)	−2.131	1.267
14	LLLG (0.5Ω)	−2.207	1.734
15	LLLG (30Ω)	−2.345	1.227

Determining the fault direction usually takes less computation time because it simply needs to determine the sign $(+/-)$ of the SPSIA. Because of its simplicity, the technique can be applied in relays with greater accuracy. Table 8.2 lists SPSIA angles in radians for various unsymmetrical/symmetrical faults of MG. SPSIA is always positive for F_U faults and negative for F_D faults, confirming the accuracy of the suggested approach.

8.3.1   Pole-ground (L-G) fault

Most of the power system faults are L-G faults. These faults should be immediately cleared in order to safeguard the power system. To analyze the performance of SPSIA, the pole-ground (L-G) fault is created by varying fault resistance. The L-G fault is also created at different locations of the suggested MG. For the forward fault/downstream (F_D) SPSIA is always calculated to be *negative (−ve)*. At the same for all unbalanced and balanced faults, the SPSIA for all reverse fault/ upstream $((F_U)$ is *positive (+ve)*. From the above observation, if SPSIA is +ve, the fault is in the reverse direction of the relay location and if SPSIA is −ve, the fault is in the forward direction of the relay location. As any small deviation from zero value can predict the direction accurately, the technique can be suitably implemented for physical relays.

A L-G type unsymmetrical fault is created in R_{Phase} and the ground with a fault resistance of 0.1 Ω. The waveform during the fault in (F_U) and (F_D) is shown in Figure 8.12. The fault is incepted at 0.2 s, and the system operates for 1 s in the RTDS platform. The value of SPSIA is calculated for the $L - G$ fault by using Equation (8.11). The angle of SPSIA, as referred to in Figure 8.13 is 2.863 radian for (F_U) and −1.147 radian for (F_D), which indicates the accuracy.

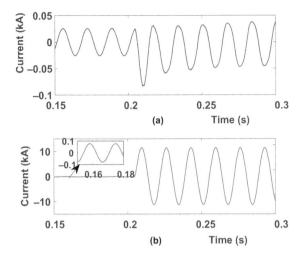

Figure 8.12   L-G fault current waveform on (a) F_U and (b) F_D side

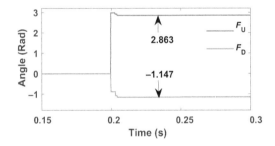

Figure 8.13   SPSIA for LG fault on F_U and F_D side

8.3.2   Pole-pole (L-L) fault

The unsymmetrical Line (L)-Line (L) fault is created between the 'R' and 'Y' with a fault resistance of 0.1 Ω. The fault is created at different locations with variations of fault impedance. The L-L fault current waveform is depicted in Figure 8.14. Different fault resistances are introduced during the double-line fault to examine the value of SPSIA. The fault is triggered at 0.2 s time instant. The real-time RTDS simulation is compiled for 1 s. The SPSIA is calculated for (F_U) and (F_D) as provided in Figure 8.15. For forward (F_D) and reverse (F_U) faults the value of SPSIA is −0.385 and 1.229 radians respectively, i.e., during the downstream fault, SPSIA is −ve and for the upstream fault SPSIA is +ve. As any small change in the angle sign indicates the direction of the fault, the direction can be detected accurately in the shortest possible time.

8.3.3   Line-line-ground (L-L-G) fault

The Pole-Pole-Ground fault is performed between 'R'-'Y'-G of the suggested MG. The fault is incepted at 0.2 s and allowed for 0.15 s. Initially, the resistance of fault is kept at

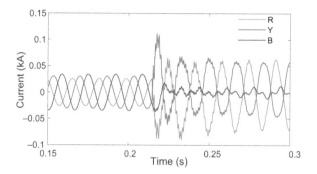

Figure 8.14   L-L fault at F_U location

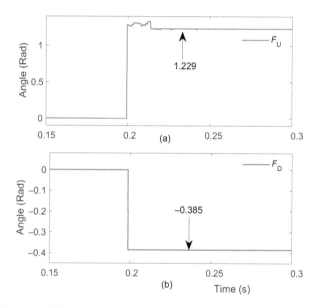

Figure 8.15   SPSIA for L-L fault on (a) F_U and (b) F_D side

0.1 Ω. The measurements are taken at the forward and backward direction of the relay position. The value of SPSIA is calculated for both directions of faults for 0.1 Ω fault resistance. The value of SPSIA for (F_U) and (F_D) is found to be 0.242 and −3.031 radian. It is confirmed from the result that during upstream F_U, the fault value of SPSIA is positive and negative for downstream (F_D) as represented in Figure 8.16. The suggested technique was found to be efficient in determining the direction of L-L-G fault.

8.3.4   *Pole (L)-pole (L)-pole (L) and pole (L)-pole (L)-pole (L)-ground (G)/(L-L-L and L-L-L-G) faults*

The SPSIA is versatile and can discern the fault current direction of symmetrical faults. This capability enhances the advantages of the fault direction estimation technique in a

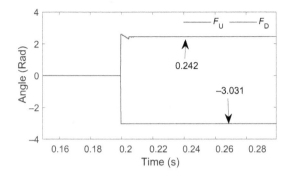

Figure 8.16 SPSIA for L-L-G fault

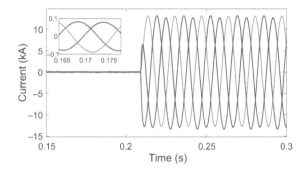

Figure 8.17 LLL current on F_U side for 0.1 Ω resistance

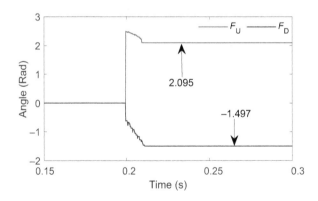

Figure 8.18 SPSIA for L-L-L fault F_U and F_D side

microgrid. The $3 - \Phi$ current waveform during fault is shown in Figure 8.17. To verify the suggested approach, L-L-L fault is created in the forward and reverse direction of the relay location by shorting the $3 - \phi$ with a resistance of 0.1 Ω. The SPSIA is calculated at the relay point. The value of SPSIA for (F_U) and (F_D) for balanced $3 - \Phi$ fault is 2.095 and -1.497 radian respectively as shown in Figure 8.18.

The suggested technique is also analyzed for pole-pole-pole-ground (L-L-L-G) fault by shorting the $3 - \Phi$ and ground with 0.1Ω resistance. The value of SPSIA for (F_D) and (F_U) is found to be -2.356 and 1.57 radians as mentioned in Figure 8.19. The results confirm that the suggested technique can effectively determine the direction of the symmetrical fault, where many techniques fail to work efficiently for balanced and unbalanced faults.

8.3.5   *Effect of fault resistance on the proposed technique*

The magnitude of the fault current reduces for high resistance faults. Many techniques fail to detect the minor magnitude fault current due to high resistance. The suggested approach is verified by varying the fault resistance. The fault resistance is varied from 0.1 to 20 Ω in steps. With variations of fault resistance, different types of faults are simulated. The most common fault (L-G fault) is simulated with 10 Ω fault resistance. The SPSIA is 2.375 and -1.144 radian for F_U and F_D of the relay location as shown in Figure 8.20. To test its validity for the most severe fault (L-L-L) fault, the $3 - \Phi$ is shorted with 10 Ω resistance. The value of SPSIA is calculated and found to be 1.57 and -2.356 radian for F_U and F_D point of the relay, which confirms the validity of the proposed approach (Figure 8.21).

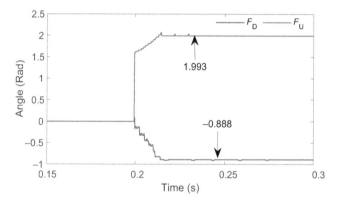

Figure 8.19   SPSIA for L-L-L-G fault on F_U and F_D side

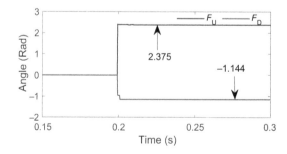

Figure 8.20   SPSIA for F_U and F_D side for 10 Ω resistance

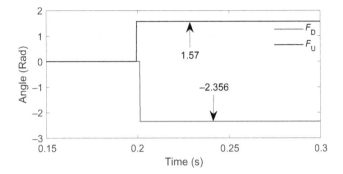

Figure 8.21   SPSIA LLL on F_U side for 10 Ω resistance

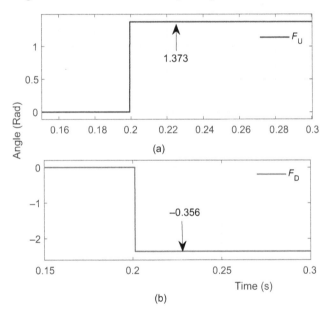

Figure 8.22   SPSIA for L-L-L fault on (a) FU & (b) FD side at reduced PV2 output

8.3.6   Insecure DER implications

Solar energy production is inherently intermittent because of weather, cloud cover, and variations in daylight. Seasonal variations exist in solar generation. Winter's fewer daylight hours impact solar output. It is difficult to predict solar generation accurately because cloud cover and atmospheric conditions are unpredictable, which causes uncertainty in output forecasts. Aging, dirt, and technical issues all impact solar panels output. Geographical location also affects solar radiation. Addressing these uncertainties is crucial for effective solar energy planning, grid integration, and overall reliability in energy systems. For reliable operation of the MG, the fault detection and direction prediction technique must work efficiently during DER uncertainty conditions.

The IEEE 34-bus system used to develop an efficient direction estimation scheme uses two solar PV numbers, each with a 1.6 MW rating at bus: 890 (PV_1) and 860 (PV_2). The PV operates at a temperature of 25 °C with an insolation of 1,000 W/m^2. The PV_1 output is reduced to 1.3 MW, and the system is allowed to operate in GCM. Several faults studies like L-G, L-L, and L-L-L are analyzed at reduced output of DER. In each case, SPSIA was calculated, and the results were analyzed. SPSIA is $+ve$ F_U (upstream) and $-ve$ for F_D (downstream) in all the cases. Again, the output of PV_2 is reduced to 1.4 *MW*, and PV_1 is allowed to operate at 1.6 MW. The SPSIA to observed for balanced $(L - L - L)$ fault as shown in Figure 8.22.

8.4   Conclusion and future scope

Smart Grid signifies the revolutionary influence on conventional grid architecture through advanced technologies and dedicated communication networks to enhance sustainability, reliability, dependability, efficiency, and high consumer satisfaction. Microgrid is an integral part of the modern smart grid. To make the microgrid more efficient, the fault must be cleared within the minimum time of its inception. Protection professionals find it more challenging to detect the fault in the microgrid because of bidirectional fault current, false tripping, and variation in fault current in different modes of operation. To determine the fault current accurately, a conventional protection scheme must be accompanied by an efficient directional estimation algorithm for MG protection. This research suggests an efficient direction estimation algorithm for AC MG based on the dynamic value of Superimposed Positive sequence impedance Angel (SPSIA), which can determine the direction of symmetrical and unsymmetrical faults in MG. The positive and negative values of SPSIA assess the direction of fault. For upstream faults (F_U), SPSIA is positive and negative for all downstream faults (F_D). The novelty of the suggested approach is verified with different adverse conditions, such as high fault resistance variation in DER output. The novel technique is also demonstrated with different fault resistance. The time taken to determine the fault is much less, and the least communication network is used, so the method can be efficiently implemented in modern MG protection applications for the physical relays.

Further research can be conducted in the following areas:

• Create more advanced and flexible control schemes for microgrids to maximize energy production, distribution, and storage instantly. Use cutting-edge algorithms and machine learning strategies to improve decision-making.
• Development of a novel scheme considering the effect of noise, pole tripping, and DER output variations.
• Microgrid cybersecurity measures should be strengthened to defend against possible cyberattacks.
• Strive towards system and component compatibility and standardization in microgrids. This would make integrating various technologies from many manufacturers easier, encouraging scalability and broad usage.

- To improve local energy resilience, investigate the idea of microgrids at the community and neighborhood levels. Examine governance frameworks, strategies of community engagement, and financial sustainability.
- Evaluate and reduce the impact of microgrid technology on the environment. Life cycle analyses, carbon footprint reduction, and sustainable practices in microgrid construction and operation should be the main research areas
- Look for ways to make microgrids more capable of responding to demand and to involve users in actively controlling how much energy they use. Provide enticing interfaces and reward systems to entice customers to participate
- Investigate the creation of peer-to-peer energy trading—enabling decentralized energy markets within microgrids. Examine how blockchain technology can be used to facilitate safe and open transactions.
- Examine how microgrids and the wider power grid interact with each other efficiently. Provide technologies and protocols that facilitate the smooth transition between islanded and grid-connected modes, guaranteeing coordination and compatibility.

References

[1]   Central Electricity Authority, Ministry of Power, Government of India; [cited 2023 December 03]. Available from: https://cea.nic.in/?lang=en.

[2]   Panda SK, Subudhi B. A review on robust and adaptive control schemes for microgrid. *Journal of Modern Power Systems and Clean Energy*. 2023;11 (4):1027–1040.

[3]   Pradhan R, and Jena P. Advanced fault detection technique for AC microgrid protection. In: *2023 IEEE 3rd International Conference on Sustainable Energy and Future Electric Transportation (SEFET)*; 2023. p. 1–6.

[4]   Saleki G, Samet H, and Ghanbari T. High-speed directional protection based on cross correlation of Fourier transform components of voltage and current. *IET Science, Measurement & Technology*. 2016;10(4):275–287.

[5]   Liang Y, Li W, and Lu Z. Effect of inverter-interfaced renewable energy power plants on negative-sequence directional relays and a solution. *IEEE Transactions on Power Delivery*. 2020;36(2):554–565.

[6]   Jalilian A, Hagh MT, and Hashemi S. An innovative directional relaying scheme based on postfault current. *IEEE Transactions on Power Delivery*. 2014;29(6):2640–2647.

[7]   Jena P, and Pradhan AK. Directional relaying during single-pole tripping using phase change in negative-sequence current. *IEEE Transactions on Power Delivery*. 2013;28(3):1548–1557.

[8]   Mahamedi B, Zhu JG, Eskandari M, *et al*. Protection of inverter-based microgrids from ground faults by an innovative directional element. *IET Generation, Transmission & Distribution*. 2018;12(22):5918–5927.

[9]   Pradhan R, and Jena P. An innovative fault direction estimation technique for AC microgrid. *Electric Power Systems Research*. 2023;215:108997.

Chapter 9

Communication and computations for smart grid protection

Udit Prasad[1] and Soumya R. Mohanty[1]

Effective communication is a cornerstone in the realm of Power Systems Protection Applications, playing a pivotal role in ensuring the reliable and secure operation of power infrastructure. The fundamental guidelines governing the implementation of communication-based protection include:

- Implementing established protective techniques within communication-based protection systems. In addition to that, evaluating potential protection scheme failure modes to reduce the likelihood of failures.
- Continuously monitoring communication channels. Incorporating protocol following retransmission (e.g., GOOSE messages) for exchanging messages to further enhance system security.
- Designing an Ethernet-based substation LAN to support multiple services, such as protection, control, monitoring, and engineering access. This involves implementing VLAN segregation and message prioritization to ensure secure transmission of protection-related data.
- Calculating real-time protection bandwidth allocation for the high-priority operation during the design phase and validating it rigorously. Using message delivery failure alarms to ensure continuous availability of adequate real-time bandwidth further enhances reliability.
- Paying attention carefully to message delivery timing, jitter requirements, and network failure recovery intervals.

Keeping the abovementioned points into consideration, this chapter embarks on a journey into the intricate world of communication within power systems, delving into various facets from performance considerations to cybersecurity strategies. To provide a comprehensive understanding, we will commence by exploring the prerequisites necessary for the communication system to support the delicate protection infrastructure. As we progress through this chapter, we will trace the application of communication across different domains of power systems protection. The spotlight then turns to the choice of Communication Medium, where we

[1]Department of Electrical Engineering, Indian Institute of Technology (BHU) Varanasi, India

analyze Communication Channel Requirements and their associated properties, followed by a comprehensive comparison of various Communication Channels.

Our discussion extends to the exploration of widely used Communication Protocols, and we navigate through different Communication Networks, comparing their applications. However, it is important to acknowledge that while communication technology brings numerous advantages, it also reveals vulnerabilities and cyber threats. We will delve into the intricacies of these vulnerabilities in protocols and network architecture. Moreover, in our quest to build a resilient and secure infrastructure, we will explore various countermeasures and proposed solutions.

9.1 Performance considerations of communication in power systems protection applications

This section provides a comprehensive overview of protection system reliability, availability, redundancy, and timing considerations, emphasizing the critical role of communication systems in ensuring the seamless operation of protection mechanisms within the power grid. Understanding these concepts is fundamental to designing and maintaining robust and dependable protection systems in complex electrical infrastructures. Furthermore, we endeavor to provide a distinct delineation of the performance metrics pertaining to the communication components and the relaying components.

9.1.1 Availability

It is defined as the proportion of the duration in which a protective operation occurred meeting or exceeding its specified performance standards compared to the total duration within a specified timeframe. To compute availability, one can subtract unavailability from 1 and express the result as a percentage. Unavailability, in this context, is the proportion of time when a protective service is not accessible due to various factors, including failures, and it is contingent on Mean Time Between Failure (MTBF) and Mean Time to Repair (MTTR) [1]. Unavailability encompasses unexpected outages, malfunction in both hardware and software, errors in procedures, and potentially any additional criteria specified by the utilities. Interested readers may refer to [2] for a detailed background on availability and unavailability. However, standard values of protection system unavailability for standard relays and relays equipped with self-test capabilities are 9.4×10^{-2} and 1×10^{-4}, respectively [3]. The availability of protection systems is a composite of the communication component and the relay element. It is imperative that communication systems are capable of functioning even in the presence of transmission disruptions, such as ground/earth potential rise, which are likely to coincide with faults in the power system.

9.1.2 Protection system redundancy

Redundancy entails duplicating essential components within the protection infrastructure in order to enhance reliability by mitigating or removing individual points of vulnerability. Moreover, redundancy permits certain elements to undergo maintenance without compromising the performance of the protective system. Forecasting the availability of a system is a multifaceted task, with an inherent potential for inaccuracies. In some cases, not all components possess redundancy, and degrees of redundancy exist. An availability analysis, a mandatory component of Reliability, Availability, and Survivability evaluation, involves mathematical calculations that combine the Mean Time Between Failure (MTBF), typically provided by the manufacturer, and the Mean Time to Repair (MTTR) of each component, factoring in the system's structural layout in terms of physical connections.

While redundant components may not give complete independence, it is vital to recognize that they do provide some independence. The extent of independence among system components is substantially influenced by the physical and electrical isolation between them, hence it is important to pay attention to this separation.

9.1.3   Reliability

The Relay Work Group has outlined two essential aspects for evaluating the reliability of a protection system: dependability and security. The assessment of reliability encompasses all components within the protection infrastructure, including relays, current transformers (CTs), voltage transformers (VTs), communication channels, supply and control connections, and sub-station DC auxiliary supply. It is important to note that "communication channels" refer to all necessary communication equipment required to facilitate information transfer between initiating relays and receiving relays at different locations. There are two primary failure modes to consider: the failure to operate, which represents dependability, and unnecessary operation, which characterizes security. Dependability ensures that the protection system responds to faults or conditions within its designated protective zone, while the operational function indicates an inability to clear a fault as expected.

The performance of a protection infrastructure's dependability can be impacted by its communication components. For example, a Permissive Overreaching Transfer Trip (POTT) scheme is used in high-speed transmission line protection along its entire length [1]. If there is an equipment failure due to the communication interface at one end of the line, the affected terminal will not receive the POTT signal from the remote terminal in the case of a fault lying inside the zone. This can result in a failure of high-speed protection for the entire faulty line. On the other hand, security ensures that the protection system can withstand faults or conditions occurring outside its intended protective zone. It is important to clarify that, in this context, "security" does not pertain to physical or cybersecurity. Unnecessary relay operations can be divided into two categories:

(i) Operation for a non-fault condition
(ii) Operation for an external fault, i.e., a fault occurring outside the primary protection zone

The integrity of a protection infrastructure can also be influenced by the communication components in the system. Consider the POTT scheme mentioned earlier, functioning correctly in all aspects except for a communication circuit loopback condition at one terminal of the transmission line. In such a scenario, a fault lying outside the protection zone could be detected by this terminal would erroneously trigger a POTT action on itself, leading to an erroneous tripping of the associated circuit breakers. Hence, the reliability of the communications system is a comprehensive measure of overall reliability.

9.1.4   Temporal assessment

Relaying operations in a power system requires the observation of a situation in one place and subsequently initiating a response in another location. The protection system introduces inherent time delays at several junctures. As depicted in Figure 9.1, the total delay time, also referred to as "End-to-End Delay Time," pertains to the interval commencing with the detection of an event at a specific point, such as A, and concluding with the initiation of a control response at a different point, for example, B. It is important to note that the end-to-end delay time does not encompass the relay detection time at A, or the time required for the breaker to operate at B. Designers of protection systems establish a unique "clearing time" for fault occurrences, which varies for each transmission line and relies on a multitude of factors.

The various time delays within the protection system include:

Detection time: The duration needed for the relay at the monitoring site to identify an abnormal event.

Message encoding time: Duration required for the relaying component/device to produce a state change message intended for transmission to the communication interfacing component/device.

Packet transfer or frame transmission time: Duration required to transmit the information packet conveying the state change message from the initial relay component to the communication interface component.

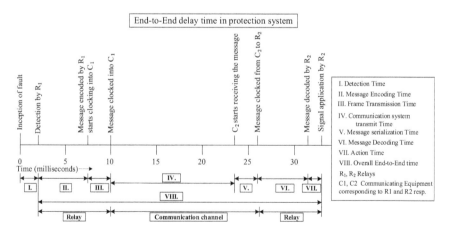

Figure 9.1   Message processing and transmission time

Propagation time: The duration needed for the communication system to transmit the information to the control center, also known as transmit time in the communication system.

Packet transfer or message serialization time: The duration needed to transmit the data packet with the state change information from the communication interface component to the relaying component at the destination.

Message decoding time: Duration needed for the relay system to comprehend the control commands/status message and decide on a course of action.

Action time: Duration needed by the protective relay to initiate the suitable control action at the control site.

Breaker operation time: Duration it takes for the circuit breaker to activate typically exceeds two cycles.

It is important to mention that certain delays consist of elements that are both constant and subject to change. For instance, the duration required for message decryption encompasses a predetermined element, allowing the receiving relay to determine the significance of the received message, but it might be extended to allow for the reception of three consecutive "take action" messages before initiating control action. In analog systems, the initiation of a trip operation could occur following a sequence of 50 ms with a valid guard zone, followed by 5 ms of a valid trip zone, further increasing the end-to-end delay further increasing the end-to-end delay while reducing the likelihood of a false control action. The time it takes for communication systems to transmit data serves as the reference used to assess the performance of the communication infrastructure. Importantly, even if transit time in communications were reduced to zero, the end-to-end delay would still persist.

9.1.5   Reliance of protection schemes on communication systems

Communication-dependent schemes rely on functional communication channels and cease to operate in the absence of communication system functionality. Notable examples of such schemes encompass line differential protection, line phase comparison protection, direct transfer trip (DTT) schemes, and distributed bus differential protection.

- *Line differential protection:* Relays transmit current information via a digital channel with a minimum bandwidth of 56 kbps. These systems are responsive to discrepancies and imbalances in communication channel propagation delay, which can be addressed by employing timestamping or external clock synchronization techniques.
- *DTT:* It necessitates secure communication because it relies on binary instructions with limited informational content. Ensuring strong coding and error detection mechanisms becomes crucial to avert operational errors resulting from message corruption, leading to bandwidth demands comparable to other communication-oriented approaches.

9.2   Communication system in power systems application

The communication infrastructure associated with protection systems is diligently designed to facilitate remote control and provide real-time indication of substation and field devices. These devices transmit crucial status updates, alarm notifications, and operational data, all contributing to the robustness of power systems. Contemporary communication networks have evolved into expansive entities, closely resembling the intricate connectivity of the foundational power networks. Safeguarding, supervising, and managing the power grid entails the exchange of locally generated data. However, the absence of an appropriate communication conduit poses a significant drawback to power system protection, rendering it incapable of precise fault discrimination. In the domain of communication-based power systems, there are various functions, each with its unique operational requirements. These applications include:

(i) *Energy management systems (EMS) and Supervisory Control and Data Acquisition (SCADA):*
    SCADA and EMS systems have traditionally been responsible for overseeing expansive geographical regions and have relied on a combination of direct serial connections, modems, radio/microwave links, and SONET/Ethernet networks to adhere to stringent availability requirements. These SCADA networks are frequently managed independently and separately from other communication resources. Notable SCADA protocols encompass:
    • Legacy protocols designed for low-speed serial communication channels, which include PG&E 2179, Modbus, and Harris.
    • Standardized, low-speed serial protocols, such as DNP3, TEC 60870-5-101, and 60870-5-103.
    • Ethernet-based protocols like Modbus-IP, DNP3-IP, and IEC 60870-104.
      The SCADA system's adaptability, driven by parameters in its database, reduces costs, enhances reliability, and promotes safety during implementation. It also serves as a foundation for future advancements, ensuring a dynamic and responsive power systems protection infrastructure. One essential component of the communication network is the Remote Terminal Unit (RTU), typically installed within substations. The RTU is equipped with digital and analog input/output (I/O) interfaces, which play pivotal roles in interfacing with substation devices. Digital I/O channels are responsible for displaying the operational statuses of field equipment, including critical information such as the state of circuit breakers—whether open or closed. On the other hand, analog I/O channels provide real-time data on electrical quantities, enabling the monitoring of parameters like current flow through a breaker or voltage levels at a bus bar. These analog inputs serve as the eyes and ears of the power system, providing essential information for timely decision-making and ensuring the integrity of the entire network.

(ii) *Engineering access and maintenance:*
    Enabling communication in engineering grants the capability for remote supervision and adjustment of substation IEDs. This technology empowers protection

engineers to remotely retrieve oscillograph-based fault records, sequential events records, and device configurations. Additionally, they can conduct firmware upgrades and carry out various management tasks. Access to this engineering feature is typically restricted to a select group of highly trained engineers who are directly accountable for the operation of the protection system.

(iii) *Pilot Protection:*

Pilot protection, also known as tele-protection, encompasses directional comparison, differential methods, and phase comparison techniques. This form of protection necessitates relay-to-relay communication and typically functions within a few electrical cycles. Relay-to-relay communication protocols employed for this purpose encompass IEC 61850 Sampled Values (SV) and IEC 61850 Generic Object-Oriented Substation Events (GOOSE).

(iv) *Security:*

The integrity of power system infrastructure is of utmost importance, as it needs to function without interruption despite various potential challenges, including equipment malfunctions, weather-related disruptions, human errors, and deliberate attacks. Contemporary communication systems offer rapid solutions to address these issues, ensuring a dependable supply of power to affected regions. Emerging applications within this realm encompass wide area protection strategies, synchrophasor measurements, real-time fault identification, Geographic Information System (GIS) outage management, and video surveillance. These innovations demand additional bandwidth and necessitate harmonious integration with existing communication services. The increased reliance on network-based communication creates a fresh vulnerability that requires safeguarding against external threats. In Section 9.6, a concise overview is provided on information security measures and methodologies employed to safeguard critical power infrastructure.

(v) *Substation and distribution automation:*

Power system communication has witnessed significant growth in recent times, with the emergence of autonomous devices that possess the ability to oversee, reconfigure, and enhance the operation of power systems. These devices encompass a range of technologies, including faulted circuit indicators, recloser controls, regulator controls, programmable logic controllers, and robust substation computing platforms. Section 9.5 comprehensively addresses substation automation systems.

(vi) *Wide-area protection and control:*

Wide-area protection and control systems play a crucial role in safeguarding extensive geographical regions through communication-based mechanisms. While conventional pilot protection systems primarily focus on safeguarding individual transmission lines, wide-area protection and control systems have a broader objective: ensuring the overall stability and resilience of the power system. Historically, these systems were prohibitively expensive and primarily deployed for the most critical and complex power system challenges. However, recent advancements in digital communication, precise time synchronization through technologies like GPS and IEEE Standard 1588, and measurement technologies like synchrophasor have significantly reduced the costs associated with wide-area systems. The integration of synchrophasor measurement functionality into protective relays has led to a substantial increase in the number of synchrophasor measurement points available, promising near-complete system coverage. The main obstacle in achieving this goal lies in establishing reliable communication channels capable of transmitting synchrophasor data across the power system. Notably, the advancements and cost reductions associated with fiber-optic-based communication are revolutionizing the approach to power system protection, control, and monitoring as a whole.

The importance of wide-area protection and control capabilities is magnified as the integration of intermittent renewable resources, such as wind and solar, continues to grow. The increasing penetration of these resources introduces new challenges to the reliability and stability of the system, necessitating enhanced situational awareness for system operators.

Table 9.1 provides a tabulation of the customary response times and event durations corresponding to various applications.

Figure 9.2 depicts a comprehensive representation of fundamental components within communication systems, as well as the diverse array of communication mediums encountered in practical installations. This visual representation elucidates the utilization of telephone networks (Public Switched Telephone Network or PSTN) for engineering purposes, the deployment of dedicated optical fibers cables, radio communication technologies, and leased analog lines, as well as the

Table 9.1   Response times and event durations for different power system communications applications

Application	Involvement	Time requirement	Event duration
Pilot protection scheme	Communication between IEDs	1–30 ms	Less than 100 ms
Substation automation	Communication between IEDs and Automation Controllers	30 ms–1 s	30 ms to several minutes
Wide-area protection and control	Communication between IEDs and Wide-Area Controllers	100 ms–1 s	100 ms to several minutes
SCADA	Communication between IEDs/RTUs and Master Station/System Operator	≤ 5 s	Seconds to hours
EMS	Operator/dispatcher communication	≤ 5 s	Minutes to hours
Security	Various	≤ 5 s	Minutes to days
Engineering access and routine maintenance	Engineers remotely communicating with IEDs/RTUs	≤ 5 s	Hours to months

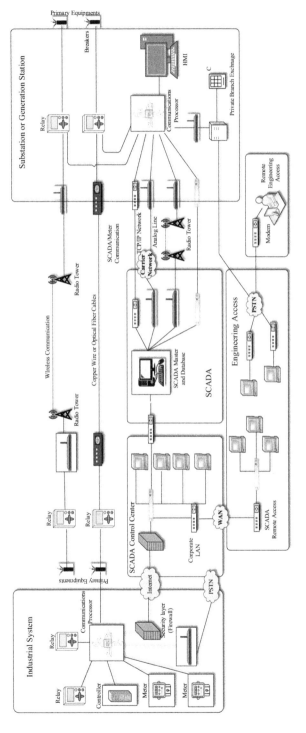

Figure 9.2 Overview of communication in power system

integration of Transmission Control Protocol/Internet Protocol (TCP/IP) networks for SCADA accessibility. Furthermore, the figure serves to highlight the optimized relay-to-relay communication protocols employed for protective measures. In addition to these specific elements, the figure also serves as a graphical testament to the multifaceted and intricate nature inherent in large-scale systems that have evolved gradually, continually adapting to user requirements over time.

9.3   Communication medium

For several decades, communication systems have been employed to augment the operational efficiency of power systems. In its initial stages, copper conductors played a pivotal role, encompassing privately operated pilot wire channels, exclusive telephone circuits, and power line carrier channels. As the 1970s dawned, utility companies started replacing copper wiring within transmission systems with their microwave links. Supervisory Control and Data Acquisition (SCADA), which has become the cornerstone of modern power systems, relies on a simple yet powerful principle—the ability to control and monitor remote assets from strategically located control centers. This entails the need for one or more strategically positioned control centers, with the primary control point often referred to as the Network Control Centre (NCC). It is here that the SCADA master station resides, acting as the nexus for communication with Remote Terminal Units (RTUs) situated in substations and field equipment, such as pole mount Auto reclosers. The interaction between the SCADA master station and RTUs occurs over a dedicated communications network. To facilitate this intricate web of communication, various communication networks come into play. These networks serve as the conduits through which data flows, connecting the SCADA master station with the RTUs. Additionally, licensed radio transceivers expanded the reach of SCADA systems into medium-voltage distribution circuits during this period. The mediums employed for these communication networks are diverse, with radio, pilot, supervisory wire, and fiber-optic being the most prevalent options. Presently, evolving communication relies on optical fiber technology coupled with digital data transmission. Fiber-optic channels are renowned for their impressive capacity and reliability, boasting attributes such as minimal noise, immunity to interference, and enhanced safety. Moreover, the utilization of spread-spectrum radio communication offers an economically viable means of extending communication to the farthest reaches of the power grid. The choice of medium depends on the specific requirements of the power system and the desired level of reliability. In this regard, the next section emphasizes communication channel requirements in the context of power systems protection infrastructure.

9.3.1   Communications channel requirements

In the realm of communication channels, we find the pivotal task of transporting information from one geographical point to another. Generally, this information falls into one of two categories: analog, depicted using a modulated sinusoidal signal of some kind, or it can take on a digital form, conveyed through electrical or optical means using the binary system of 1 and 0. This section delves into the essential attributes defining a communication channel. The following subsections aim to expound upon the fundamental characteristics of communication channels embedded within protection systems. It is imperative that, for each specific application, all these characteristics find a clear definition and consensus among the responsible parties overseeing the relay and communication aspects.

9.3.1.1 Interface types

Interface types pertain to the physical and electrical (or optical) connection points where the components of relaying and communication converge. Naturally, these interfaces must align harmoniously with one another. In certain instances, an adapter serves as the bridge to reconcile disparities between device interfaces. Some prevalent interfaces include the four-wire audio, employing screw terminals; digital RS-232, employing a DB-9 connector; IEEE C37.94, utilizing multimode fiber and STC connectors; and direct Ethernet, which can operate over either fiber or copper mediums [1].

9.3.1.2 Bandwidth

Bandwidth, the channel's traffic-carrying capacity, holds considerable significance. In the realm of protective relaying, the demand for high bandwidth is typically minimal. Modem-transmitted audio signals usually traverse voice-grade circuits, characterized by a frequency response ranging from 300 to 3,000 Hz. This range amply supports reliable communication at speeds up to 19.2 kbps. Digital channels dedicated to relaying typically operate at bit rates of 19.2 kbps or lower. Meanwhile, Ethernet interfaces, now increasingly prevalent, boast speeds of 10/100/1,000 Mbps.

9.3.1.3 Quality

The quality of the signal transmitted from the relay at Location A to its destination, Location B, is of paramount importance. It is imperative that no information is lost during this transmission, ensuring that the relay at Location B can accurately decipher the received information. Analog channels are assessed based on parameters such as frequency response, signal-to-noise ratio, total harmonic distortion, and phase noise. On the other hand, digital channels rely on metrics like packet loss rate (PLR), bit error rate (BER), errored seconds (ES), severely errored seconds (SES), unavailable seconds (UAS), and jitter (bit timing variations). The IEC Standard 60834-1[4] provides detailed guidelines concerning the susceptibility of noise bursts in pilot protection schemes, encompassing blocking, permissive tripping, and direct tripping. Table 9.2 presents the minimum number of noise bursts

Table 9.2 Noise burst sensitivity according to IEC standard 60834-1

Pilot scheme type	Security (bursts/undetected error)
Blocking	10^4
Permissive tripping	10^7
Direct tripping	10^8

needed to generate an undesirable output. Furthermore, IEC Standard 60834-1 specifies that the likelihood of a protection command not being received within a 10 ms timeframe should be below 10^{-4}.

9.3.1.4   Latency/delay

Latency or delay represents the time it takes for signals to traverse the entire length of the communication channel, from one end to the other. Various elements of latency resulting from communication transit times play a crucial role in protective relaying applications. It is imperative to thoroughly examine these factors for each protective application:

Magnitude: The extent of the delay involves various factors such as signal transmission, information handling, and buffering within all communication elements. The ultimate permissible delay within any protective system is determined by the stability needs of the power grid, a value that protection engineers are responsible for providing.

Variability: Communication transit times may not remain constant over time, with fluctuations arising from routing, processing, or switching by equipment. Solid-state relays (SSR) can be sensitive to varying latency in communication, which may lead to maloperation. Some modern microprocessor relays adopt a digital communication system to assess latency during communication, blocking relay operations during channel dropouts until the delay is determined. These relays can withstand variations in delay, but they do not tolerate asymmetric delays, as elaborated later in this discussion.

Asymmetry: Communication transit times may differ in each direction, as is often the case with Synchronous Optical Network (SONET) rings or networked communication systems. Certain contemporary microprocessor relays utilize a digital channel-oriented communication system for assessing the End-to-End latency in data transmission or employing time-based synchronization for data synchronization purposes. These relays have the capability to inhibit relay functions when there are interruptions in the communication channel, and they will resume normal operation either when the delay is assessed or upon receipt of a predefined quantity of valid messages. Nonetheless, in the context of channel-based communication, there is an underlying assumption of symmetry. If the End-to-End delay were to be reduced by half compared to the measured loop delay, these relays would endeavor to synchronize remote terminal data with local terminal data in a corresponding manner. Any deviation from this symmetry could lead to a malfunction in the relay. For time synchronization-based relays, they rely on high-precision clocks for absolute time alignment, and their clocks must be phase-locked to external time sources. Some relays allow for time-of-day-based synchronization for data alignment, emphasizing the importance of verifying specific requirements.

Table 9.3 presents the latency times for data transmission in both wireless and optical fiber communication channels.

Table 9.3   Propagation delay of communication paths as a function of path length

Path length (in km)	Propagation delay	
	Wireless	Optical fiber
1	3.3 μs	4.9 μs
20	66.7 μs	97.8 μs
50	166.7 μs	244.6 μs
100	333.3 μs	489.2 μs
250	833.3 μs	1.223 ms
500	1.666 ms	2.446 ms

9.3.1.5 Availability and redundancy

Availability, expressed as the ratio of time during which the communication channel performs at or above its required performance levels to the total time within a specified period, carries significant implications. This calculation excludes planned maintenance or construction outages from the assessment period. The concept of unavailable time encompasses unforeseen disruptions caused by issues like failure of communication and power systems, radio signal weakening, fiber optic cable damage, software glitches, and procedural errors, such as mistakes made by workers. It is important to note that scheduled downtime is not considered when calculating unavailable time.

The communication system's availability must align with the specific requirements of the protection application. In some instances, there may be a need for multiple relay-to-relay connections. These parallel communication channels might share the same infrastructure in some cases, while others require entirely independent communication paths, often referred to as "fully diverse" as discussed in Section 9.5. To qualify as fully diverse, the communication system design must ensure that no single failure can disrupt the protection mechanism. In situations where redundant paths are in use, the absence of the protective function is only taken into account when both the main and backup channels become unavailable at the same time.

9.3.1.6 Channel capacity

Channel capacity refers to the maximum information transmission capability of a communication channel. Shannon's work [5] demonstrated that a noisy channel can reliably transmit data when the transmission rate is kept low, as described by Equation (9.1), where the logarithmic base depends on the symbol alphabet size.

$$C = BW \log_2 \left(\frac{P}{N} + 1 \right) \tag{9.1}$$

BW, P, and *N* are the channel bandwidth (Hz), signal power (W), and noise power (W) respectively. Improved data transmission speeds can be achieved through channels exhibiting higher signal-to-noise ratios (SNR) but achieving Shannon's limits in practice is challenging. Engineers use encoding and modulation methods to get closer to these limits while balancing data rate and reliability for specific applications. Empirical evidence demonstrates that an enhanced communication channel can convey a greater amount of data within a given bandwidth unit.

9.3.2   Comparison of different communication channel

A typical communication system comprises a transmitter, a receiver, and various communication pathways. Different communication media and network configurations in the field of communications offer various prospects for enhancing the speed, security, reliability, and protective capabilities of relay systems. Communication media encompass options like microwave, radio systems, and fiber optics, each with their unique characteristics. A comprehensive exploration of the strengths and weaknesses of currently employed communication media, encompassing both analog and digital technologies, has been provided in Table 9.4.

Table 9.4   Analysis of strengths and weaknesses in analog and digital communication media for power system applications

Media	Advantages	Disadvantages
Fiber optic	• Economical • Ample data transfer capacity (BW) • Faster data transfer speed • Resistant to disruptions caused by electromagnetic interference • Application: Telecommunication, SCADA, data transfer including message, video, voice, etc.	• Costly testing apparatus • Identifying failures can pose a challenging task • Prone to experiencing damage
Leased phone	• Efficiency is achieved when a robust connection is needed to access a site through telephone service	• Costly over an extended period • Not preferred for multi-channel applications
Microwave	• Economical • Reliable • Suitable for establishing backbone communication infrastructure • Ample data transfer capacity (BW) • Faster data transfer speed	• Exorbitant expenditure for upkeep • Specialized testing apparatus and the requirement for proficient technicians • Clearance of the line of sight is necessary • Signal degradation due to fading and the effects of multipath propagation
Radio system	• Appropriate for establishing communication in remote regions • Mobile applications	• Alterations in channel speed, and channel transition while data is being transferred • Noise • Limitations in signal strength • Lack of security • Interference from neighboring channels

(Continues)

Table 9.4 (Continued)

Media	Advantages	Disadvantages
Satellite communication	• Extensive geographic reach • Low error rates • Appropriate for establishing communication in remote regions • The expense remains unaffected by the distance	• Less control over the transmission, continual leasing cost • Subject to eavesdropping (tapping) • Complete reliance on distant areas and remote locations • Delays spanning 250 ms from start to finish, eliminate the feasibility of a majority of protective relay uses, as indicated in reference [26] • Susceptible to unauthorized monitoring (tapping)
Power line carriers	• Economical • Appliances placed within an area belonging to the utility company • Appropriate for communicating between different stations	• Restricted range of coverage • Low bandwidth • Inherently limited number of channels available • Exposed to public access

9.3.3 Fiber-optic-based communication

Fiber-optic communication has become the favored option for the evolving communication system in power systems due to its high bandwidth, reliability, SNR, resistance to electromagnetic interference, and safety from electric hazards.

Attributes:

- Optical fiber cable functions as a dielectric medium for waves, employing total internal reflection to convey light along its lengthwise direction.
- It consists of a high refractive index core (e.g., silica glass) surrounded by cladding with a lower refractive index.
- Buffer jackets offer an additional protection layer, reinforcing mechanical strength, and making it resistant to water infiltration.
- Extremely pure core, typically composed of Germania and Silica, reduces the extent of attenuation during optical transmission,
- There are two categories of optical fibers: multimode (MM) and single-mode (SM). Single-mode (SM) fibers exhibit reduced attenuation, decreased dispersion, and increased bandwidth compared to multimode fibers (MM).
- The size of a fiber-optic cable is typically represented using a pair of numerical values, which indicate the core diameter and the outer diameter of the cladding. For instance, in the case of multimode fiber, it might be expressed as 50/125.

Advantages:

- Dedicated fiber-optic channels offer reliability, protection, fast operation, and ease of use

- Transceivers can be powered by the relay, eliminating the need for a separate power source.
- Certain transceivers have the capability to connect directly to the relay, thus eliminating the need for metallic cables and reducing EMI susceptibility.
- Multiple fiber connections are cost-effective for short communication paths.
- Multiplexers increase data capacity for longer paths, but dedicated fiber pairs offer simplicity and reliability.
- Dedicated fiber-optic links enhance overall system reliability.

9.3.4   Ethernet communication

Ethernet is a versatile communication technology that has evolved significantly over time. Originally, it used tapped coaxial cables and employed Carrier-Sense Multiple Access with Collision Detection (CSMA/CD) to manage simultaneous access. Today, it employs dedicated twisted-pair cables or dual optical fibers, enabling full-duplex communication, and has seen a substantial increase in speed, reaching 10 Gbps. Additionally, it supports features like VLANs and multiple priority levels to ensure efficient and secure data transmission.

Attributes:

- *Ethernet Port Speed and Fiber-Optic Interface:* Ethernet port speeds have advanced from 10 Mbps to 100 Mbps and even 1 Gbps. Modern fiber-optic transceivers prefer 100 Mbps and 1 Gbps options, making the 10 Mbps interface nearly obsolete. Linking network segments or sub-networks with various speeds necessitates the use of bridging, which, in turn, can introduce latency.
- *Full-Duplex:* Modern switches use full-duplex interfaces, effectively doubling available bandwidth and eliminating local packet collisions. Switches store traffic, ensuring optimal scheduling and maximizing data throughput. Ethernet switches establish connections based on MAC addresses, allowing unicast packets to travel directly between communicating ports.
- *IEEE 802.3 Flow Control:* When a switch's queue is full, IEEE Standard 802.3 flow control can pause traffic at the source.
- *Priority Queuing and VLAN Support:* VLANs and class of service (CoS) play a crucial role in segmenting and giving priority to Ethernet traffic of similar types as networks grow. VLANs create separate broadcast domains, enhancing security. Managed switches are required to configure VLANs, and handle tagged traffic. Priority queuing mechanisms use a three-bit quality-of-service field to manage traffic with varying delivery time requirements.
- *Loss-of-Link Management:* To ensure high reliability, relay manufacturers offer products with redundant Ethernet ports. These ports can detect incoming fiber loss, and some switches employ FEFI to disable outgoing fiber upon incoming fiber failure.
- *Remote Monitoring, Port Mirroring, and Diagnostics:* Managing and diagnosing issues in modern Ethernet networks can be challenging. Tools like port mirroring allow remote monitoring of specific ports for troubleshooting. Vendors often provide their network management tools in addition to standard solutions like SNMP.

- *Network Protocols for LAN:* Ethernet is compatible with widely used SCADA protocols such as Modbus RTU and TCP, DNP3, and IEC 60870-5-101/104. It also offers services like DHCP, NTP, FTP, and Telnet for data transfer and remote engineering access.
- *Ethernet-Based Protection Message Standards:* In power system protection, the IEC 61850 standards introduce specialized messaging protocols, including Generic Substation Event (GSE), GOOSE, and IEC 61850-9-2 Sampled Values (SV) compliant information, tailored for protection applications.

9.3.5 Wireless systems

Wireless systems hold significant appeal due to their cost-effectiveness and broad accessibility. The following subsection briefly delves into the key components of wireless systems:

- *Microwave Systems:* Microwave technology plays a pivotal role in wireless systems by enabling direct relay-to-relay communication, impervious to power system glitches. Typically, electric utilities oversee microwave infrastructure, ensuring added control over performance and reliability. However, vulnerabilities persist, stemming from issues in multiplexers, radio equipment, cabling, and antenna alignment.
- *Narrow-band VHF/UHF Radio Systems:* In the past, narrow-band VHF/UHF radios served as the backbone for SCADA (Supervisory Control and Data Acquisition) and dedicated pilot channels. These systems utilized analog (voice-grade) channels, making them susceptible to interference. Although radio channels remained resilient to power system faults, legacy radio technology faced vulnerabilities related to antenna misalignment and adverse weather conditions. Modern systems have transitioned to digital radios with direct digital communication, utilizing an on/off-keyed tone interface connecting the radio to the relay contact input/output.
- *Spread-spectrum Radio Systems:* The realm of spread-spectrum radio communication presents several advantages, including enhanced communications security, robust resistance to interference, low probability of detection, and cost-effectiveness. Initially used for secure government communication, commercial applications have expanded following approval by the U.S. Federal Communications Commission for license-free operation under specified conditions. In the context of power system protection, spread-spectrum radio channels offer benefits such as extended range, congestion avoidance, and exemption from licensing requirements. Signal spreading in the frequency domain can be achieved through two predominant methods: frequency hopping and direct-sequence spreading. Frequency hopping disseminates the signal across a spectrum comprising numerous discrete frequencies, sequentially occupied in a pseudorandom sequence. In contrast, direct-sequence spreading multiplies the information bit stream using a considerably faster PRBS (Pseudorandom Binary Sequence), resulting in a wider bandwidth. However, it is essential to note that spreading, despreading, synchronization, and forward error correction (FEC) introduce time

delays. Depending on the chosen scheme, the radio channel may exhibit insufficient speed for pilot protection. Most contemporary radios rely on data terminal equipment to negotiate a half-duplex channel. Nevertheless, some radio models can swiftly switch their half-duplex channel, simulating full-duplex transmission at speeds of up to 19,200 bps, with a forward delay of approximately 8 ms.

Evolving modern microprocessor-based relays that are compliant with communication protocols (as explained in the subsequent Section 9.4) and that utilize communication media discussed above, such as utility-owned fiber-optic and spread-spectrum radio channels, enable communication-based protection at all voltage levels of the power system. The emergence of digital network technologies with sufficient bandwidth to support utility requirements is driving the adoption of these protective strategies.

9.4 Communication protocol

A protocol entails a prescribed framework of guidelines essential for facilitating structured communication among multiple interacting entities. Their roles encompass enabling and overseeing network communication, specifying data representation, signaling mechanisms, error detection, and authentication of networked computing devices. The challenge of establishing seamless communication between IEDs originating from diverse manufacturers often arises from the independent evolution of data processing and data communication methodologies. Consequently, this situation frequently leads to the development of intricate and costly interfaces. The International Standards Organization (ISO) model, often referred to as the Open Systems Interconnection (OSI) model, has introduced a solution to this issue. This model segregates the communication process into seven fundamental layers, as illustrated in Figure 9.3. These layers delineate the precise mechanisms governing the transmission of data between endpoints within a communication network. Effective communication between two devices is contingent upon the alignment of each layer in the model at the transmitting device with the corresponding layer in the model at the receiving device [6–8].

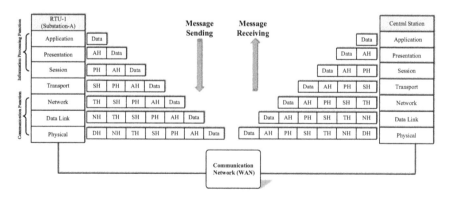

Figure 9.3   OSI model flow

In the context of communication in power systems, since the early 90s, there has been a recognized need to standardize communication protocols for integrating protection, control, and data acquisition within a LAN of a substation. Various organizations like IEC, CIGRE, and IEEE have focused on defining peer-to-peer communication specifications among Intelligent Electronic Devices (IEDs) to ensure interoperability across different vendor products. CIGRE Study Committee B5 (formerly 34) covers Power System Protection and Automation, addressing system protection, substation control, remote control, and metering. WG 07 within this committee focuses on substation control, automation, and communication standards. It highlights the importance of automating substations in the evolving power industry. Substation automation typically includes centralized or distributed architectures, and the integration of protection and control devices is crucial. Recent technological advancements emphasize the need for IEDs with standardized communication protocols and object models. UCA 2.0 and IEC 61850 are significant standards in this context, with IEC 61850 being a broader standard since its publication in 2003. Choosing the right communication protocol involves considering three key factors:

(i) *Performance:* The primary performance criteria involves ensuring monitoring signals and protective control commands are transmitted between IEDs incorporating different functionalities in less than 4 ms over the LAN. This necessitates a peer-to-peer communication protocol in conjunction with either a shared Ethernet setup with a speed of 100 Mbps or a switched Ethernet setup with a speed of 10 Mbps.
(ii) *Interoperability:* When the LAN is used for control and monitoring, interoperability becomes crucial. A well-defined data model specifying the syntax and semantics of exchanged information is necessary for achieving interoperability. This requirement applies to all substations, both new and retrofitted.
(iii) *Maturity:* Assessing the maturity of a communication architecture is essential to mitigate cost and schedule risks. This involves evaluating installed user bases for the chosen protocol version and ensuring alignment with the utility's vision for substation automation.

Gateways may be required in cases where different communication protocols are used within and outside the substation, among LAN segments, or for streaming data from the substation yard. Proper integrated system testing is recommended to ensure the correct operation of peer-to-peer communication with foreign protocols.

These communication protocols can be broadly classified into two categories: (i) physical-based protocols and (ii) layered-based protocols, which will be briefly elucidated in this section.

9.4.1 Physical-based protocols

The Physical-based protocols have emerged to guarantee the harmonious interoperability of devices from various producers and to facilitate effective data transmission across defined distances and/or data transfer rates. The Electronics Industry

Association (EIA) has introduced protocols like RS232, RS422, RS423, and RS485 for the management of data communication. Furthermore, these protocols are integral components of the "Physical Layer" within the OSI model, a concept elucidated in the subsequent section dedicated to Layered-based protocols [9].

• The RS232 Protocol serves as a fundamental communication protocol delineating the parameters for inter-device communication. This mode of communication can manifest in three distinct configurations: simplex, wherein one device assumes the transmitter role while the other functions as the receiver, allowing unidirectional data flow from the transmitter to the receiver; half duplex, where either device can alternate between transmitting and receiving but not simultaneously; and full duplex, permitting either device to concurrently transmit and receive data. An essential component of this protocol entails a single twisted pair connection bridging the two devices.

• *RS 485 Protocols:* This protocol bears a resemblance to the RS232 protocol, enabling concurrent communication among multiple relays (up to 32) in a half-duplex fashion. In this half-duplex configuration, just one relay can either send or receive command data at any given time. This necessitates a polling/responding mechanism for information exchange. The communication is perpetually instigated by the "Master unit" (host), while the "Slave units" (relays) remain passive, transmitting data only upon receiving a request from the "Master unit" and lacking inter-communication capability. The RS485 protocol encompasses two distinct communication modes: (i) Unicast mode and (ii) Broadcast mode as depicted in the figure. In the unicast mode, the "Master unit" issues polling commands, and only a designated "Slave unit" (identified by a unique address) responds accordingly. The "Master unit" awaits a response from the selected "Slave unit" or terminates the request if a predefined time limit elapses. In the broadcast mode, the "Master unit" disseminates messages to all "Slave units" [9].

9.4.2   Layer-based protocols

The class of proto under consideration in this context has been devised in accordance with the OSI model, as documented in reference [10]. The OSI model is the outcome of collaborative efforts undertaken by the International Organization for Standardization within the framework of the OSI initiative. This model delineates a communication system into multiple layers, with each layer comprising a set of analogous functions responsible for rendering services to the layer situated above it while simultaneously receiving services from the layer beneath it. At each layer, an instance is tasked with delivering services to instances at the higher layer and soliciting services from the layer below. During the transmission of data from one device to another, each layer appends pertinent information to the "headers," and this information is subsequently decrypted at the receiving endpoint. A visual representation of data communication utilizing the OSI model is provided in Figure 9.3, where "H" signifies "headers" [9].

Table 9.5 presents a comprehensive elucidation of the functions associated with each layer.

Table 9.5 OSI model function

Layers	Function
Application	Enables user interaction with the software by attaching an application header to data, specifying the requested application type. Various standards, like HTTP and FTP, govern this layer.
Presentation	This function performs data format conversion, data compression, and decompression for network communication. It also includes a presentation header with data format and encryption details in the application data unit.
Session	Facilitates user interaction, ensures fault handling and crash recovery, and creates session data units by incorporating session headers into presentation data units.
Transport	This layer manages data packets, TCP and UDP are two transport protocols here. TCP ensures reliability and order, making it stateful, while UDP offers transmission with minimal overhead and reduced error verification
Network	Manages package routing and addressing, guiding packets along the quickest route while appending a network header containing the Network Address to the Transport Data Unit.
Data link	This layer defines the MAC Address and incorporates features such as error handling and retransmission. It appends a Data Link Header to the Network Data Unit, encompassing the Physical Address, resulting in a data link data unit.
Physical	Assesses the physical medium's electrical, mechanical, functional, and procedural characteristics.

9.4.2.1 DNP-3.0 [11]

Distributed Network Protocol (DNP)-3.0 is a communication protocol designed for interoperability among substation computers. It utilizes layers 1, 2, and 7 from the OSI model, and optionally incorporates a pseudo-transport layer for message segmentation, known as the Enhanced Performance Architecture (EPA) model. DNP-3 is mainly employed for communication between master stations in SCADA systems, RTUs, and IEDs in the electric utility sector. Unlike TCP/IP, it does not wait for delayed packets due to embedded time synchronization (time tag) with millisecond accuracy, enabling asynchronous message exchange at a typical processing throughput rate of 20 ms.

DNP-3 is a vital part of SCADA systems, facilitating communication among their components. It supports various features, including multiplexing, data fragmentation, error checking, prioritization, time synchronization, and user-definable objects. Additionally, DNP-3 is used in electrical and water industries for utility purposes.

The protocol addresses the need for standardized communication between SCADA components from different vendors, offering an open protocol based on IEC 60870-5. DNP-3 ensures reliable communication in challenging environments, resisting distortion from EMI and legacy components, but it lacks security measures, which should be considered when planning SCADA systems. Implementing a network alarm monitoring system can enhance the security and reliability of DNP-3 communications in SCADA systems.

9.4.2.2   ModBus [12]

ModBus is a three-layer protocol utilizing a "master–slave" communication technique. In this approach, the master device initiates transactions, and the slave devices respond with requested data or actions. Unlike DNP 3.0, ModBus lacks embedded time synchronization but can employ external sources like GPS or IRIG for synchronization. The choice of synchronization protocol depends on device quantity, type, and physical arrangement. ModBus typically operates with an 8-ms throughput rate.

ModBus features three frame formats: ASCII, RTU, and TCP/IP, as shown in Figure 9.4. ASCII and RTU are for serial communication, with ASCII using two ASCII characters per 8-bit byte and RTU using two 4-bit hexadecimal characters (or 8-bit binary). ASCII allows longer time intervals between characters but has lower data throughput compared to RTU. ModBus TCP/IP is a modification with Ethernet-TCP/IP, replacing the Address Field with a ModBus Application (MBAP) Header. ModBus and DNP 3.0 differ in communication purposes. ModBus suits substation communication for protection, control, and metering devices, while DNP 3.0 is designed for data communication from substations to master control centers, offering more flexibility, reliability, and security due to its specific data objects.

9.4.2.3   IEC 61850

IEC 61850, a standard pertaining to electrical substations established by the International Electrotechnical Commission (IEC), provides flexible data models that can be customized for different communication protocols. An instance of this is the Generic Object-Oriented Substation Events (GOOSE) protocol, which facilitates the seamless exchange of analog and digital data between peers. Figure 9.5 depicts the message structure of GOOSE data packets. It incorporates time tags and

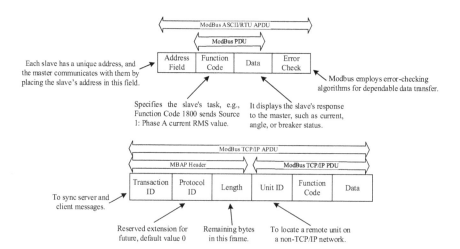

Figure 9.4   Components of ModBus PDU/APDU for ASCII, RTU, and TCP/IP

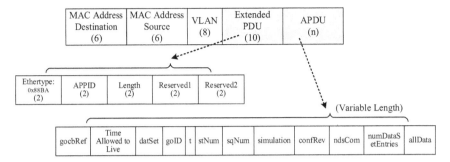

MAC Address Destination (6)	MAC Address Source (6)	VLAN (8)	Extended PDU (10)	APDU (n)

Ethertype: 0x88BA (2)	APPID (2)	Length (2)	Reserved1 (2)	Reserved2 (2)

(Variable Length)

gocbRef	Time Allowed to Live	datSet	goID	t	stNum	sqNum	simulation	confRev	ndsCom	numDataS etEntries	allData

Figure 9.5   Structure of IEC 61850 GOOSE protocol-compliant message

asynchronous message exchange with a typical processing rate of 12 ms [13]. IEC 61850 boasts advantages such as independent programming from wiring, enhanced data exchange performance, and redundancy to prevent data loss [14,15]. The IEC 61850 standard-based protocol can be further divided into client–server and multicast communication protocols.

- *Client–server communication protocols* refer to the interactions that occur between servers, serving as protection and control devices, and clients, such as SCADA and gateways. These exchanges employ the MMS protocol (Manufacturing Message Specification), a standard outlined in ISO 9506. Originally designed for industrial automation, this protocol stands out as one of the pioneering systems that employ hierarchical naming conventions for data. This communication adheres to the OSI layered model and operates over TCP/IP. A distinct TCP channel is established for each client–server pair. Through these channels, clients can access data, modify configurations/settings, receive unscheduled updates or request control commands. In a SCADA system, the number of open TCP/IP channels corresponds to the servers being monitored.
- The IEC 61850 standard introduces efficient *point-to-multipoint message exchange protocols* known as GOOSE (Generic Object-Oriented Substation Event) messages. These communications utilize ethernet multicast, incorporating quality of service and a high-priority approach to reduce latency during switching in substations and distribution to subscribers, with response times of either 3 ms or 20 ms based on specific needs. Additionally, IEC 61850 includes a mechanism for GOOSE message repetition to verify sender status, a valuable improvement over traditional copper wiring, which may not readily detect cable faults.

Furthermore, the digitalization of secondary measurements in Current and Voltage Transformers enables data transmission via optical fibers from switchyards to control cabinets. This innovation has led to the development of Merging Units responsible for analog-to-digital conversion and transmission. The IEC 61850-9-2 specification streamlines information transmission from merging units, enhancing interoperability. It mandates either 80 or 256 samples per cycle and defines the specific data to be included in sampled values messages, encompassing four voltage and four current measurements per frame. Figure 9.6 illustrates the structure of IEC 61850-9-2 SV message.

The standard's primary aim is to ensure interoperability among Intelligent Electronic Devices (IEDs) from different manufacturers, addressing the need for device interchangeability without system alterations. IEC 61850 leverages existing standards and widely accepted communication principles, enabling seamless information exchange among IEDs. However, it does not standardize substation operation functions or their allocation within automation systems. Substation automation systems typically consist of three levels: station (level-2), bay (level-1), and process (level-0), managing high-voltage equipment and the grid. Logical interface mappings between these levels form the basis for the IEC 61850 standard series. According to IEC 61850, real-time services are efficiently linked to the Data Link layer (layer-2) to reduce protocol overhead and attain swift, low-latency message delivery [16]. It is stipulated in Table 9.6 that the total time for message generation, transmission, reception, and processing should not exceed 3 ms.

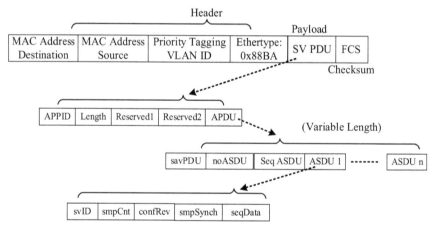

Figure 9.6   IEC 61850-9-2 Sampled Value (SV) structure

Table 9.6   IEC 61850 time requirements for various message types

Application	Protocols	Time
Fast Message	GOOSE	≤3 ms
Raw Data	SV	≤3 ms
Medium Speed Messages	MMS	≤100 ms
Low Speed Messages	MMS	≤500 ms
Time Synchronization	IEEE 1588	>500 ms
File Transfer	MMS	>500 ms

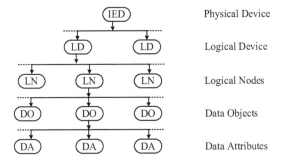

Figure 9.7   Hierarchy in IEC 61850

IEC 61850 establishes an information model hierarchy comprising several key levels, namely IEDs, LD (Logical Device), LN (Logical Node), DO (Data Object), DA (Data Attribute), and Basic Type, as depicted in Figure 9.7. LDs serve as organizational containers within an IED, categorizing its information into various segments. Vendors often assign logical device names, such as PROT (protection), CTRL (control), and REC (Recorder), to classify their data. These names are not standardized and can vary between vendors. IEC 61850 categorizes all known substation automation system functions into LNs, which are sub-functions located in physical nodes for data exchange. It is denoted by four-letter codes, representing the automated functions or components within the system. LNs are grouped by application area, described briefly, and may have device function numbers. These functions encompass control, protection, measurement, supervision, and more. For instance, "PIOC" stands for "Protection Instantaneous Over Current." This approach decouples applications from communication, allowing communication through different protocols. Vendors and utilities maintain application functions optimized for specific needs with high maturity and quality. Each LN includes a set of mandatory and optional data objects to fulfill its designated tasks. These data objects have the capability to communicate status updates, measurements, pre-defined values, adjustable parameters, or descriptive information.

With plug-and-play capabilities and successful pilot projects, IEC 61850 promises to revolutionize substation automation systems worldwide. It fulfills the long-standing promise of automation and integration in utilities [17]. A non-proprietary, high-speed protocol, without the need for protocol converters, has become essential for robust, integrated substation communication networks. The introduction of IEC 61850 and the Utility Communications Architecture has enabled the integration of station IEDs by means of standardization. By using standardized high-speed communication, utility engineers can replace costly stand-alone devices with sophisticated functionality and utilize available data to its fullest extent.

The proliferation of Ethernet network communication and the adoption of IEC 61850-9-2 SV service have rendered distributed bus differential schemes, which were formerly costly and dependent on proprietary communication methods, increasingly cost-effective and applicable across a wide range of voltage levels and utility scenarios. In addition to that, Contemporary advancements in technology such as programmable logic, IEC 61850 GOOSE messaging, and synchrophasor technology have facilitated the extensive adoption of communication-centric protection, control, and monitoring systems. This adaptability facilitates the creation of specialized protection strategies designed to meet the precise demands of power system configurations.

9.5   Communication network

Manufacturers nowadays typically design automation systems with a common architecture, with slight variations. These architectures involve a centralized server linked to decentralized clients/systems, protective IEDs, synchronization, and communication components. A LAN enables the functioning of a system by providing an HMI for controlling and monitoring the system.

This section focuses on exploring communication network specifics within automation systems. These networks play a crucial role in adapting automation systems to SCADA protocols and various IED communication protocols. Presently, communication network advancements center on standards and, notably, enhancing the compatibility between automated systems and equipment connected to them. While some projects achieve extensive interoperability, it is commercially unfeasible for a central computer from one manufacturer to coexist with decentralized modules from another.

9.5.1   Substation architecture

The term power system refers to the physical infrastructure responsible for generating, transmitting, and distributing power. In contrast, the protection and control system oversees and safeguards this power system [18–22]. This system is structured into four levels, as illustrated in Figure 9.8:

1. **Process bus:** This is the foundational layer of measurement and control equipment, installed at the process level. These devices are physically connected to the power system, continuously monitoring its status. Components at this level encompass resistance thermal detectors (RTDs) for temperature sensing, voltage transformers (VTs) for voltage monitoring, current transformers (CTs) for current measurement, and various other sensory devices. Transducers, meanwhile, convert the output from these sensors to different levels.

2. **Bay level:** The bay level comprises IEDs responsible for gathering sensor data, transforming it into actionable information, and responding accordingly. A bay denotes a specific area housing a power system device (e.g., a feeder breaker) along with all associated measurement and control equipment. IEDs at this level encompass protective equipment (relays), measurement devices (meters), event recorders, load tap changers, VAR controllers, RTUs, and PLCs.

3. **Station level:** Substation controllers handle data acquisition and control of IEDs at this level, containing local I/O capabilities. They house data pertaining

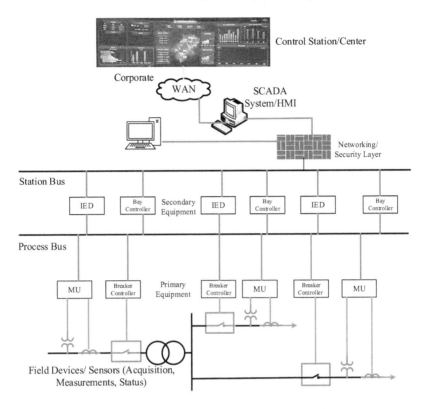

Control Station/Center

Corporate

WAN

SCADA System/HMI

Networking/ Security Layer

Station Bus

| IED | Bay Controller | Secondary Equipment | IED | Bay Controller | IED | Bay Controller |

Process Bus

| MU | Breaker Controller | Primary Equipment | Breaker Controller | MU | Breaker Controller | MU |

Field Devices/ Sensors (Acquisition, Measurements, Status)

Figure 9.8   Chronological structure of digital substation

to the entire substation. Possible substation controllers include RTUs, PLCs, bay controllers, and HMI software tools installed on personal computers.

4. **Enterprise level:** This term encompasses all individuals and customers who utilize power system information, whether they are located inside or outside of the substation. These software programs collect data from devices at both the station level and the unit level. For instance, a utility's system integration task serves the following threefold purpose: transferring sensor measurements and data-created information between IEDs, between IEDs and a substation controller, and directly to end-user clients from both IEDs and the substation controller.

9.5.2   Different communication network topologies

This section explores various LAN configurations, including star, multidrop, and ring topologies, that can be formed by utilizing two types of IED port connections: point-to-point and multidrop. In a point-to-point connection, exemplified in Figure 9.9, data flows directly between two interconnected devices. The medium for data transfer can be metallic, wireless, or optical fiber, with each device featuring distinct transmit-and-receive connections. This arrangement, limited to two

Figure 9.9   Point-to-point connection

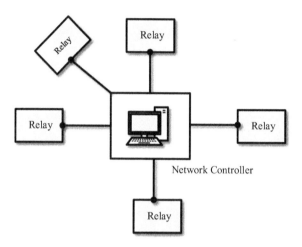

Figure 9.10   Star connection

devices per connection, allows each device to maintain control over its transmission and knowledge of its connected peer. Point-to-point communication offers a distinct advantage for applications necessitating a rapid exchange of data between two specific nodes. However, it comes with the drawback that any disruption in the communication channel results in the complete loss of information exchange.

However, larger networks with multiple coordinated PCM (Pulse Code Modulation) applications reliant on distributed communication necessitate a network controller to oversee message exchanges among IEDs. In this context, the term network controller encompasses one of several functions within a multifunction device.

The establishment of point-to-point connections between each IED and the network controller results in a star network configuration, as illustrated in Figure 9.10. A star layout enables the network controller to employ unique protocols and data rates tailored to individual IEDs. Furthermore, the failure of one communication link does not disrupt the entire network, as data between two IEDs follows a path from the first IED to the network controller and then to the second IED. Benefits of star communication further include the ease of adding or removing nodes, straightforward management and monitoring, and the ability to ensure that a node breakdown does not disrupt the overall system. On the downside, the entire network's reliability hinges on the failure of a single hub.

Figure 9.10 illustrates how a star configuration delivers a straightforward and dependable LAN setup. Conversely, Figure 9.11 depicts a straightforward multidrop network, where all devices connect to a shared physical communication link or bus, taking turns to communicate. Typically, client devices sequentially query each IED to retrieve data, permitting only one client to poll the IEDs at any given time. However, engineering analysis suggests that multidrop networks are insufficient for robust LAN construction. Advantages of utilizing multidrop communication include the ability to redistribute bandwidth from unaffected channels in the event of a drop in a specific communication channel. On the flip side, one drawback is the absence of channel redundancy in cases of fiber or equipment failures.

In some multidrop network diagrams, Ethernet switches between the network controller and the IEDs may not be depicted, aiming to display only logical connections. Figure 9.11 serves as an example of such a diagram that, when describing an Ethernet network, could lead to potential confusion, as the network controller and IEDs are physically connected in a star fashion to an Ethernet switch between them.

Despite Ethernet's origin for flexible, non-deterministic message communication to all nodes on a network, it has gained popularity in PCM networks. Thus far, its use has primarily been restricted to asynchronous client/server applications, such as SCADA and engineering access. However, as protection and control engineers seek more deterministic and robust network designs for peer-to-peer messaging, Ethernet switch and router vendors have struggled to develop suitable technologies for high-reliability, mission-critical PCM network applications. As a result, PCM manufacturers have been compelled to incorporate networking technology within the IEDs, including multiple physical Ethernet ports and Ethernet switching and message routing. These multiple ports on IEDs can operate in primary-and-failover or switch mode simultaneously.

Figures 9.12 and 9.13 showcase IEDs with ports in switch mode forming a ring topology, where each IED handles messages, generates and transmits them, and can even relay messages intended for other IEDs on the LAN. In this network, data between two IEDs traverse a point-to-point path around the ring from the sending IED to the receiving IED. All network traffic enters and exits each IED, resulting in a higher traffic processing burden compared to star networks. In the context of ring-based communication in the SONET system, there exist two configurations,

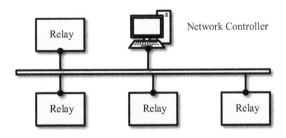

Figure 9.11   Multidrop connection

each with its own set of advantages and disadvantages. The first configuration is known as the SONET Path Switched Ring [9], consisting of two distinct optical fiber links that interconnect all nodes in a counter-rotating pattern, as depicted in Figure 9.12. In typical scenarios, data travels from point A to C through the outer ring, marked as the primary route (depicted in the left diagram). However, in the event of a channel failure, data switches to the inner ring, serving as the secondary route (as shown in the right diagram). One of its benefits lies in its redundancy, ensuring that communication remains unaffected by channel failures. Conversely, a potential drawback emerges in the form of uneven time delays between transmitters and receivers, which could lead to the false operation of protective relays when transitioning from the primary to the secondary route during channel failures.

On the other hand, the alternative SONET ring communication mode is the SONET Line Switched Ring [9], sharing a similar structure with the SONET Path but featuring one active path and one reserved path, as shown in Figure 9.13. Under normal circumstances, data is transmitted through the active path using the outer ring (as illustrated in the left diagram). However, in the event of a channel failure, the inner ring is activated to reverse the data transmission direction (as shown in the

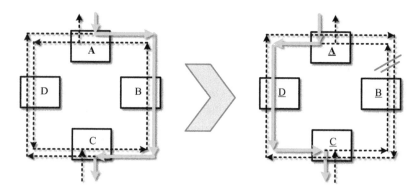

Figure 9.12   Ring topology: case-1

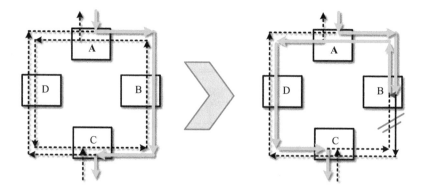

Figure 9.13   Ring topology: case-2

right diagram). One advantage of this configuration is its more efficient utilization of fiber optics for certain applications. However, a notable disadvantage is its unsuitability for tele-protection applications due to the complex handshaking (synchronization) requirements, which introduce a 60 ms delay.

While engineering analysis favors star connections, well-constructed ring topologies using rugged components can suffice for non-mission-critical PCM applications. Ethernet ring networks channel all traffic in the same direction, and in case of a failure, traffic begins traveling in both directions away from the failure. Present-day Ethernet systems predominantly adopt star or ring topologies, with support for both star and multidrop logical connections. The LAN topologies depicted above accommodate both TDM (Time Division Multiplexing)-based and packet-based communication. When utilizing packet-based messaging, the network controller collects and manages packet distribution based on packet header information and network bandwidth provisioning methods. Consequently, in-service performance relies largely on engineering design and implementation, rather than just the selected IEDs and communication technology. The network controller's responsibilities include collecting, managing, prioritizing, and dispersing all packets, making it possible for IEDs to receive unwanted or out-of-order packets or to miss expected ones.

In TDM-based communication scenarios, data travels directly between two end devices, either between an IED and the network controller or through the network controller, for instance, between a client and an IED. In either TDM example, the message consumer utilizes and passes on the content without involving the message or packet itself.

Communication-enhanced protection schemes entail the dissemination of an internal operational status among various relay devices. Illustrative instances encompass directional comparison techniques for the protection of transmission lines, such as permissive tripping and blocking mechanisms, as well as zone-interlocked methods for the protection of electrical substations. These methodologies employ sophisticated digital communication modalities to facilitate the exchange of directional parameter information amongst relays and to maintain uninterrupted vigilance over the communication channel, thereby ensuring its reliability.

When making a decision regarding the choice of a pilot protection system, it is imperative to take into account the potential consequences of channel failures on the reliability and security of the system. Blocking schemes have the propensity to trigger in response to internal faults, even in scenarios where the channel is not operational, and are particularly applicable when multiple channels share the same pathway with the protected line. Conversely, permissive tripping schemes are better aligned with situations in which channels remain unaffected by line faults.

9.6 Cyber security

When the initial communication networks were established for electric power systems, they operated in a different threat landscape, devoid of many current risks. Back then, critical communications were conducted within relatively isolated networks, shielded from electronic attacks. Nowadays, we have interconnected these networks to enhance system operation, protection, and automation, forming vital electronic connections responsible for transmitting various data types, from emails to synchrophasor data. Safeguarding this intricate infrastructure in an increasingly hostile environment is imperative. To effectively protect these vulnerable system components, a comprehensive understanding of potential threats is essential. This section offers an overview of these threats, outlining potential attackers and the tools and techniques they may employ.

The most effective approach to ensure the security of these vulnerable communication links is through the application of security technologies. Modern cryptographic solutions and advanced features found in microprocessor-based IEDs offer robust electronic security. This chapter details strategies and technologies for defending against electronic attacks, including:

- Utilizing modern cryptographic security modules to safeguard sensitive data and block unauthorized access to critical networks.
- Employing robust access control features within power system protection, control, and monitoring equipment to prevent unauthorized entry.
- Leveraging security status monitoring features in these devices for rapid detection and response to electronic threats.

9.6.1 Important security tips

To shield critical networks from electronic attacks, consider following key power system security recommendations:

- Gain a comprehensive understanding of all communication paths within and between networked critical assets, creating a detailed communication network diagram.
- Utilize strong passwords, avoiding weak predictable passwords.
- Effectively manage passwords by avoiding default ones, changing them periodically, and controlling access to them.
- Encrypt and authenticate critical communications across various systems.
- Implement need-to-know access principles, limiting system details access to authorized personnel.
- Establish multiple secure communication paths to key assets to mitigate denial-of-service attacks and enable security alarms through redundant paths.
- Take proactive security measures without waiting for government mandates or attacks.
- Continuously review access activity and monitor security alarms.
- Embrace security in-depth, covering physical security, cybersecurity, training, and organizational culture.
- Safeguard access tools, including computers, passwords, encryption equipment, system documentation, and software.

9.6.2 Background and intentions of attackers

Historically, major power outages have primarily resulted from natural disasters and human errors. However, intentional acts of sabotage or malicious interference with power system equipment can also trigger local or widespread disruptions. Coordinated manipulation of circuit breakers or deliberate adjustments to protection settings can destabilize the power grid. Individuals and groups with malicious intentions are aware of society's heavy reliance on electricity and may aim to target power infrastructure.

9.6.2.1 Electronic attack

There are various ways motivated individuals could harm power infrastructure. Using electronic attack methods, such as compromising networked devices, offers advantages over physical methods like explosives or sabotage. Exploiting network connections enables attackers to spread their actions to remote devices. Additionally, electronic attacks can be more practical when targeting multiple locations separated by long distances, allowing coordinated actions from a remote, safe location with the potential for anonymity and global reach.

9.6.2.2 Threat actors targeting power infrastructure

Recent attention to vulnerabilities in critical infrastructure networks highlights the susceptibility of global economies to electronic attacks. These attacks are facilitated by readily available hacking tools and precompiled software that exploit existing weaknesses. This suggests that governments, organizations, and individuals worldwide possess the capacity and technical knowledge to launch cyberattacks on infrastructure.

- *Government-Sponsored Information Warfare:* Hostile governments can allocate significant resources to conduct information warfare against enemy governments and civilian infrastructure networks. Electronic information warfare can level the playing field against stronger adversaries, enabling foreign governments to map vulnerabilities in power infrastructure discreetly.
- *Hostile Organizations:* Hostile organizations, driven by ideological motivations, pose a distinct threat. Unlike governments, they may not be deterred by global consequences and may even seek attention through successful attacks.
- *Insiders:* Insiders with access to sensitive information or equipment can also pose a threat. Malicious insiders may act out of personal motivations like greed or revenge, while innocent employees could be manipulated into compromising network security through social engineering techniques.
- *Hackers:* Hackers, motivated by financial gain or notoriety, continually challenge network defenses. The vast number of Internet-connected devices presents numerous potential targets. Companies that connect their critical networks to the Internet expose valuable data and intellectual property to hackers, making hacking a profitable and persistent endeavor.

9.6.3 Attack methods and tools

The availability of pre-made attack tools and instructional materials for download can enhance the efficiency of electronic attacks. To defend against such attacks, understanding the techniques employed by potential adversaries is crucial. Most electronic attacks follow a common pattern:

- *Network reconnaissance:* Attackers discreetly gather information about the target.
- *Active scanning:* Attackers send messages to identify vulnerabilities.
- *Exploiting vulnerabilities:* Attackers leverage electronic weaknesses to gain control or cause harm.

Understanding these stages reveals the skill and commitment of potential adversaries. For further details, refer to [15,16].

9.6.3.1   Network reconnaissance

Electronic attacks often commence with passive information gathering, which generates minimal network activity. Information such as employee data, company locations, IP addresses, and phone numbers aids in focusing on an electronic attack. Attackers can conduct reconnaissance globally using internet resources. Tools like Sam Spade facilitate this process. However, minimizing sensitive information availability is essential. Information sources may include company dumpsters, corporate websites, and job search sites.

9.6.3.2   Active scanning

Once reconnaissance provides initial insights, attackers actively probe the target's defenses. This involves TCP/IP scans, telephone number sweeps for dial-up modems, and social engineering attempts. Attackers worldwide can exploit accessible vulnerabilities.

Tools like Nmap quickly identify live hosts and potential vulnerabilities, while dial-up modem scanners automate modem discovery. Additionally, commercial telephone scanners like Phone Sweep identify various devices and perform password-guessing attacks.

9.6.3.3   Exploiting vulnerabilities

The ultimate goal is to access networked assets and disrupt systems. Attackers seek electronic vulnerabilities in power system communications, such as SCADA networks and digital substations.

- Exploiting SCADA network vulnerabilities: Attackers can exploit SCADA protocols' lack of inherent security features to manipulate control points.
- Password Cracking: Attackers may intercept passwords or use online guessing techniques, aided by knowledge of the system's hardware.
- Exploiting Software Vulnerabilities: Hackers target flaws in network services, such as buffer overflow attacks. Prepackaged scripts and programs are available to exploit these vulnerabilities.
- Trojan Horse Applications: Malicious code hidden within seemingly innocent applications can compromise a system.
- Backdoors: Attackers can install backdoors, granting remote control over compromised computers. Tools like Back Orifice provide extensive control.

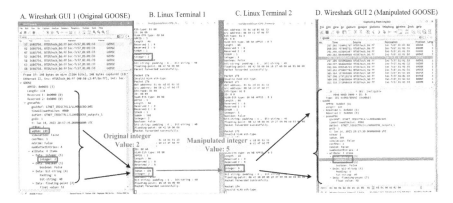

Figure 9.14   Vulnerability assessment of standard IEC 61850 GOOSE protocol

- Sabotaging the smooth functioning of digital substations through the exploitation of vulnerabilities in IEC 61850 standard-compliant protocols, like GOOSE and Sampled Values, commonly used in digital substation automation. This entails actions like issuing deceptive tripping commands, manipulating the circuit breaker statuses, and intercepting/manipulating measurements from merging units, and more.

Figure 9.14 illustrates the vulnerability assessment conducted within a laboratory environment on the IEC 61850 GOOSE protocol. The evaluation was conducted in real-time using Real-Time Digital Simulator (RTDS), with the GOOSE messages being generated through the GTNETx2 network interface card (NIC). Data packet reception and manipulation were executed on a Linux OS-based system. Within the figure, Windows A and D represent the utilization of Wireshark for validating the test by intercepting data packets at the Ethernet network interface. Terminals B and C are associated with the development of a module in the C programming language, designed for receiving and manipulating GOOSE data packets. The focus of this test was on a specific data attribute, an integer value, which underwent modification from 2 to 5 and was subsequently resent through port forwarding.

9.6.3.4   Attack propagation

Once inside a protected network, attackers can utilize the mentioned techniques against critical assets like workstations, SCADA servers, RTUs, PLCs, and protective relays. They can move from one compromised asset to another, using network connections to propagate malicious traffic.

9.6.4   *Prioritizing electronic security risks in the electric power industry*

The electric power industry relies on a complex communication infrastructure comprising various technologies, protocols, and functions. Advances in relays and related devices have led to increased electronic connectivity among them.

Figure 11.7 illustrates common power system communication links. Figure 11.7 reveals that real-time protection communications primarily utilize secure point-to-point connections like fiber-optic, copper wires, or wireless links, minimizing electronic vulnerability. However, SCADA and engineering access links are more susceptible to electronic intrusion due to less secure communication methods such as radio channels, dial-up modems, leased lines, and internet connections. The figure also illustrates an indirect connection between a SCADA network and the internet, posing a significant security concern.

Moreover, many power system PCM equipment vendors now offer engineering functionality via dedicated SCADA interfaces, blurring the lines between the two. This growing electronic exposure underscores the urgency of assessing network vulnerabilities and implementing appropriate electronic security measures. For further information on communication security in the electric power industry, refer to [23,24].

9.6.5   *Defensive technologies and strategies*

In this section, the focus is on safeguarding networks against electronic attacks. An illustration in Figure 11.8 depicts an attacker attempting to compromise a relay through a vulnerable communications link. To thwart such actions, three protective barriers must be overcome, as represented by brick walls in the figure:

(i) *Access to the Communications Channel:* The attacker needs to read and/or write data to the communications channel.
(ii) *Bypassing Link-Security Mechanisms:* Defensive technologies protecting the network's perimeter must be defeated.
(iii) *Bypassing Local Electronic Access Control Mechanisms:* The attacker must overcome access-control measures for the target device.

Strengthening these defensive barriers is crucial for securing critical networks and complying with cybersecurity regulations [25]. Link security mechanisms safeguard data during transit over insecure media and prevent unauthorized access from such media into protected networks. These mechanisms include cryptographic modules and TCP/IP traffic-filtering firewalls. Electronic access controls prevent unauthorized entry into protected devices, employing methods like password entry and cryptographic authentication. The following sections delve deeper into defensive strategies.

*   *Defining the Electronic Security Perimeter:* The initial step in securing a communication infrastructure is defining logical electronic security perimeters (ESPs) to counter electronic attacks. This involves identifying critical electronic assets, creating a network diagram detailing connectivity, and grouping equipment into logical subnetworks (subnets) with similar security needs. Subnets should be arranged to minimize internal threats and prioritize external threats. For example, assets in remote substations are grouped together, assuming external threats are more likely. Network connections entering ESPs act as electronic access points into protected subnets and must be secured using appropriate link security and electronic access control technologies.

- *Limiting Access to Protected Networks:* Reducing outside attackers' access to communication channels at electronic access points is the primary defense. Implement secure networking to minimize malicious traffic on protected subnets.
  1. *Choosing Secure Communications Technologies:* Selecting secure communication links is vital. Security often correlates inversely with cost. Options range from economical but risky internet connections to costly dedicated lines offering higher security. A wholly owned network infrastructure provides maximum security but at a high cost.
  2. *Implementing Restrictive Traffic Filtering:* Properly segmenting TCP/IP networks into subnets and controlling traffic flow through routers and firewalls is crucial. Routers forward traffic between LAN subnets, and advanced routers offer firewall capabilities. Filtering rules based on properties like IP addresses and port numbers should only allow essential traffic through.
- *Implementing Strong Cryptographic Link Security:* Insecure WAN links serve as potential electronic attack entry points into protected networks. Link security technologies use modern cryptography to ensure confidentiality, integrity, and authentication of data in transit. These cryptographic protections rely on secret key values rather than keeping the algorithms themselves secret. IEC 62351-6 addresses security considerations pertaining to the IEC 61850-compliant protocol by incorporating authentication and encryption features, which were introduced in the revised 2020 edition. Figure 9.15 illustrates the process of enhancing the standard GOOSE data packet by appending an authentication tag corresponding to the payload while simultaneously integrating encryption. Additionally, Table 9.7 presents an overview of the various security options put forth in IEC 62351-6.
- *Implementing Strong Local Electronic Access Controls in Critical Devices:* While link security defends against external threats, internal threats must be addressed separately. Strong electronic access controls within ESPs are necessary to prevent attacks originating from within.
- *Securing Personal Computers:* Securing all PCs on critical networks is essential. This involves configuring PCs securely, applying vendor-issued patches promptly, and using antivirus software to guard against vulnerabilities and malware threats.

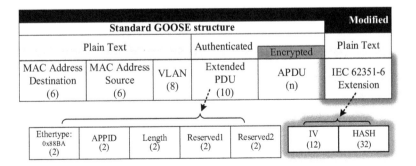

Figure 9.15 GOOSE message structure with security features

Table 9.7   Security provisions in the revised 2020 edition of IEC 62351-6 for protocols compliant with the IEC 61850 standard

Algorithms	Size (bits)	Usage	Security
SHA-256	80	Mandatory	Authentication
SHA-256	128	Mandatory	Authentication
SHA-256	256	Mandatory	Authentication
AES-GMAC	64	Mandatory	Authentication
AES-GMAC	128	Mandatory	Authentication
AES-GCM	64	Optional	Authentication and Encryption
AES-GCM	128	Optional	Authentication and Encryption

9.7   Conclusion

This chapter has delved into various aspects of communication in power systems protection applications, shedding light on the critical considerations and elements that play a pivotal role in ensuring the effectiveness and security of these systems. In the first section, we explored the performance considerations, emphasizing the importance of availability, redundancy, reliability, and temporal assessment in maintaining the integrity of power systems protection applications. These factors serve as the foundation for a robust communication infrastructure. The subsequent section focused on the communication systems used in power systems applications, encompassing energy management systems (EMS), SCADA, pilot protection, security, substation, distribution automation, and wide-area protection and control. Understanding these systems' roles and interplay is crucial for optimizing power system performance.

Moving on to the communication medium, we discussed the various requirements and compared different communication channels, including fiber-optic-based communication, Ethernet, and wireless systems. The choice of communication medium should align with the specific needs and constraints of each application. Communication protocols were explored in detail, from physical-based protocols like RS232 and RS485 to layer-based protocols such as DNP-3.0, ModBus, and IEC 61850. Selecting the appropriate protocol is essential for achieving seamless communication between power system components. The chapter further delved into communication networks, addressing substation architecture, different communication network topologies, and their implications for power systems. Understanding the network structure is vital for efficient data exchange and control.

In the context of cybersecurity, we discussed important security tips, the motives of attackers, attack methods, and tools used in the context of power infrastructure. We also delved into prioritizing electronic security risks and outlined defensive technologies and strategies to safeguard critical communication channels and devices.

In summary, effective communication in power systems protection applications is a multifaceted endeavor that demands careful consideration of performance, medium, protocol, network, and cybersecurity aspects. A comprehensive understanding of these elements is essential for building resilient and secure power systems that can withstand modern challenges and threats.

References

[1] Relay Work Group and Telecommunications Work Group, "Communications Systems Performance Guide for Electric Protection Systems," June 10, 2021.

[2] J.J. Kumm, M.S. Weber, E.O. Schweitzer, and D. Hou, "Assessing the Effectiveness of Self-Tests and Other Monitoring Means in Protective Relays," *Western Protective Relay Conference*, October 1994.

[3] WECC Telecommunications Work Group, "Guidelines for the Design of Critical Communications Circuits," March 10, 2016.

[4] Tele protection Equipment of Power Systems – Performance and Testing – Part 1: Command Systems, IEC Standard 60834-1

[5] C.E. Shannon, "A mathematical theory of communication," *The Bell System Technical Journal*, vol. 27, no. 3, pp. 379–423, 1948, doi: 10.1002/j.1538-7305.1948.tb01338.x.

[6] R. Lai and A. Jirachiefpattana, *Communication Protocol Specification and Verification*. New York: Kluwer Academic Publishers, 1998, pp. 5–25.

[7] F.F. Driscoll, *Data Communications*. International ed., Orlando, FL: Harcourt Brace Jovanovich Publishers, 1992, pp. 233–35.

[8] G.J. Holzmann, *Design and Validation of Computer Protocols*, 2nd ed., Englewood Cliffs, NJ: Prentice-Hall, 1991, pp. 27–30.

[9] R. Leelaruji, and L. Vanfretti, "State-of-the-art in the industrial implementation of protective relay functions, communication mechanism and synchronized phasor capabilities for electric power systems protection," *Renewable and Sustainable Energy Reviews*, vol. 16, no. 7, 2012, pp. 4385–4395, doi: 10.1016/j.rser.2012.04.043. [online]

[10] C. Strauss, *Practical Electrical Network Automation and Communication Systems*. Amsterdam: Elsevier, 2003.

[11] J. Beaupre, M. Lehoux, and P.-A. Berger, "Advanced monitoring technologies for substations," in *2000 IEEE ESMO – 2000 IEEE 9th International Conference*, August 2000, pp. 287–292.

[12] MODICON, Inc., Industrial Automation Systems. Modicon Modbus Protocol Reference Guide. [Online]. Available: https://www.modbus.org/docs/PI_MBUS_300.pdf

[13] E. Schweitzer, and D. Whitehead, "Real-time power system control using synchrophasors," in *61st Annual Conference for Protective Relay Engineers*, 2008, pp. 78–88.

[14] SIEMENS. IEC 61850 V Legacy Protocols. [Online]. Available: https://library.e.abb.com/public/811733b652456305c2256db40046851e/SPAcommprot_EN_C.pdf

[15] Kalkitech Intelligent Energy Systems. IEC 61850. [Online]. Available: http://www.kalkitech.com/offerings/solutions-iec61850offerings-iec61850overview/

[16] IEC, "IEC 61850-6: Configuration description language for communication in electrical substations related to IEDs," 2004.

[17] M.C. Janssen, and C.G.A. Koreman, "Substation Components Plug and Play Instead of Plug and Pray" The http://www.Nettedautomation.com/standardization/IEC–TC57/WG10–12/index.html impact of IEC 61850. Kema T&D Power, Netherlands. [Online] Avail Amsterdam.

[18] C. Ozansoy, A. Zayegh, and A. Kalam, "Communications for substation automation and integration," *Presented at the 2002 AUPEC Conference*, Melbourne, Australia, 2002.

[19] O.K. Sahingoz, and N. Erdogan, "Agvent agent systems", *Proc. of 4th Int. ICSC Symp. On Engineering of Intelligent Systems (EIS 2004)*, Madeira, Portugal, 2004, pp. 548–554. [Online] Available: http://www.cs.itu.edu.tr/~erdogan/sahingoz-erdogan-EIS2004-Agv.pdf.

[20] Bolton, F. 2001. *Pure CORBA*. Indianapolis, IN: Sams Publishing.

[21] Object Management Group, "Real-Time CORBA Specification," Version 1.1, OMG, August 2002.

[22] M. Henning, and S. Vinoski, 2000, *Advanced CORBA Programming with C++*. Boston, MA: Addison-Wesley.

[23] General requirements for the competence of testing and calibration laboratories, ISO/IEC Standard 17025, 2017.

[24] Acceptability of Electronic Assemblies, IPC Standard A-610, 2017.

[25] J.J. Kumm, M.S. Weber, E.O. Schweitzer, III, and D. Hou, "Philosophies for testing protective relays," *Proceedings of the 48th Annual Georgia Tech Protective Relaying Conference*, Atlanta, GA, May 1994.

[26] D.G. Fink and H. Beaty, *Standard Handbook for Electrical Engineers*, 15th ed. New York: McGraw-Hill, 2006

[27] Power Systems Management and Associated Information Exchange—Data and Communications Security—Part 6: Security for IEC 61850, document IEC 62351-6, 2020.

Chapter 10

Holonic architecture for cyber-power distribution grid monitoring and operation

S. Basumallik[1], F. Rafy[1], S. Konar[2] and A.K. Srivastava[1]

The underlying cyber-infrastructure of the modern smart distribution grid improves system resiliency, efficiency, reliability, and security through real-time communication and monitoring, automated control, and flexible data management. With the increase in distributed energy resources (DERs) and the widespread adoption of Internet of Things (IoT) devices in power systems, the number of cyber components has drastically increased. For improved situational awareness and decision support, there is a need for automated control architecture that takes into account uncertainty under increased penetration of DERs in the distribution system. This chapter introduces holonic control architecture, a smart, flexible, adaptive control system that is capable of seamlessly adjusting to the evolving operational conditions of the distribution grid. The fundamental unit of the holonic architecture is a holon, an autonomous and self-governing unit that is "both a whole and a part," and can operate independently as well as cooperate within a larger, interconnected system, to achieve common system goals. In particular, the holonic architecture has multiple modes of operations, i.e., centralized, decentralized, distributed, and local. For each control architecture, we discuss the different cyber needs in terms of frequency of communication, frequency of model exchange, database management, and associated cyber vulnerabilities. Further, the holons can collectively determine a lead holon when one lead node fails. The overall versatility of holonic systems lies in their ability to seamlessly transition between these control modes, adapting to the challenges of diverse operational conditions and the specific needs of autonomy and cooperation. We then present two cases of holonic control—the volt–watt control and service restoration. We describe in detail all related centralized and distributed optimization techniques such as primal-dual methods, alternating direction method of multipliers, and proximal atomic coordination. For each control optimization technique, we illustrate the related cyber-physical communication architectures and demonstrate the actual implementation on a real-time hardware-in-the-loop testbed. Lastly, this chapter highlights future needs of holonic

[1]Smart Grid Resiliency and Analytics Lab, West Virginia University, USA
[2]Smart Grid Demonstration and Research Investigation Lab, Washington State University, USA

cyber architecture in terms of computational needs and networking, and discusses future applications of delay-aware robust control and integrated electrical-transport-information networks.

10.1   Introduction

10.1.1   Evolving cyber-power distribution grid

The modern cyber-power active distribution system leverages advanced communication and automated controls to integrate a growing number of Distributed Energy Resources (DER), Electric Vehicles (EV), Internet of Things (IoT) devices, and competitive markets [1,2], shown in Figure 10.1. This increases the complexity of the grid and expands ownership. Nevertheless, with the proliferation in the number of edge devices and increasing uncertainty with renewables, the performance requirements of the power systems still remain the same—the grid needs to be *reliable, efficient, sustainable, and secure.* Reliability demands the distribution system's ability to deliver consistent and uninterrupted electricity to meet the ever-growing demands of users. Efficiency emphasizes the need for optimal energy utilization and minimal losses during generation, transmission, and distribution. Sustainability

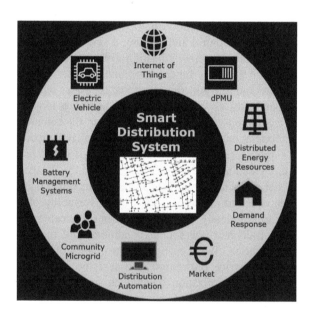

Figure 10.1   The advent of DER, advanced sensing using synchrophasor technology, coupled with the dynamics of a competitive market, the influence of the Internet of Things (IoT), implementation of Demand Response (DR), Electric Vehicles (EV), community microgrids, battery management systems have ushered in a new era for modern distribution grids

considerations involve integrating renewable energy sources to minimize environmental impact. Finally, traditional system security includes both steady state and transient state security that focuses on stability under normal operating conditions, and swift and effective response to sudden disturbances, such as faults to prevent widespread outages. The overall performance of the distribution power system thus effectively relies on design, planning, operation, and well-informed policies.

To ensure the optimal performance of the distribution system, monitoring plays an important role by providing real-time insights that allow operators to perform local and remote sensing, control, and operation of substation equipment, identify potential issues, and swiftly respond to dynamic changes. Monitoring and situational awareness are gained through high-speed data from remote terminal units (RTU) which takes a few minutes, Supervisory Control and Data Acquisition (SCADA) systems which take a few seconds, and micro phasor measurement unit (μ-PMU) which provide 60/120 samples per second. The incorporation of DERs and IoT, along with the utilization of advanced measurement units and intelligent computing devices, has resulted in a proliferation of these digital devices that process an ever-growing volume of data [3]. This has allowed easier control of the power distribution system for demand response mechanisms, load balancing, voltage control, fault detection and isolation, islanding detection, substation automation, power quality monitoring, and outage management. While the underlying cyber layer has simplified control and operations, this tight coupling between distribution and cyber systems, embedded communication, and information technologies has nevertheless resulted in a highly complex cyber-power distribution system [4].

10.1.2 Overview of existing distribution grid control architectures

Typically, automated control systems are of six different types: (a) local, (b) centralized, (c) decentralized, (d) distributed, (e) hierarchical, and (f) hybrid controls [5], as shown in Figure 10.2.

Central approach: In central control systems, measurement data from various locations in the field are aggregated at a central controller such as the advanced distribution management system (ADMS). This centralized control structure allows for efficient communication and coordination of control actions from the central control center to the field equipment. The centralized control in a complex system is often slow, has limited scalability, and is more susceptible to failures [1]. Thus, there is a need for the centralized control to be resilient to missing sensor data and unresponsive meters. Nevertheless, the solution of the central controller is always optimal.

Decentralized approach: In decentralized architecture, the power network is strategically segregated into multiple clusters, each functioning as a self-contained unit. Within each cluster, the sensors play a crucial role in both sensing and decision-making processes. They communicate directly with the lead controller specific to their cluster for decision-making [5]. To enhance coordination and scalability, all these clusters can be further organized into a hierarchical framework through a central coordinator. Clusters can also collaborate at the same level,

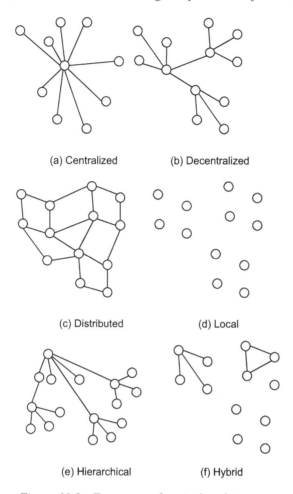

Figure 10.2　Taxonomy of control architectures

engaging in the exchange of boundary information. The weak coupling of these clusters implies that fast communication is not a prerequisite, contributing to a more resilient and adaptable network. By distributing tasks across clusters, the communication load is significantly reduced, addressing concerns related to potential single points of failure in the system. Decentralized approaches naturally exhibit hierarchy, with the upper layer implicitly responsible for clustering, while the lower layer actively engages in decision-making. These decentralized approaches can potentially devolve into local approaches if the control area is limited to a single node [6].

Distributed approach: Distributed control architecture is a system design that allocates decision-making processes across multiple interconnected components or nodes. In this approach, each node possesses a level of autonomy, enabling it to make local decisions based on its specific inputs and conditions. These nodes then communicate and coordinate with each other, forming a networked system without a centralized authority [1]. The nodes or computing agents, associated with individual controllers, collectively employ the measure–communicate–compute–deploy approach to address control problems [5]. These agents communicate with their neighboring controllers, involving the exchange of primal and dual variables, and aim to reach a consensus. Formulated by decomposing the original centralized optimization problem, these solutions can be proven to be globally optimal, however may have slow convergence rates [6]. This decentralized structure enhances system resilience, scalability, and adaptability, as failures in one node do not necessarily impact the entire system.

Local approach: Local control systems typically incorporate local measurements and utilize physical relationships among these measurements and control variables available within a single sub-station to make decisions. Control actions may include the opening/closing of circuit breakers to reroute power, the adjustment of transformer taps in response to changes in terminal voltage, the activation of capacitors or reactors to modify power flow, or the modification of generator mechanical input to control output power. The control systems depend on the "measure-compute-deploy" approach and do not consider system states in other locations when making control decisions [5]. This speeds up the control performance, however, the absence of coordination often prevents these approaches from attaining performance optimality. These local controllers can often be hard-coded, and lack robustness to fault tolerance [1]. Moreover, the lack of coordination can give rise to conflicting operations among various regulating devices or give rise to racing conditions. As a result, the implementation of local algorithms may lead to an unstable system operation, requiring parameter tuning to prevent such instability [6].

Hierarchical approach: This architecture is a combination of centralized and decentralized approaches. It operates on the principle of organizing control into multiple layers. At the top level, a central controller oversees global decision-making and coordinates activities across the entire system to meet system-wide objectives. Simultaneously, lower levels of the hierarchy implement decentralized control, allowing for local autonomy and quick responses to specific conditions or challenges.

Hybrid approach: In this approach, certain nodes may operate with local control, while other nodes may have decentralized and distributed approaches.

10.2 Cyber requirements for distribution grid monitoring and operation

10.2.1 Cyber-infrastructure supporting the distribution grid

The cyber layer of the smart distribution system consists of the communication network layer and the control layer. Communication networks link substations to both centralized and distributed control centers. The tools operating in the Energy Management System (EMS) at the control center require data input from RTUs,

SCADA, and μ-PMUs, characterized by acceptable latency and reliability. The effectiveness of this input significantly hinges on the underlying communication network [7]. The control layer includes EMS which determines and executes control actions for the distribution system operations. The control room operator continuously monitors the power system's real-time operating state through a Human Machine Interface (HMI). Real-time measurements obtained from μ-PMUs and RTUs in the power layer are used for this assessment. The HMI is part of the data storage and interface module within the control layer, providing storage capabilities for general system data and acting as the interface between human operators and the system. Databases are employed to store sensor and communication network data [8]. The control room connects to substations via fiber optic or dial-up modems.

While the modern distribution grid has evolved with an increased integration of cyber components, yet the fundamental performance requirements for the power system persist unchanged. Cyber elements now play a crucial role in supporting infrastructure, enabling the power grid to meet its performance objectives efficiently. As discussed before, this involves utilizing cyber capabilities to enable real-time monitoring and control of the grid, optimizing energy distribution, and implementing intelligent automation. To this end, establishing basic security requirements includes confidentiality, integrity, and availability. Confidentiality safeguards against unauthorized users accessing private information. Integrity protects information from being altered by unauthorized users or attackers. Availability ensures that a resource is accessible to legitimate users whenever needed. The various cyber needs for the power distribution system include:

1. Redundant Communication Paths: Incorporate redundant communication paths to enhance reliability, ensuring continuous data flow even in the face of network disruptions or failures.
2. Scalability: Allow for seamless integration of additional devices and components as the power distribution system expands.
3. Edge Computing Capabilities: Integrate edge computing capabilities to process data closer to its source, reducing latency and enhancing real-time decision-making.
4. Adaptive Control Architecture: Implement an adaptive control architecture for decision-making processes for flexibility and responsiveness to dynamic conditions of the grid.
5. Data Storage and Retrieval Systems: Design efficient data storage and retrieval systems to manage the large volumes of data generated by sensors and devices.
6. Remote Monitoring and Management: Integrate remote monitoring and management capabilities to facilitate centralized oversight and control, allowing administrators to efficiently monitor and manage the entire power distribution system.

While good situational awareness and decision support, enabled by cyber systems, are pivotal in the effective management of power distribution systems, the complexity and potential for errors in dynamic environments necessitate the reliance on automated control systems.

10.2.2   Cyber requirements for existing control architectures

The integration of modern cyber-infrastructure has a huge influence on the evolution of the smart grid technology. In modern power systems, cyber networks and communication networks are closely related, though cyber has a broader perspective in comparison [9]. Whereas, communication networks are only concerned with data exchange, cyber encompasses data computing, processing, and security in the network. It enhances system reliability, efficiency, security, and resiliency through real-time communication, flexible data management, and automated control. The reliability of the grid is characterized by the consistency and dependability of the delivery of electricity to consumers. Real-time communication improves reliability by enabling continuous monitoring of grid performance, quick responses to potential issues, and improved resilience to withstand cyber threats. Efficient communication technology further assists in maintaining grid stability, balancing demand-supply, and guaranteeing minimal interruptions in power supply. Efficiency on the other hand pertains to the optimal use of available resources to achieve the desired output with marginal losses. This involves a structured distribution of electricity, reduced energy losses in the distribution network, and optimal usage of renewable sources. Advanced data management and automation assist in achieving this efficiency to empower precise control over power flow and resource allocation. The protection of the distribution grid and its infrastructures from threats and natural events is crucial for maintaining grid security. Even after experiencing substantial damage, the resiliency of the grid should be efficient enough to recover from unexpected events and failures. In such events, real-time monitoring and automated control systems are essential to detect, protect, respond to, and isolate problems, thereby minimizing the impact.

The advancements brought by the integration of modern cyber-infrastructure in power distribution systems have led to the development of different controller architectures. These control architectures can be primarily categorized based on the inclusion of communication technology into the distribution grid. The communication-based category includes centralized, distributed, and decentralized control strategies, whereas the other category, autonomous or local control strategies, avoids information exchange between the components [10]. Earlier stages of development in large-scale industrial systems like power distribution grids predominantly relied on centralized operation for ease of management and maintainability. However, the complexities of such systems, dictated by the high dimension and volume of data, significant delays in inherent communication networks, management of remotely located components, and integration of independent distributed energy resources, posed substantial challenges. These challenges have been pivotal in driving the transition from centralized architecture to more dynamically adaptive mechanisms, such as distributed, decentralized, and local architectures. These four architectures are major categories of decision-making and resource allocation processes within the underlying cyber infrastructures. Intelligent electronic devices (IEDs) play a crucial role in the bidirectional flow of power and transmission of data in all those different coordinated networks to facilitate grid modernization with the help of cyber-empowered communication architecture. In general, Field Area

Networks (FAN) or Neighborhood Area Networks (NAN), which range from 100 to a couple of kilometers, are employed for the underlying communication network. IEDs, communicating with each other or the central unit depending on the coordination mechanism, can utilize wired Power Line Communication (PLC) or high bandwidth Wi-Fi to support different applications in NAN or FAN [11]. PLC leverages the existing power line infrastructure for effective communication and hence minimizes redundancy in wiring in NAN or smaller networks.

10.2.2.1   Centralized controller architecture

A centralized controller, as the name suggests, follows centrality in decision-making to coordinate the operations in the network. The associated cyber layer follows the same mechanism as this architecture. In this coordination mechanism, a single controller operates as the central processing unit, coordinating with the associated nodes and other elements in the network. This controller gathers and processes data from smart electronic devices linked to power-generating components, such as thermal, wind, or energy storage systems over the communication system. The central unit leverages this data to make strategic decisions aimed at enhancing the system's overall performance. Efficient applications and tools such as optimization algorithms are also used by the central unit to achieve performance goals, like reducing energy consumption, maximizing renewable energy production, and improving grid resilience. IEDs like sensors, smart meters, and relays, communicate real-time data about power usage, grid conditions, and potential faults back to the central control system. Similar to the SCADA, the central controller oversees the monitoring, controlling, and decision-making process of the grid. All major operational decisions and actions, like providing set points to the edge devices, are highly reliant on the single controller unit in this architecture. From the cyber perspective, a server acts as the sole coordinator in the cyber physical system (CPS) network while connected through ethernet channels. As a core of the computerized control mechanism, servers process data from intelligent sensors through wired or wireless technology and distribute actuation tasks among the IEDs [12]. As shown in Figure 10.3, the centralized controller is the heart of the system. It collects data from the sensors or Intelligent Electronic Devices in the network. Throughout the process, the IEDs across the network send real-time data—voltage levels, load data, fault indicators, etc. — to the central server or controller. Based on this data, the controller makes informed decisions to optimize the performance of the CPS to achieve global optimization. If needed, it also makes adjustments, like tap settings adjustments or rerouting power flows for the connected components in the network.

Central management of control decisions makes this design provide high scalability in the power distribution network to avoid significant structural changes. It is easier to expand the system according to CPS production capabilities. Thus, adding new components to the existing network without major changes in this architecture is beneficial to cut costs during industrial operations and expansions. The cost-effectiveness of this design relatively outperforms other controller architectures in terms of affordability to set up and maintain a centralized control system. The standard types of equipment and associated software make it further inexpensive to

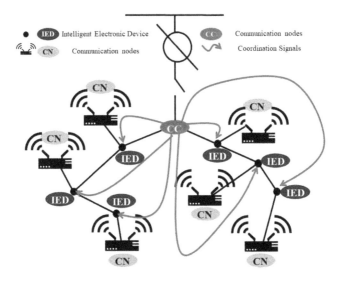

Figure 10.3   Data flow and generalized design of centralized controller

manage eliminating the need for highly specialized experts. This ease of management in the system allows for easy adjustments to how the CPS operates. Changes can be implemented using industry-standard software from a single location, which can help in prompt response to new production needs or challenges [12].

Unlike other controller systems, the information in this design flows only towards and from the centralized controller to the IEDs through suitable communication mediums. This centrality in the design presents several challenges though. One of the major challenges arises with the need for accurate data collection and dissemination across a range of resources. As the central controller is responsible for the operational management, real-time monitoring and control become difficult to maintain with large-scale systems. The growing number of renewable energy resources adds to this complexity due to their operational diversity and unpredictability in cyber-physical systems. Often massive computing power is necessary to collect and analyze data from different edge devices in the network. This high computing requirement and control can introduce inevitable delays in control actions.

10.2.2.2   Distributed controller architecture

In modern power distribution systems, the adoption of operational strategies that utilize inter-controller communication for distribution optimization marks a significant shift towards enhancing overall system performance. This approach is fundamentally different from traditional centralized systems, offering several advantages in terms of scalability, computational efficiency, and system resilience. The cyber layer of the distribution network is configured to support distributed operability. The network is designed to permit multiple controllers within the system to connect and communicate with each other. Each controller of the network has a specific role in the system. There are dedicated controllers for sensors, which collect data from the network; computing devices, which process the data; and cyber components, which manage the security and resilience of the network [1]. The role of SCADA in this control architecture is more like overseer rather than sole decision-maker. SCADA may still provide central points for data aggregation and higher-level decision-making but leaves the local control and processing to the controllers.

As an independent part, the controllers are capable of sharing information with neighboring devices. The inter-controller communication is essential for coordinating actions and distributing informed decisions across the entire distribution network. Figure 10.4 shows how the distributed controller communicates with multiple edge-based computing nodes. Bidirectional communication occurs between neighboring IEDs to exchange information such as local voltages, current flows, power quality metrics, or other relevant measurements. If DERs are connected to the system, then the operational status of those DERs including data on current generation levels, capacity, and any operational constraints or issues is also essential information to share in this architecture. Also, nodes in this system often share control commands or set point data with the adjacent nodes to align the local operation of grid-connected DER nodes. For example, a node might communicate its planned reactive power output to help neighboring nodes adjust their own outputs for optimal voltage regulation. The shared data also may include predictive data based on local forecasts like expected solar generation based on weather forecasts or any alerts triggered due to disturbances to assist in adjusting control strategy proactively. IEDs in distributed control cooperate with each other with the help of connected communication mediums and devices. Unlike the central command in the centralized controller, the IEDs collaborate to make joint decisions. The decision is based on predefined goals set by either the grid operator or the end-user. Since, in this design, communication occurs between immediate neighbors, there is no need to know the entire grid's state to make decisions. This means that

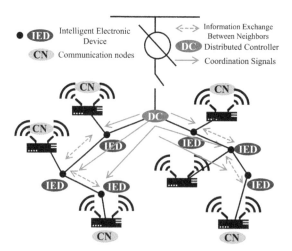

Figure 10.4 Data flow and generalized design of distributed controller

local interactions in this case are effective ways to bypass the interference with the grid and enhance efficiency in distributed operation. It allows the system to have the plug-and-play capability to easily integrate and coordinate various components in the network [10,13]. A key benefit of this approach is scalability, especially when distributed energy resources (DERs), like solar arrays, wind turbines, and battery energy systems are integrated from time to time. Interfacing new DERs in the system can be done efficiently as this architecture is capable of managing the complexities associated with it. Unlike centralized systems, where the single entity is burdened with the entire computational load, the distributed strategy spreads this burden across multiple independent controllers. In the distributed approach, the computational task is divided into smaller sub-problems, managed by different controllers. This subdivision makes it easier for the controllers to manage and focus on specific problems. This not only prevents a single point of failure but also improves the performance of the system to handle complex operations. With control and monitoring functions spread across various points in the network, identifying and responding to faults becomes more flexible and efficient. Each controller can monitor its segment of the network, quickly detecting and addressing issues.

Nevertheless, to ensure consistent and secure communication in various nodes, this controller architecture is intricately designed like the structure of the mesh networks. Managing this increased complexity is a challenging issue with the distributed strategy in the power grid. This is particularly evident in the continuous integration of DERs with the constant demand in the grid. These systems must process data rapidly, make decisions locally, and coordinate the decisions across the network to maintain grid stability and efficiency. Despite their inherent robustness to failure due to their distributed nature, the main challenge is to ensure the reliability and flexibility of each controller. The underlying communication network must be robust to withstand failures in individual components without compromising overall performance. Furthermore, the widespread use of communication technology and the intra-communication of this system increases their vulnerability to cyber threats. Another significant challenge lies in integrating distributed control architectures into existing power systems, which often contain legacy components and technologies. This integration demands meticulous planning and management to guarantee compatibility and ensure the smooth operation of the electrical system [14].

10.2.2.3 Decentralized controller architecture

The architecture of a decentralized controller integrates elements of both centralized and distributed control systems. It encompasses similar aspects of functionality, the flow of commands, and coordination mechanisms. A distinctive characteristic of this mechanism is the separation of the entire network into multiple zonal areas that manage several inclusive IEDs. In a decentralized control setup, the controllers work together but not as closely as in other systems. This is like a distributed system, where each controller is connected but operates independently. However, like in a centralized system, each controller in its own area (or zone) makes important decisions using information and commands from smart devices known as IEDs. This allows for both independent and coordinated control across the network [10].

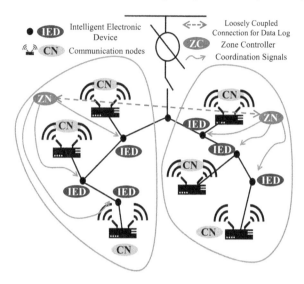

Figure 10.5   Data flow and generalized design of decentralized controller

Communication in a decentralized controller system takes place to facilitate coordination among local networks to achieve local optimization while sharing logs of each zone to achieve somewhat global optimization. However, due to the loose coupling between zones, the global optimization is rather less significant than the local one. By facilitating zonal partitioning into the network, each zone coordinator focuses on the optimization of their respective area to manage voltage levels, balance loads, integrate local DERs, etc. The goal is to obtain optimal operation based on local conditions and requirements. Hence, communication is limited within a zone between a zone coordinator and the associated IEDs in the zonal network. Every zone also can communicate and can be loosely coupled with each other as shown in Figure 10.5 to achieve broader network goals, such as overall system stability, efficient distribution of power, and minimizing network losses. The circulated information mostly consists of status updates or performance metrics rather than command or coordination signals [10].

As this architecture employs autonomous controllers at each node, the need for hierarchical control is minimized. In addition, communication is more localized, which reduces the dependency on a centralized network. Even though the nodal autonomy introduces complexity, minimal communication and independent operability of nodes contributed to high resiliency during system failures. Fault location identification, isolation, and event response are comparatively easier because of the network segmentation through zones. Also, due to the independent characteristics, the failure of one node does not necessarily affect the others in the system.

10.2.2.4   Local controller architecture

The cyber layer associated with the local controller system is useful in managing localized energy resources and responding promptly to local grid disturbances. However,

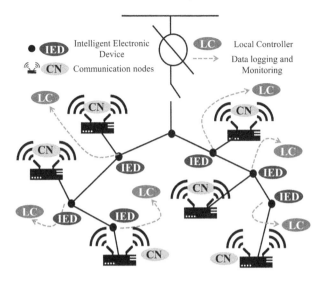

Figure 10.6   Data flow and generalized design of local controller

since there is no intra-communication between the IEDs or controllers, coordination is not available in this strategy. Instead, the controllers utilize the measurements at the point of common coupling (PCC) without any other external information exchange inside the network. The PCC, in this case, is a point of contact where the electric utilities are interfaced with the consumer market. The lack of intra-network communication allows this architecture to be fast in response and resilient to communication failures and other complexities. Yet, it invokes limitations in such a system, as the minimal use of information may therefore lead to sub-optimal or even non-optimal control solutions. [10,15]. Communication protocols in the local controller system are often considered for an immediate environment where prioritizing swift response and processing data locally is the sole responsibility. Even if a limited communication setup and sensors are present, it is used for monitoring purposes. The data from that limited setup is typically logged into the databases instead of generating commands and setpoints for the controllers as seen in other controller architectures [15]. As shown in Figure 10.6, the local controllers are connected to each of the IEDs in the network in a one-to-one mapping format. Even though the coordination of network devices is absent in this architecture, the local controller may receive performance logs through the communication nodes.

There exist two other controllers, (a) hierarchical and (b) hybrid, which employ similar cyber characteristics. Details are omitted here.

10.3   Introduction to holonic control for distribution grid

10.3.1   *Defining holons and holonic control*

Holons are characterized as self-sufficient and self-governing entities that capture the dual characteristic concept of being ''both a whole and a part'' [16].

This hierarchical arrangement mirrors the self-similarity found in fractals, and much like the self-repeating patterns in fractals, holonic control exhibits a recursive nature. The holonic architecture is characterized by this unique ombination of independence and cooperation. Operating autonomously, holons collaborate within a broader, interconnected system known as a holarchy, synergizing their efforts to achieve common system goals. As these entities work independently yet collectively, the resulting holonic architecture becomes a dynamic framework capable of responding effectively to the complexities of ever-changing operational landscapes. This dual nature, balancing autonomy and collaboration, makes the holonic systems agile and responsive. Some existing holonic architectures of smart power systems applications that exhibit such "control within control" include reactive power control [17], smart grid IT information exchange [18], operation of distribution systems under normal and emergency modes [19], and IEC 61850 and IEC 61499-based system automation [20].

10.3.2   Holonic control architecture

Depending on system conditions and requirements, holonic architecture can switch between different modes of operations—centralized, decentralized, distributed, hierarchical, or local. For example, holons can work collaboratively to establish a centralized control architecture. This centralized holarchy operates with the primary objective of optimally allocating resources from a central locus, facilitating more uniform decision-making processes. Simultaneously, it maintains a comprehensive and vigilant oversight of the entire system. Under normal conditions, this is similar to existing Advanced Distribution Management Systems (ADMS), which collect real-time data from various sensors and monitoring devices and run optimization algorithms centrally for operations and control. It can also act as a hierarchal controller where the central controller communicates with lower-level substation controllers which further talk to other nodes such as DERs to request information, for broader communication and control.

In the event of a breakdown in the communication system or network, these holons have the ability to adapt and reconfigure their arrangement to ensure maximum nodes are working. A set of holons can talk with other holons and execute computations autonomously and in a decentralized manner. This decentralized holarchy avoids overwhelming the central controller, enhances system scalability, resource utilization, and handling of data processing requirements. It also helps in cutting down on communication issues and lowers the chances of single points of failure. Even if the electrical grid is isolated, the primary holons in each isolated segment can communicate with the central controller. Should communication with the central controller break down, the decentralized holons can still sustain communication among themselves.

However, when communication failures lead to a holon being isolated from other holon groups, the holon can independently operate in a decentralized fashion, specifically within local contexts for making decisions. In such situations, the disconnected local holon utilizes its autonomous capabilities to navigate and address tasks within its immediate environment, ensuring continued functionality even when communication with the broader holarchy is disrupted. An example is shown in Figure 10.7.

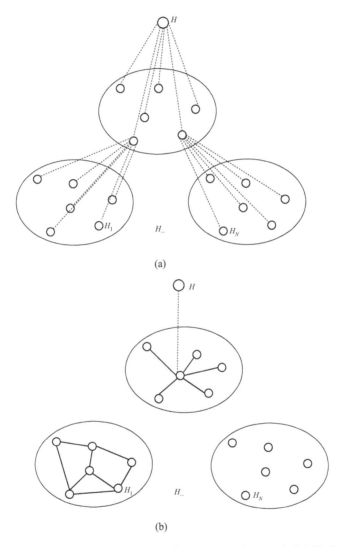

(a)

(b)

Figure 10.7   An overview of holonic architecture and control. (a) Under normal circumstances, holonic control operates in a centralized hierarchical manner for efficiency and optimization. (b) However, during emergencies, the system seamlessly decomposes into subsystems, each acting as a resilient copy of the whole system, ensuring adaptability and continuity.

10.3.2.1   Requirements of holonic architecture

The requirements for holonic control are shown in Figure 10.8. Flexibility is needed in multiple layers which include holonic data layer, computing layer, security layer, control layer, communication layer, and policy layer.

1.   Holonic data layer—This encompasses both centralized and federated data management approaches. Holonic databases can have centralized repositories where commonly used data is stored. These central data sources serve as reference points within the larger holonic structure. They ensure consistency, reliability, and a single source of truth for specific types of information. Simultaneously, the holonic data layer embraces federated principles. This means that data is distributed across different entities or nodes, allowing for decentralized decision-making and local autonomy. To enable holonic data layer, dynamic database management is employed. Dynamic database management gives holons the ability to adapt and respond to changing conditions, requirements, or data patterns in a database system. Unlike static database management, where the structure and configurations are mostly fixed, dynamic database management systems offer flexibility and responsiveness to handle evolving data needs that come with dynamic clustering of the system. Dynamic database management supports real-time data integration, allowing seamless incorporation of streaming or real-time data into the database, allows for data repartitioning, can dynamically adjust indexing strategies based on query

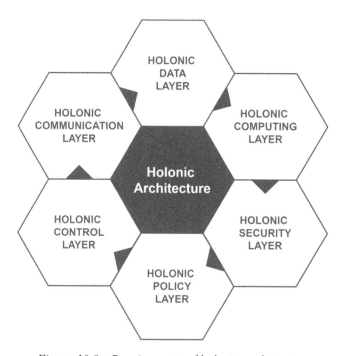

Figure 10.8   Requirements of holonic architecture

patterns and the nature of the data, can automatically scale in response to changing data volumes or processing demands, and allows for changes to the database schema without requiring a complete shutdown or significant disruption. Dynamic database management is especially relevant in holonic environments where data requirements are subject to change, and the ability to quickly adapt to these changes is crucial for maintaining optimal performance and efficiency.

2. Holonic communication layer—Communication between nodes occurs through various modes: (a) certain critical information is exchanged through centralized hubs, ensuring consistency and reliability across the system, (b) nodes in a decentralized system have the autonomy to communicate and make decisions independently, (c) communication is distributed across multiple nodes, fostering collaboration and information sharing, and (d) nodes interact locally, making decisions based on immediate surroundings and needs, allowing quick responses to specific local conditions.

3. Holonic control layer—The computing layer combines both centralized and distributed optimization strategies. Certain computational tasks are centrally optimized which involves consolidating resources and leveraging centralized decision-making to enhance efficiency, while tasks can be distributed across multiple nodes, allowing for parallel processing and decentralized decision-making under emergency situations. Before addressing the challenges associated with decentralized optimization, a critical preliminary step involves the initiation of a leader election process within the holonic architecture. By designating a leader, the system introduces a focal point for decision-making and coordination among the participating entities. This selected leader assumes the responsibility of guiding the collective toward the successful resolution of the decentralized optimization task. This approach enhances the efficiency of the holonic system by preventing conflicts, streamlining communication, and enabling a more organized execution of the optimization task. The leader election from multiple lead nodes is implemented on the Resilient Information Architecture Platform for Smart Grid (RIAPS) platform [21,22].

4. Holonic computing layer—This integrates centralized and parallel/distributed computing paradigms. Centralized computing concentrates the computing resources and decision-making in a central hub or node ensuring privacy, efficient utilization, and data security. Distributed computation involves the simultaneous execution of computational tasks across multiple nodes or processors, improving performance, processing speed, and efficiency.

5. Holonic security layer—This involves the implementation of centralized and decentralized defense layers within the system. The centralized cyber defense layer consolidates security measures and decision-making, providing a unified and coordinated approach to protect against cyber threats within a system or network. In the decentralized defense approach, instead of relying solely on a centralized security model, individual nodes or components within the system have autonomous security measures—security entities or nodes can interact and share threat intelligence. This collaborative approach strengthens the

overall security posture, as nodes work together to identify and mitigate potential risks.

6. Holonic policy layer—This layer orchestrates changes in system architecture, facilitating dynamic clustering and seamless reconfiguration of networks in power systems. Numerous techniques exist for the formation of dynamic subsystems or clusters, each offering distinct approaches to achieve effective grouping. These methods include epsilon-partitioning which involves defining partitions based on a specified threshold parameter, adaptive zone divisions which dynamically adjust cluster boundaries based on physical relationships, and others such as sensitivity-based approach, electrical distance, modularity index, eigen-value-based decomposition. These dynamic clustering facilitates the seamless reconfiguration of holonic structures, enabling the system to quickly adapt to unforeseen events, disturbances, or changes in operational objectives. This flexibility facilitates self-organization capability is essential for maintaining optimal performance, resilience, and responsiveness. Holonic policy also ensures adaptive power system protection settings, responding to evolving conditions through a holonic approach to policy management.

10.4   Use cases of holonic architecture

10.4.1   *Volt–watt control in DER-rich distribution grid*

The first application of holonic control is the volt–watt control. The purpose of the volt–watt control function is to manage feeder voltage and mitigate voltage violations. This function restricts the active power generation from DERs based on the voltage levels, activating when the voltage surpasses a predefined threshold. This becomes increasingly important as the utilization of renewable energy resources rises. While it is crucial to operate these resources at their maximum power point, relying solely on VAR control may prove insufficient to maintain voltages within acceptable limits. Consequently, there may be a need to curtail active power via volt–watt control to avert potential issues.

The holonic architecture described in this chapter has the capability to support both the distributed controller and centralized controller frameworks. In the distributed setting, nodes communicate exclusively among neighboring DER nodes, and the distributed coordinator application facilitates this communication upon detecting changes in the network topology. In a distributed framework, communication among neighboring nodes is necessary only for updating primal variables. Updating dual variables operates independently of measurements at neighboring nodes. This approach reduces the need for exchanging voltage and maximum power points among the DER nodes. In the centralized setting, the algorithm exclusively collaborates with the centralized controller to determine setpoints for the DERs. The model focuses solely on communication between the central unit and the DER nodes. In a centralized framework, both primal and secondary variables are computed using the same algorithm within the central unit.

The optimization of the volt–watt problem is formulated as follows,

$$\min_{\mathbf{x}} \sum_{\forall i} f_i(x_i) \tag{10.1a}$$

$$\text{s.t.} \quad \underline{y}_i \leq y_i(x_i) \leq \bar{y}_i \tag{10.1b}$$

$$\underline{x}_i \leq x_i \leq \bar{x}_i \tag{10.1c}$$

Here, the objective function ($f_i : \mathbb{R} \to \mathbb{R}$) must satisfy μ-strong convexity and *l*-smoothness criteria. The problem exhibits complete decomposability with respect to x_i, where these variables are interconnected only through the function y_i, which may inherently be non-linear. A linear relationship between x_i and y_i is identified for updating primal variables in the optimization problem solution. The plant is used to determine $y_i(x_i)$, which is then employed in updating the dual variables, making our approach feedback-based or dynamic. The iterative use of this feedback-based methodology serves to progressively mitigate model inaccuracies stemming from the approximated relationship.

We assume a linear power flow approximation of the distribution system. Consider an unbalanced radial distribution network with three phases and $N + 1$ nodes, where the node set is defined as $\mathscr{N} = \{0, 1...., N\}$. This simplified model linearizes the power flow equations, omitting line losses and assuming balanced node voltages. The linear approximation of the power flow equations discussed is,

$$\tilde{v} = \bar{Z}^P \tilde{P} + \bar{Z}^Q \tilde{Q} + v_0 1_{3N} \tag{10.2}$$

where the network voltage vector is $\tilde{v} = [v_1 ..... v_N]^T$ with $v_j = [|V_j^a|^2, |V_j^b|^2, |V_j^c|^2]^T$, and $\forall j \in \mathscr{N}$, active power injection vector is $\tilde{P} = [p_1 ..... p_N]^T$ where $p_j = [p_j^a, p_j^b, p_j^c]^T, \forall j \in \mathscr{N}$, reactive power injection vector is $\tilde{Q} = [q_1 ..... q_N]^T$ where $q_j = [q_j^a, q_j^b, q_j^c]^T, \forall j \in \mathscr{N}$, substation end voltages are $v_0 1\_3N$ and resistances and reactances of the distribution network are $3 - \phi$ matrices $\bar{Z}^P$ and $\bar{Z}^Q$ [23].

The active power injection in (10.2) is separated into curtailable $\tilde{P}^C$ and fixed $\tilde{P}^F$ components, where, $\tilde{P} = \tilde{P}^F + \tilde{P}^C$. The network-wide voltage profile is given as,

$$\tilde{v}(\tilde{P}^F) = \bar{Z}^P \tilde{P}^C + \tilde{v}^{unc} \tag{10.3a}$$

$$\tilde{v}^{unc} = \bar{Z}^P \tilde{P}^F + \bar{Z}^Q \tilde{Q} + v_0 1_{3N} \tag{10.3b}$$

where $\tilde{v}^{unc}$ is a factor dependent on $\tilde{P}^F$ and $\tilde{Q}$. Here, $\tilde{v}(\tilde{P}^F)$ characterizes the input-output relationship between active power curtailment and the network voltage profile. At time *t*, if the active power curtailment vector throughout the network is denoted as $\tilde{P}^F(t)$ and, following its implementation, the measured voltage is $\mathbf{v}(t)$, the resultant Lagrangian multiplier dictates the control action $\tilde{P}^F(t + 1)$.

10.4.1.1   Distributed approach

The distributed approach for solving the volt–watt problem is shown in Figure 10.9. The lagrangian multiplier of the optimization problem (10.1) is,

$$\mathcal{L}(\hat{p}, \xi, \lambda) = \sum_{\forall i} f_i(\hat{p}) + \underline{\lambda}^T (\underline{v} - v(\hat{p})) + \bar{\lambda}^T (v(\hat{p}) - \bar{v})$$
$$+ \sum_{\forall i} K_i(\hat{p}_i, \xi_i) \tag{10.4}$$

Here, $\hat{p}$ denotes the required active power curtailment and $K_i(\hat{p}_i, \xi_i)$ is a quadratic penalty function designed to expedite the convergence of the objective function,

$$K_i(\hat{p}_i, \xi_i) = \begin{cases} \xi_i \left(\hat{p}_i - \underline{p}\right) + \dfrac{c}{2}\left(\hat{p}_i - \underline{p}\right)^2 & \hat{p}_i + \dfrac{\xi_i}{c} < \underline{p} \\[2mm] -\dfrac{\xi_i^2}{2c} & \underline{p} \le \hat{p}_i + \dfrac{\xi_i}{c} \le \bar{p} \\[2mm] \xi_i \left(\hat{p}_i - \bar{p}\right) + \dfrac{c}{2}\left(\hat{p}_i - \bar{p}\right)^2 & \bar{p} < \hat{p}_i + \dfrac{\xi_i}{c} \end{cases} \tag{10.5}$$

For the primal-dual algorithm, the primal updates are as follows,

$$\hat{p}_i(t+1) = \hat{p}_i(t) - \alpha \left\{ \left(\bar{\lambda}_i - \underline{\lambda}_i\right) + \sum_{\forall j \in \mathcal{N}} \left[\bar{Z}^P\right]^{-1} \left[f_{i'}(x_i) + \mathrm{ST}_{\underline{cp}_i}^{c\bar{p}_i}(\xi_i + c\hat{p}_i)\right] \right\} \tag{10.6}$$

Cyber-Physical System Workflow for Distributed VWATT control

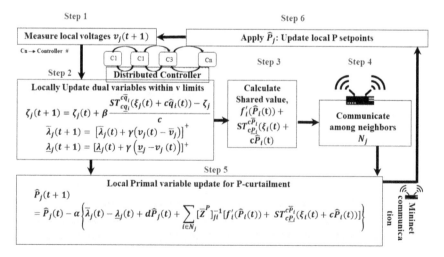

Figure 10.9   Distributed volt–watt control application

For any $e_1 < e_2$, the soft-thresholding function $\mathrm{ST}_{e_1}^{e_2}(\cdot)$ in (10.6) is $\mathrm{ST}_{e_1}^{e_2}(z) = \max(\min(z - e_1, 0), z - e_2)$. The distributed nature of the method is achieved by scaling the gradient for updating the primal variable by $\left[\bar{Z}^P\right]^{-1}$. However, increased modeling accuracy makes $[\bar{Z}^P]^{-1}$ lose positive semi-definite properties, thus not guaranteeing asymptotic convergence. To this end, the linearized 3-phase network with no cross-coupling among phase resistances and reactances is employed.

The dual update is given as,

$$\xi_i(t+1) = \xi_i(t) + \beta \frac{\mathrm{ST}_{\underline{cp}_i}^{\overline{cp}_i}(\xi_i + c\hat{p}_i) - \xi_i(t)}{c} \tag{10.7}$$

$$\overline{\lambda}_i(t+1) = \overline{\lambda}_i(t) + \gamma \left[\left(v_i^{meas}(t) - \overline{v_i}\right)\right]^+ \tag{10.8}$$

$$\underline{\lambda}_i(t+1) = \underline{\lambda}_i(t) + \gamma \left[\left(\underline{v_i} - v_i^{meas}(t)\right)\right]^+ \tag{10.9}$$

Here, $[\cdot]^+$ is a projection operator onto the non-negative orthant. The use of voltage measurements $v_i^{meas}(t)$ eliminates the need to calculate $\tilde{v}^{unc}$, which is dependent on system-wide loading conditions and the non-linearity of AC power flow equations. The tuple $[\underline{p_i}, \overline{p_i}]$ is non-zero only for DER nodes. If P_i^{DER} is the maximum active power injection capability of DER node i, the corresponding set-points for the DER inverter are,

$$p_i^{inj}(t+1) = P_i^{DER}(t) + [\hat{p}_i(t+1)]_{\underline{p_i}}^{\overline{p_i}} \tag{10.10}$$

Here, $[\cdot]_{\underline{p_i}}^{\overline{p_i}}$ denotes projection onto the set $[\underline{p_i}, \overline{p_i}]$. To determine DER set-points, both the maximum power point and curtailable part must be calculated. Given that most DERs are renewable, they are expected to operate at their maximum power point (MPP) level, $p_i^{mpp}(t)$, at any time t. The active power injection set-points are updated as,

$$p_i^{inj}(t+1) = \left[p_i^{mpp}(t+1) + [\hat{p}_i(t+1)]_{-p_i^{mpp}(t)}^{0}\right]_0^{p_i^{mpp}(t+1)} \tag{10.11}$$

The intermediate primal and dual variables are updated using $p_i^{mpp}(t)$. Controller set-point computation utilizes the primal auxiliary variable $\hat{p}_i$ and dual auxiliary variables ξ_i, $\overline{\lambda}_i$, and $\underline{\lambda}_i$ (each as vectors for all three phases) at each node. Voltage controllers are assumed to be present for all phases at time t, with constraints indicating the absence of DERs in specific phases. Voltages at DER nodes $v_i^{meas}(t)$, and active power capability based on MPP, $p_i^{mpp}(t)$, are measured during this time. Auxiliary variables, including $\hat{p}_i(t)$, $\xi_i(t)$, $\overline{\lambda}_i(t)$, and $\underline{\lambda}_i(t)$, are updated using available values and exchanged information from neighboring DERs. Local active power injection set-points $p_i^{inj}(t+1)$, are computed and applied thereafter.

For the distributed framework, each DER controller for a given node j ($j \in \mathcal{N}$) has four steps.

Step 1: Measure All local phase voltages $v_j(t)$ are measured at all available phases.

Step 2: Calculate, $\hat{p}_j(t+1)$, $\xi_j(t+1)$, $\bar{\lambda}_j(t+1)$, $\underline{\lambda}_j(t+1)$, using the following equations:

$$
\hat{p}_j(t+1) = \hat{p}_j(t) - \alpha \left\{ \left(\bar{\lambda}_j(t) - \underline{\lambda}_j(t) \right) \right.
$$

$$
\left. + \sum_{\forall i \in \mathcal{N}_j} \left[\bar{Z}^P \right]_{ji}^{-1} \left[f_{i'}(\hat{p}_i(t)) + ST^0_{-cp_i^{mpp}(t)}(\xi_i(t) + c\hat{p}_i(t)) \right] \right\}
$$

(10.12a)

$$
\xi_j(t+1) = \xi_j(t) + \beta \frac{ST^0_{-cp_i^{mpp}(t)}\left(\xi_j(t) + c\hat{p}_j(t) \right) - \xi_j(t)}{c}
$$

(10.12b)

$$
\bar{\lambda}_j(t+1) = \bar{\lambda}_j(t) + \gamma \left[\left(v_j^{meas}(t) - \bar{v}_j \right) \right]^+
$$

(10.12c)

$$
\underline{\lambda}_j(t+1) = \underline{\lambda}_j(t) + \gamma \left[\left(\underline{v}_j - v_j^{meas}(t) \right) \right]^+
$$

(10.12d)

where $\mathcal{N}_j$ represents the set of all neighboring nodes of node j for every $j \in \mathcal{N}$.

Step 3: (Active Power Set-Point Deployment): The active power injection set-point at time $t+1$ is determined as,

$$
p_j^{inj}(t+1) = \left[p_j^{mpp}(t+1) + \left[\hat{p}_j(t+1) \right]_{-p_j^{mpp}(t)}^0 \right]_0^{p_j^{mpp}(t+1)}
$$

(10.13)

Step 4: (Communication): Neighboring DER nodes receive values $f_{j'}(\hat{p}_j(t+1)) + ST^0_{-cp_j^{mpp}(t+1)}\left(\xi_j(t+1) + c\hat{p}_j(t+1) \right)$.

Figure 10.10 shows the results of the distributed approach.

10.4.1.2 Centralized approach

The centralized approach for solving the volt–watt problem is shown in Figure 10.11.

Setpoint computation for DERs in the system in the centralized controller scenario involves exclusive collaboration with the centralized controller. The communication modeled here is solely between the central unit and the DER nodes. In the centralized framework, both primal and secondary variables are computed using a unified algorithm within the central unit. Figure 10.12 shows the results of the centralized approach.

To perform a comparative analysis of the centralized and distributed controller architectures, a CPS testbed with a modified IEEE-123 bus as a use case

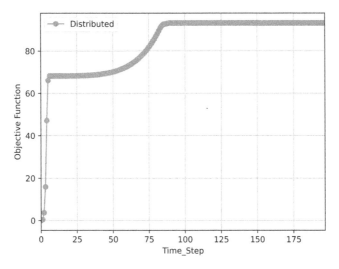

Figure 10.10   Objective function plot for distributed volt–watt control application

Cyber-Physical System Workflow for Centralized VWATT control

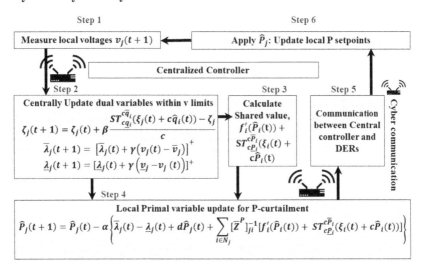

Figure 10.11   Centralized volt–watt control application

has been developed in our earlier work. The CPS testbed assists us in under-standing how those different control strategies function differently in real-world scenarios through emulated environments. The volt–watt algorithm has been incorporated into the testbed to control voltage levels and facilitate power flow management. As shown in Figure 10.10, the objective function, as described in

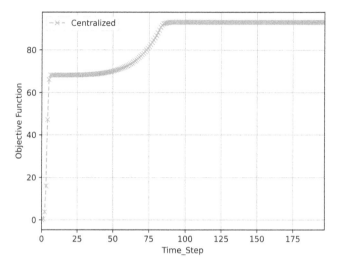

Figure 10.12   Objective function plot for centralized volt–watt control application

Section 10.4.1.1, shows stable power flow in the system after a certain number of adjustments to voltages in connected distributed controller nodes. To allow these adjustments, the algorithm maintains the voltage stability by curtailing the calculated amount of active power from specific controller nodes. As shown in Figure 10.12, the system reaches the optimal solution after a number of iterations where the voltage control algorithm adjusts the amount of active power curtailment to reach global minima.

A centralized controller requires information from the IEDs in the network to obtain global optimization. In this process, the controller requests relevant information from the sensors through different communication mediums and protocols. In our testbed, we demonstrated the communication between those connected IEDS and the central controller using Mininet as an emulated network environment. Figure 10.13 shows how the volume of data transfer between the sensor devices and the controller varies between the two distinctive voltage control strategies. It justifies how a centralized controller requires less number of instances to exchange information compared to distributed ones. Because the distributed controllers communicate with each other to reach local optimization that further contributes to the global optimization of the network. However, the volume of data exchange in each instance is comparatively much lower than the contemporary one in distributed control as the neighboring controllers require less amount of information from a selective number of neighbors to perform power flow calculations in the local controllers.

In our CPS testbed, we closely monitored the time both centralized and distributed controllers took for optimization and data exchange in a co-simulated 123-bus test system. As shown in Figure 10.14, the time difference between these two strategies is minimal. This minimal variation can be partly attributed to the scale of

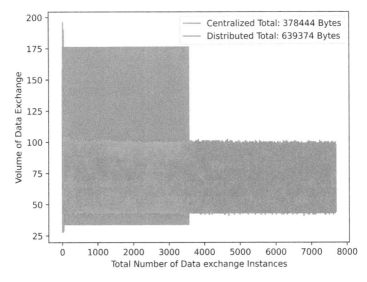

Figure 10.13   Volume of data exchange between IEDs and controllers

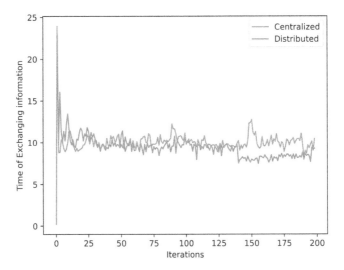

Figure 10.14   Total of optimization and communication in CPS testbed

our test system, which only considers a number of IEDs and their inter-communication due to computational constraints. In real-world power networks, particularly larger systems with many IEDs, distributed control is often faster due to its parallel processing and localized decision-making capabilities. On the contrary, the centralized system's potential delay-causing factors, like a single decision-making point, do not significantly impact our relatively simple model.

10.4.2   Distribution system restoration

The second application of holonic control is the restoration of the distribution grid. First, we introduce the centralized portion of the holonic control. Next, we briefly discuss the consensus-based distributed ADMM algorithm. Following ADMM, we present the recent improvements in ADMM to solve mixed integer problems, especially the penalty-driven ADMM (PD-ADMM) algorithm. Finally, we provide numerical examples of centralized and distributed restoration of the distribution grid using the algorithms discussed.

10.4.2.1   Centralized approach

As stated in the previous sections in a centralized approach measurements from all the sensors carrying information like location of faulted lines, load demand, and voltage profile of the system, are communicated to the control center. The centralized controller (P1) optimizes an objective function (10.14a) by satisfying constraints (10.14b) and (10.14c). Due to the presence of integer variables for example status of line switches, load switches, buses, etc., the restoration problem becomes mixed-integer in nature.

$$P1 := \min_{X, \mathcal{Z}} g(X, \mathcal{Z}) \tag{10.14a}$$

$$\text{Subject to}: h_1(X, \mathcal{Z}) = 0 \tag{10.14b}$$

$$h_2(X, \mathcal{Z}) \leq 0 \tag{10.14c}$$

Here, $X \in \mathbb{R}^p$ and $\mathcal{Z} \in \mathbb{Z}^q$ are the global continuous and integer decision variables, respectively. Therefore, distribution grid restoration problem (P1) is often formulated as mixed integer linear programming, mixed integer quadratic programming, and mixed integer second-order cone programming problem [24].

10.4.2.2   Distributed approach

In this section distributed consensus ADMM algorithm is discussed. For simplicity of discussion, we assume P1 (10.14) is convex and a function of X which is a vector of continuous decision variables.

ADMM-based distributed optimization: For distributed computation, we split the decision vector X and assume the objective function is also separable across this splitting. Let us assume, a distribution network is represented by a connected and directed graph $G(\mathcal{N}, \mathcal{E})$. For distributed computation, the network is decomposed/split into L subsystems $\mathcal{H}^l, l \in 1, 2, \dots L$. A subsystem $\mathcal{H}^l$ is a graph with vertices $\mathscr{A}^l \subseteq \mathcal{N}$ and edges $\mathscr{E}^l \subseteq \mathscr{E}$. The function $\zeta(lj)\{\mathcal{H}^l \times \mathcal{H}^j\} \to \mathscr{E}$ identifies overlap region(s) or line(s) between any two subsystems, $\mathcal{H}^l$ and $\mathcal{H}^j$ (See Figure 10.15). $\mathcal{N}^l$ denotes the set of overlapping regions of the lth subsystem. Let us assume $X_{\zeta(lj)}^l$ represent vectors associated to $\mathcal{H}^l$, collecting variables related to the lines connecting $\mathcal{H}^l$ and $\mathcal{H}^j$, respectively.

Following the decomposition, each subsystem ($\mathcal{H}^l$) solves a very similar problem (10.14) but is written for the subsystem under the influence of its distributed

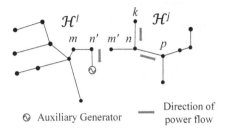

Figure 10.15   Area-level decomposition

controller (l^{th}) only [25]. The decomposition of the function (10.14) results in

$$P2 : \min_{X^l} \sum_{l=1}^{L} g_l(X^l) \tag{10.15a}$$

Subject to : $h_1(X^l) = 0, \forall l \in L$ (10.15b)

$h_2(X^l) \leq 0, \forall l \in L$ (10.15c)

$X^l_{\zeta(lj)} = X^j_{\zeta(lj)}$ (10.15d)

Here, $g_l(X^l)$ is the objective function of $\mathcal{H}^l$ obtained in accordance with the decomposition of the function (10.14). Constraints (10.15d) require $\mathcal{H}^l$ and $\mathcal{H}^j$ consent on the values of the continuous variables of lines $\zeta(lj)$.

Distributed controller responsible for computing X^l for $\mathcal{H}^l$ requires information to imitate the rest of the system to achieve globally optimal solutions. This is realized by introducing auxiliary variables at the lines $\zeta(lj)$ of $\mathcal{H}^l$.

Let us assume mn is a line* identified by $\zeta(lj)$ between $\mathcal{H}^l$ and $\mathcal{H}^j$ (See Figure 10.15). Here, n' and m' are auxiliary nodes introduced for the decomposition. In the decomposition approach, the active power ($P_{mn'}$) and reactive power ($Q_{mn'}$) flowing through the boundary node n' in the neighboring subsystem(s) $\mathcal{H}^j$ are represented by virtual entities or auxiliary generators connected to the node n' of $\mathcal{H}^l$ (See Figure 10.15). Here, $\hat{P}_{n'}$ and $\hat{Q}_{n'}$ are the copies of the active, reactive power values shared by $\mathcal{H}^j$ to $\mathcal{H}^l$ following (10.16a) and (10.16b). $\hat{P}_{n'}$ and $\hat{Q}_{n'}$ represent $\mathcal{H}^j$ at node n' of $\mathcal{H}^l$. These values are updated at each inter-subsystem iteration (k) of ADMM. This representation (10.16) is repeated for all lines identified by ζ.

$$\hat{P}_{n'} = P_{np} - P_{kn} \tag{10.16a}$$

$$\hat{Q}_{n'} = Q_{np} - Q_{kn} \tag{10.16b}$$

---

*Multiple lines can exist between any two subsystems.

In Figure 10.15, m and n' are neighboring nodes. Therefore, the coupling variables of $\mathcal{H}^l$ (X^l) and the copies shared by $\mathcal{H}^j$ ($\hat{X}^j$) are as follows

$$X^l_{\zeta(lj)} = [P_{mn'}, Q_{mn'}, V_m, V_{n'}, \hat{P}_{n'}, \hat{Q}_{n'}] \tag{10.17a}$$

$$\hat{X}^j_{\zeta(lj)} = [P_{m'n}, Q_{m'n}, V_{m'}, V_n, (P_{np} - P_{kn}), (Q_{np} - Q_{kn})] \tag{10.17b}$$

The steps of the ADMM algorithm discussed above are provided in Algorithm 10.1.

Algorithm 10.1: ADMM algorithm

Initialization: $\psi^l_{\zeta(lj)}(0), \mu^l_{\zeta(lj)}(0), \rho(0)$

Step 1: Solve

$$X^l(k+1) := \operatorname{argmin}_{X^l} \sum_{l=1}^{L} g_{l(X^l)} + \sum_{l=1}^{L} \sum_{j \in \mathcal{N}^l} [\frac{\rho}{2} \|X^l_{\zeta_{(lj)}} - \psi^l_{\zeta_{(lj)}} + \mu^l_{\zeta_{(lj)}}\|_2^2 - \frac{\rho}{2} \|\mu^l_{\zeta_{(lj)}}\|_2^2] \tag{10.18a}$$

$$\text{Subject to} : h_1(X^l, Y^l) = 0, \forall l \in L \tag{10.18b}$$

$$h_2(X^l, Y^l) \le 0, \forall l \in L \tag{10.18c}$$

Step 2: Update $\psi^l_{\zeta(lj)}$:

$$\psi^l_{\zeta_{(lj)}}(k+1) = \frac{1}{2}(X^l_{\zeta_{(lj)}}(k+1) + \mu^l_{\zeta_{(lj)}}(k) + \hat{X}^j_{\zeta_{(lj)}}(k+1) + \mu^j_{\zeta_{(lj)}}(k)), \\ \forall \mathscr{E} \in \zeta(lj) \tag{10.18d}$$

Step 3: Update $\mu^l_{\zeta(lj)}$:

$$\mu^l_{\zeta_{(lj)}}(k+1) = \mu^l_{\zeta_{(lj)}}(k) + X^l_{\zeta_{(lj)}}(k+1) - \psi^l_{\zeta_{(lj)}}(k+1) \tag{10.18e}$$

Here $\rho > 0$ is known as the penalty coefficient or augmented Lagrangian parameter. In ADMM primal (X^l) and global auxiliary variables (ψ^l) are updated in an alternating or sequential fashion [26]. Here ADMM is written in a slightly different form by combining linear and quadratic terms of the augmented Lagrangian and scaling the dual variables. μ^l is the vector of scaled dual variables for l^{th} subsystem. In Algorithm 10.1, Step 1 (10.18a, 10.18b, and 10.18c) computes primal variables from the augmented Lagrangian of the decomposed problem (10.15). The global auxiliary variables of a subsystem (l) are updated in Step 2 (10.18d) by its own ($X^l_{\zeta_{(lj)}}$) coupling primal variables, shared ($\hat{X}^j_{\zeta_{(lj)}}$) coupling primal variables by the adjacent subsystems of lth subsystem and associated scaled dual variables. The scaled dual variables are updated in Step 3 (10.18e). These steps are repeated till convergence criteria are satisfied.

ADMM-based mixed integer distributed optimization: As stated before, restoration of the distribution grid is a mixed integer programming problem. Exact optimization techniques for mixed integer programming (MIP) problems discussed in the literature [27] are neither scalable nor decomposable and often suffer from high convergence time [28–31]. A popular distributed optimization approach to solve MIP problems is to use a primal-dual iterative approach where integer variables are temporarily relaxed and projected onto integer space. Primal-dual-based optimization approach guarantees optimality at zero duality gap [32]. Relaxation of integer variables transforms the non-convex problem with a non-zero duality gap into a convex problem [29,30]. As ADMM is a dual ascent optimization approach also capable of parallel computation, it has been used to solve distributed MIP problems in many recent works. Integer relaxation and projection of corresponding auxiliary variables onto integer space can obtain solutions close to primal optimum [28,33,34]. In this section, PD-ADMM algorithm proposed in [35] and applied for distributed restoration is discussed.

The steps for the distributed PD-ADMM algorithm for solving mixed integer problem (P1) with L subsystems are provided in Algorithm 10.2.

In this section, we assume every subsystem (l) is solving a mixed integer problem. Here X^l is continuous variables and Y^l is the continuous *relaxation* of integer variables ($\mathcal{Z}^l$) associated with problem solved by subsystem l. The PD-ADMM algorithm consists of optimization jointly over state variables X^l and Y^l. The projection of a continuous variable y onto the integer set $\mathbb{Z}$ is denoted by $\Pi_{\mathbb{Z}}(y)$. This projection/rounding off technique is embedded into the regularization terms by rounding off the copies of global state vectors associated to $Y^l_{\xi_{(jl)}}$, i.e., $\psi^l_{2_{\xi_{(jl)}}}$ to its nearest integer (10.19a). This approach will obtain integer solutions for integer variables located at the overlap regions between l^{th} and adjacent subsystem (e.g., j^{th}). To achieve integer solutions for relaxed integer variables located inside a subsystem (l^{th}) penalty function (ϕ^l) is defined and appended to the augmented Lagrangian of every subsystem (l) (10.19a). Here, x_s is the vector of relaxed integer variables of a subsystem excluding the boundary integer variables (10.19d). The projection of a continuous variable x_s onto the integer set $\mathbb{Z}$ is denoted by $\Pi_{\mathbb{Z}}(x_s)$.

In Step 2, to achieve a fully distributed solution, the global state vectors are updated using the average of the most recently computed local state vectors of the adjacent subsystems (10.19e) and (10.19f). The dual variables are updated as per (10.19g) and (10.19h) in Step 3. Primal and dual residuals (10.19i) and (10.19j) are used to define stopping criteria (10.19k) for PD-ADMM iterations in Step 4. The penalty coefficient is updated following the residual balancing technique (10.19l) [26] in Step 5. Here, γ_1 and γ_2 are algorithmic parameters. In (10.19m), τ is a penalty coefficient of the penalty function which is updated iteratively. The penalty function adds a significant penalty value to the objective function at the non-integer solutions of x_s to drive it close to its corresponding integer values $\Pi_{\mathbb{Z}}(x_s)$.

Algorithm 10.2: PD-ADMM algorithm

Define function: ϕ^l, $\forall l \in L$

Initialization : $\psi^l_{1_{\zeta(lj)}}(0)$, $\psi^l_{2_{\zeta(lj)}}(0)$, $\mu^l_{1_{\zeta(lj)}}(0)$, $\mu^l_{2_{\zeta(lj)}}(0)$, $\rho(0)$

while $(\delta(k) \geq \varepsilon, k \leq k_{max})$ **do**

Step 1: Solve

$$X^l(k+1), Y^l(k+1) = \text{argmin}_{X^l, Y^l} \sum_{l=1}^{L} g_l(X^l, Y^l)$$

$$+ \sum_{l=1}^{L} \sum_{j \in \mathcal{N}^l} \left[\frac{\rho}{2} \|X^l_{\zeta(lj)} - \psi^l_{1_{\zeta(lj)}} + \mu^l_{1_{\zeta(lj)}} \|_2^2 - \frac{\rho}{2} \|\mu^l_{1_{\zeta(lj)}}\|_2^2 \right] \tag{10.19a}$$

$$+ \sum_{l=1}^{L} \sum_{j \in \mathcal{N}^l} \left[\frac{\rho}{2} \|Y^l_{\zeta(lj)} - \Pi_{\mathbb{Z}}(\psi^l_{2_{\zeta(lj)}}) + \mu^l_{2_{\zeta(lj)}} \|_2^2 - \frac{\rho}{2} \|\mu^l_{2_{\zeta(lj)}}\|_2^2 \right] + \tau \phi^l(x_s)$$

Subjectto : $h_1(X^l, Y^l) = 0, \forall l \in L$ $\qquad$ (10.19b)

$h_2(X^l, Y^l) \leq 0, \forall l \in L$ $\qquad$ (10.19c)

Here,

$$\phi^l(x_s) = \|x_s - \Pi_{\mathbb{Z}}(x_s)\|_2^2 \tag{10.19d}$$

Step 2: Update $\psi^l_{1_{\zeta(lj)}}(k)$ and $\psi^l_{2_{\zeta(lj)}}(k)$

$$\psi^l_{1_{\zeta(lj)}}(k+1) = \frac{1}{2}\left(X^l_{\zeta(lj)}(k+1) + \mu^l_{1_{\zeta(lj)}}(k) + X^j_{\zeta(lj)}(k+1) + \mu^j_{1_{\zeta(lj)}}(k) \right), \tag{10.19e}$$
$$\forall \mathscr{E} \in \zeta(lj)$$

$$\psi^l_{2_{\zeta(lj)}}(k+1) = \frac{1}{2}\left(Y^l_{\zeta(lj)}(k+1) + \mu^l_{2_{\zeta(lj)}}(k) + Y^j_{\zeta(lj)}(k+1) + \mu^j_{2_{\zeta(lj)}}(k) \right), \tag{10.19f}$$
$$\forall \mathscr{E} \in \zeta(lj)$$

Step 3: Update $\mu^l_{1_{\zeta(lj)}}(k)$ and $\mu^l_{2_{\zeta(lj)}}(k)$:

$$\mu^l_{1_{\zeta(lj)}}(k+1) = \mu^l_{1_{\zeta(lj)}}(k) + X^l_{\zeta(lj)}(k+1) - \psi^l_{1_{\zeta(lj)}}(k+1) \tag{10.19g}$$

$$\mu^l_{2_{\zeta(lj)}}(k+1) = \mu^l_{2_{\zeta(lj)}}(k) + Y^l_{\zeta(lj)}(k+1) - \psi^l_{2_{\zeta(lj)}}(k+1) \tag{10.19h}$$

Step 4: Compute primal residual (r_p), dual residual (r_d) and define convergence (δ):

$$r_p(k+1) = \|X^l_{\zeta(lj)}(k+1) - \psi^l_{1_{\zeta(lj)}}(k+1)\|_2$$
$$+ \|Y^l_{\zeta(lj)}(k+1) - \psi^l_{2_{\zeta(lj)}}(k+1)\|_2 \tag{10.19i}$$

$$r_d(k+1) = \|\psi^l_{1_{\zeta(lj)}}(k+1) - \psi^l_{1_{\zeta(lj)}}(k)\|_2$$
$$+ \|\psi^l_{2_{\zeta(lj)}}(k+1) - \psi^l_{2_{\zeta(lj)}}(k)\|_2 \tag{10.19j}$$

$$\delta(k+1) = \left\| \begin{pmatrix} r_p(k+1) \\ r_d(k+1) \end{pmatrix} \right\|_{\infty} \leq \varepsilon \tag{10.19k}$$

Step 5: Update $\rho(k)$

$$\rho(k+1) = \begin{cases} \rho(k)(1+\gamma_1) & \text{if} \|r_p(k+1)\|_2 \geq \gamma_2 \|r_d(k+1)\|_2 \\ \dfrac{\rho(k)}{(1+\gamma_1)} & \text{if} \|r_d(k+1)\|_2 \geq \gamma_2 \|r_p(k+1)\|_2 \\ \rho(k) & \text{otherwise} \end{cases} \tag{10.19l}$$

Step 6: if $(\phi^l > 0)$

$$\tau(k+1) = \gamma_3 \tau(k) \tag{10.19m}$$

end while

10.4.2.3 Numerical example

Here we provided a numerical example of centralized and distributed restoration of the distribution grid. In these centralized and distributed algorithms, the restoration of the distribution grid in post-fault conditions is achieved by controlling line switches, load switches, and the generation of DERs [35–37]. The objective function $g(X)$ for restoration is (10.20).

Minimize : $g(X)$

$$= \alpha_1 \left(-\sum_{n \in \mathcal{N}} s_n P_n \right) + \alpha_2 \left(-\sum_{n \in \mathcal{N}^{Der}} P_n^{inj} \right) + \alpha_3 \left(\sum_{n \in \mathcal{N}} \|1 - V_n\|_2^2 \right) \tag{10.20}$$

The objective function tries to optimize load restoration, DER capacity utilization and tries to keep all the bus voltages close to their nominal value. The nominal value of bus voltages is 1 p.u. Here, the global state vector $X = [P_{mn}, Q_{mn}, f_{mn}, V_n, P_n^{inj}, Q_n^{inj}, z_{mn}, z_{nm}, \beta_{mn}, v_n, s_n]$. Here, $P_{mn}(Q_{mn})$ indicates active (reactive) power flow through line mn. V_n indicates the p.u. voltage of nth bus. $P_n^{inj}(Q_n^{inj})$ indicates active (reactive) power injection by the DER connected to n^{th} bus. $P_n(Q_n)$ indicates active (reactive) load connected to n^{th} bus. The status of the line switch of line mn and the status of load switch of bus n are indicated by integer variables β_{mn} and s_n, respectively. Here, α_1, α_2 and α_3 are constants.

Details of the decomposable centralized restoration algorithm of the distribution grid are discussed in [35]. The efficiency of the PD-ADMM algorithm in the restoration of the distribution grid is evaluated in IEEE 123-bus system. The rated power and rated voltage of the system are 1 MVA and 4.16 kV, respectively. The minimum and maximum voltage magnitude square limits are 0.95 and 1.1 p.u., respectively. Substation 1 and 2 are operating at 1.0 p.u. voltage (see Figure 10.16). All line switches are marked in Figure 10.16. The total number of controllable line switches present in the system is 46

322   *Control, communication, monitoring and protection of smart grids*

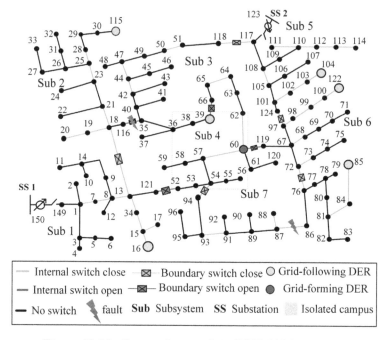

Figure 10.16   Restoration result in IEEE 123-bus system

and all loads are switchable. All the DERs connected to the network are operating in the grid forming (GFM) or grid following (GFL) mode [35–37]. DERs are connected to buses 16, 39, 57, 60, 85, 104, 115, and 122 with total MVA capacity of 1.6 MVA or nearly 40% of the total MVA demand. The GFM-DER connected to the bus 60 is operating at voltage 1.0 p.u. The GFM-DER primarily supplies power to subsystem 4 (isolated campus), however, if required the excess generation is supplied to the neighboring subsystems. For implementation of the distributed PD-ADMM algorithm, in this study IEEE 123-bus system is divided into seven subsystems/holons. Switches (13−18), (35−116), (117−118), (39−66), (52−121), (54−94), (97−124), (60−119) and (72−76) form the boundaries or overlap regions of these subsystems (see Figure 10.16). A contingency scenario with two faulted lines (86−87) and (35−116) is studied to evaluate the performance of centralized and distributed algorithms in restoration of distribution grid. The fault at the line (86−87) is internal to the subsystem 7 whereas the fault at (35−116) is an outage of a boundary switch. These two faults are cleared by opening respective lines. As shown in Figure 10.16 the network is restored in the case of both centralized and distributed approaches by closing switches (13−18), (117−118), (97−124), (72−76), and (54−94) and opening switches (39−66), (52−121), and (60−119). As per the topology of the restored network in Figure 10.16, the voltage level of the subsystem 1 and 2 are regulated by the substation 1, voltage level of subsystem 3 and 5 are regulated by substation 2 and the voltage level of subsystem 4, 6 and 7 are regulated by the GFM-DER. The convergence of the boundary integer, continuous variables, and residual term of PD-ADMM algorithm is illustrated in

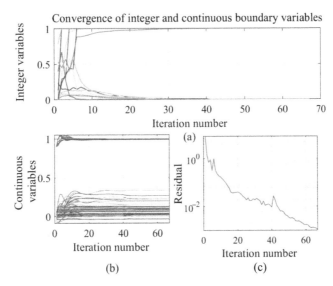

Figure 10.17 *Performance of PD-ADMM in the restoration of an unbalanced IEEE-123 bus system: convergence of a) integer variables b) continuous variables c) residual term*

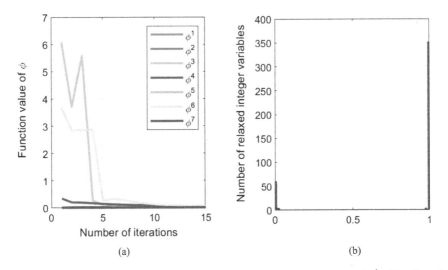

Figure 10.18 *Penalty functions of PD-ADMM: (a) convergence of ϕ^l, (b) relaxed integer variables at the point of convergence*

Figure 10.17. The convergence of penalty functions (ϕ) of all seven subsystems is illustrated in Figure 10.18a. The numerical values of relaxed integer variables at the point of convergence of PD-ADMM algorithm are illustrated in Figure 10.18b. A comparison between optimum control variables for the centralized and distributed solutions is provided in Table 10.1.

Table 10.1   Comparison results for ε = 0.001

Optimization	Outage	F^{Der}(p.u.)		Load shedding (p.u.)	Substation power (p.u.)
Centralized	35-116 86-87	$P^{inj}_a = 0.4901$	0.0		$P^{inj}_a = 0.9299$
		$P^{inj}_b = 0.5051$	0.0		$P^{inj}_b = 0.4099$
		$P^{inj}_c = 0.441$	0.0		$P^{inj}_c = 0.7149$
Distributed	35-116 86-87	$P^{inj}_a = 0.4907$	0.0		$P^{inj}_a = 0.9293$
		$P^{inj}_b = 0.5057$	0.0		$P^{inj}_b = 0.4093$
		$P^{inj}_c = 0.4407$	0.0		$P^{inj}_c = 0.7143$

In this case study the configuration of the restored network identified by both centralized and distributed restoration algorithms is the same. Every bus in the restored network is connected to a source bus and the restored network is radial. The presence of the isolated campus also shows the ability of both centralized and distributed optimization algorithms in the formation of partial islands. The power injections by DERs and substations computed by both centralized and distributed algorithms are within ±0.001. Convergence of penalty functions in the finite number of iterations shows the efficiency of PD-ADMM approach in identifying integer solutions of relaxed integer variables.

10.5   Summary and path forward

10.5.1   Summary

In the future, the continued integration of EVs and DERs into distribution grids presents challenges and opportunities for novel control architecture. Further exploration is needed to address the complex interplay between transportation and communication layers, offering solutions that enhance the seamless integration of DERs and EVs into the grid. The application of holonic control, introduced in this chapter, opens the door to a dynamic, adaptive, and resilient framework for managing the evolving challenges posed by the convergence of transportation, cyber, and power systems.

The inherent flexibility of holonic systems lies in their smooth transition among these different control modes, allowing them to adjust effortlessly to the demands of various operational conditions and meet specific requirements for autonomy and collaboration. Two distinct operational modes: "Economic," for cost reduction, and "Resilient," involving a multi-step spatial-temporal mode are needed to allow gracefully transition among control modes based on events.

Holonic architecture will be important to offer conflicting objectives of scalability, efficiency, reliability, optimally, economics, and resiliency for a DER-rich distribution grid by switching among different architectures with their own advantages and limitations. The information, sensor network, and computing need to support the holonic architecture to enable a reliable and agile distribution grid. Implementation of the holonic architecture nevertheless has several challenges as discussed in the next subsection.

10.5.2 Path forward

One crucial aspect is the need for thorough performance analysis. As holonic control systems become more complex and interconnected, assessing their performance becomes paramount. Future research should focus on developing comprehensive methodologies and metrics to evaluate the efficiency, reliability, and scalability of holonic control architectures. Understanding the performance implications of various configurations and interactions within holonic systems will contribute to their optimization and practical deployment.

Addressing cybersecurity concerns is another critical area for the future of holonic control. With the increasing integration of digital technologies, holonic systems become susceptible to cyber threats. Research efforts should focus on enhancing the security measures within holonic control architectures, ensuring robust protection against cyber attacks and unauthorized access. Developing resilient and secure communication protocols, encryption methods, and intrusion detection systems will be crucial for the successful implementation of holonic control in cyber-physical systems.

Another future need for holonic control is addressing delays in communication and decision-making processes. Real-world systems often encounter delays in information transmission and processing. Future research should explore innovative solutions to minimize latency within holonic control systems, ensuring timely and responsive decision-making.

Noise and uncertainties in the system pose additional challenges for holonic control. Future research should focus on developing strategies to handle noise and uncertainties in sensor data, communication channels, and decision-making processes. Robust control algorithms, machine learning techniques, and advanced filtering methods can help holonic systems maintain stability and performance in the presence of uncertainties.

Lastly, the future of holonic control also requires attention to the hardware needed for implementation. As holonic systems scale up and become integral parts of critical infrastructures, ensuring that the hardware infrastructure can support the computational demands and communication requirements becomes crucial. Future research should explore hardware architectures that optimize the performance of holonic control systems, considering factors such as adaptability, dynamic database management, scalability, and fault tolerance.

Acknowledgments

The authors would like to acknowledge partial support from the NSF CPS, US DOE RACER, US DOE SolarTestbed, and US DOE GRIP awards to support this work.

References

[1]   V. V. G. Krishnan, S. Gopal, Z. Nie, and A. Srivastava. Cyber-power testbed for distributed monitoring and control. In *2018 Workshop on Modeling and Simulation of Cyber-Physical Energy Systems (MSCPES)*, pp. 1–6. Piscataway, NJ: IEEE, 2018.

[2]   H. Lopes Ferreira, A. Costescu, A. L'Abbate, P. Minnebo, and G. Fulli. Distributed generation and distribution market diversity in Europe. *Energy Policy*, 39(9):5561–5571, 2011.

[3]   C. B. Vellaithurai, S. S. Biswas, R. Liu, and A. Srivastava. Real time modeling and simulation of cyber-power system. In: Khaitan, S., McCalley, J., and Liu, C. (eds), *Cyber Physical Systems Approach to Smart Electric Power Grid* (pp. 43– 74). Berlin, Heidelberg: Springer; 2015.

[4]   A. Ahmed, S. Basumallik, A. Gholami, *et al*. Spatio-temporal deep graph network for event detection, localization and classification in cyber-physical electric distribution system. *IEEE Transactions on Industrial Informatics*, 20 (2):2397–2407, 2023.

[5]   S. Majumder, N. Patari, A. K. Srivastava, P. Srivastava, and A. M. Annaswamy. Epistemology of voltage control in der-rich power system. *Electric Power Systems Research*, 214:108874, 2023.

[6]   S. Majumder, A. Vosughi, H. M. Mustafa, T. E. Warner, and A. K. Srivasava. On the cyber-physical needs of der-based voltage control/optimization algorithms in active distribution network. *IEEE Access*, 11:64397– 64429, 2023.

[7]   H. M. Mustafa, M. Bariya, K. S. Sajan, et al. RT-meter: A realtime, multi-layer cyber-power testbed for resiliency analysis. In *Proceedings of the 9th Workshop on Modeling and Simulation of Cyber-Physical Energy Systems*, pp. 1–7, 2021.

[8]   A. Srivastava, T. Morris, T. Ernster, C. Vellaithurai, S. Pan, and U. Adhikari. Modeling cyber-physical vulnerability of the smart grid with incomplete information. *IEEE Transactions on Smart Grid*, 4(1):235–244, 2013.

[9]   I. Jawhar, J. Al-Jaroodi, H. Noura, and N. Mohamed. Networking and communication in cyber physical systems. In *2017 IEEE 37th International Conference on Distributed Computing Systems Workshops (ICDCSW)*, pp. 75–82. Piscataway, NJ: IEEE, 2017.

[10] K. E. Antoniadou-Plytaria, I. N. Kouveliotis-Lysikatos, P. S. Georgilakis, and N. D. Hatziargyriou. Distributed and decentralized voltage control of smart distribution networks: Models, methods, and future research. *IEEE Transactions on Smart Grid*, 8(6):2999–3008, 2017.

[11] L. M. B. A. Dib, V. Fernandes, M. L. Filomeno, and M. V. Ribeiro. Hybrid plc/wireless communication for smart grids and Internet of Things applications. *IEEE Internet of Things Journal*, 5(2):655– 667, 2018.

[12] D. A. Zakoldaev, A. V. Gurjanov, I. O. Zharinov, and O. O. Zharinov. Centralized, decentralized, hybrid principles of cyber-physical systems

control of the industry 4.0. *Journal of Physics: Conference Series*, 1353:012143, 2019.

[13] P. S. Sarker, S. Majumder, M. F. Rafy, and A. K. Srivastava. Impact analysis of cyber-events on distributed voltage control with active power curtailment. In *2022 IEEE International Conference on Power Electronics, Drives and Energy Systems (PEDES)*, pp. 1–6, 2022.

[14] M. H. Nazari, S. Grijalva, M. Egerstedt, et al. Communication-failure-resilient distributed frequency control in smart grids: Part I: Architecture and distributed algorithms. *IEEE Transactions on Power Systems*, 35(2):1317–1326, 2019.

[15] P. Srivastava, R. Haider, V. J. Nair, V. Venkataramanan, A. M. Annaswamy, and A. K. Srivastava. Voltage regulation in distribution grids: A survey. *Annual Reviews in Control*, 55:165–181, 2023.

[16] A. Koestler. *The Ghost in the Machine*. New York: Macmillan, 1968.

[17] S. Ionita. Multi agent holonic based architecture for communication and learning about power demand in residential areas. In *2009 International Conference on Machine Learning and Applications*, pp. 644–649. Piscataway, NJ: IEEE, 2009.

[18] M. H. Moghadam and N. Mozayani. Research article a novel information exchange model in it infrastructure of smart grid. *Research Journal of Applied Sciences, Engineering and Technology*, 6(23):4399–4404, 2013.

[19] A. Pahwa, S. A. DeLoach, S. Das, *et al.* Holonic multi-agent control of power distribution systems of the future. *CIGRE Grid of the Future Symposium*, 2012. Available from: https://cigre-usnc.org/wp-content/uploads/2017/01/4-C6_Pahwa.pdf.

[20] N. Higgins, V. Vyatkin, N. K. C. Nair, and K. Schwarz. Distributed power system automation with IEC 61850, IEC 61499, and intelligent control. *IEEE Transactions on Systems, Man, and Cybernetics, Part C (Applications and Reviews)*, 41(1):81–92, 2011.

[21] Y. Du, H. Tu, S. Lukic, D. Lubkeman, A. Dubey, and G. Karsai. Implementation of a distributed microgrid controller on the resilient information architecture platform for smart systems (RIAPS). In *2017 North American Power Symposium (NAPS)*, pp. 1–6. Piscataway, NJ: IEEE, 2017.

[22] S. Eisele, I. Mardari, A. Dubey, and G. Karsai. RIAPS: Resilient information architecture platform for decentralized smart systems. In *2017 IEEE 20th International Symposium on Real-Time Distributed Computing (ISORC)*, pp. 125–132. Piscataway, NJ: IEEE, 2017.

[23] V. Kekatos, L. Zhang, G. B. Giannakis, and R. Baldick. Voltage regulation algorithms for multiphase power distribution grids. *IEEE Transactions on Power Systems*, 31(5), 3913–3923, 2016.

[24] J. A. Taylor and F. S. Hover. Convex models of distribution system reconfiguration. *IEEE Transactions on Power Systems*, 27(3):1407–1413, 2012.

[25] F. U. Nazir, B. C. Pal, and R. A. Jabr. Distributed solution of stochastic volt/var control in radial networks. *IEEE Transactions on Smart Grid*, 11 (6):5314–5324, 2020.

[26] S. Boyd, N. Parikh, E. Chu, B. Peleato, and J. Eckstein. *Distributed Optimization and Statistical Learning via the Alternating Direction Method of Multipliers*. Boston, MA: Now Publishers Inc., 2011.

[27] P. Bonami, M. Kılınç, and J. Linderoth. Algorithms and software for convex mixed integer nonlinear programs. In: Lee, J. and Leyffer, S. (eds), *Mixed Integer Nonlinear Programming. The IMA Volumes in Mathematics and its Applications*, vol. 154. New York, NY: Springer, 2009.

[28] S. Diamond, R. Takapoui, and S. Boyd. A general system for heuristic minimization of convex functions over non-convex sets. *Optimization Methods and Software*, 33(1):165–193, 2018.

[29] Z. Liu and O. Stursberg. Efficient solution of distributed MIP in control of networked systems. *Proceedings in Applied Mathematics and Mechanics*, 20 (1), e202000160, 2021.

[30] Z. Liu and O. Stursberg. Distributed solution of mixed-integer programs by admm with closed duality gap. In *2022 IEEE 61st Conference on Decision and Control (CDC)*, pp. 279–286, 2022.

[31] B. Stellato, V. V. Naik, A. Bemporad, P. Goulart, and S. Boyd. Embedded mixed-integer quadratic optimization using the OSQP solver. In *2018 European Control Conference (ECC)*, pp. 1536–1541, 2018.

[32] S. Boyd and L. Vandenberghe. *Convex Optimization*. Cambridge: Cambridge University Press, 2004.

[33] A. Alavian and M. C. Rotkowitz. Improving admm-based optimization of mixed integer objectives. In *51st Annual Conference on Information Sciences and Systems (CISS)*, pp. 1–6, 2017.

[34] R. Takapoui, N. Moehle, S. Boyd, and A. Bemporad. A simple effective heuristic for embedded mixed-integer quadratic programming. In *2016 American Control Conference (ACC)*, pp. 5619–5625, 2016.

[35] S. Konar, A. K. Srivastava, and A. Dubey. Distributed optimization for autonomous restoration in DER-rich distribution system. *IEEE Transactions on Power Delivery*, 38(5):3205–3217, 2023.

[36] S. Konar and A. K. Srivastava. Look – ahead corrective restoration for microgrids with flexible boundary. In *2021 North American Power Symposium (NAPS)*, pp. 1–6, 2021.

[37] S. Konar and A. K. Srivastava. MPC-based black start and restoration for resilient der-rich electric distribution system. *IEEE Access*, 11:69177–69189, 2023.

Chapter 11

Condition monitoring and health prognosis applications in smart grids

B. Sivaneasan[1], D.R. Thinesh[1], Z. Yunyi[1], K.T. Tan[1], A.K. Rathore[1] and K.J. Tseng[1]

The traditional power system has undergone a remarkable evolution over the years, transitioning into a smart grid that has revolutionized the way we generate, distribute, and consume electricity. The shift from a centralized and one-way flow of power to a decentralized, bidirectional network has brought about numerous benefits and opportunities for efficiency, reliability, and sustainability. One of the key drivers of this evolution is the integration of advanced technologies and digital communication systems. Smart grids leverage sophisticated sensors, meters, and control devices that enable real-time monitoring and optimization of power generation, distribution, and consumption. These technologies provide granular data and insights, allowing utilities and consumers to make informed decisions regarding energy usage.

Furthermore, the smart grid facilitates the integration of renewable energy sources (RES) into the power system. With the increasing adoption of solar panels, wind turbines, and other clean energy technologies, the traditional grid faced challenges in managing intermittent generation and balancing supply and demand. The smart grid, however, enables seamless integration of renewables by dynamically adjusting the grid's operation based on real-time data, optimizing the utilization of clean energy sources, and reducing reliance on fossil fuels.

Another significant aspect of the smart grid is its ability to empower consumers and promote energy efficiency. Advanced metering infrastructure and home energy management systems (EMS) allow consumers to monitor their energy usage in real time, enabling them to make more conscious decisions about when and how they consume electricity. This, in turn, helps reduce overall demand, lower costs, and alleviate strain on the grid during peak periods.

Moreover, the smart grid enhances the resilience and reliability of the power system. By utilizing two-way communication and automated monitoring, the grid can quickly identify and isolate faults, minimizing downtime and improving

[1]Engineering Cluster, Singapore Institute of Technology, Singapore

response times. Additionally, the ability to reroute power and balance loads dynamically ensures uninterrupted supply, even in the face of disruptions or disasters.

The global outlook for smart grid implementation is promising and shows a positive trajectory. Governments, utilities, and industry stakeholders worldwide recognize the potential of smart grids in addressing various energy challenges, enhancing efficiency, and promoting sustainability. As a result, we are witnessing an increasing number of countries and regions actively investing in and deploying smart grid technologies.

North America has been at the forefront of smart grid implementation, with significant investments and initiatives in the United States and Canada. The U.S. Department of Energy has been supporting research and development projects, and utility companies are rolling out smart meters and advanced grid infrastructure. Canada has also made significant strides in integrating smart grid technologies, particularly in provinces like Ontario and British Columbia.

Europe has been a frontrunner in promoting sustainable energy practices and has set ambitious targets for renewable energy generation. Countries like Germany, Spain, and the Netherlands have made substantial investments in smart grid infrastructure, focusing on integrating large-scale RES and improving grid efficiency. The European Union has also established regulations and funding programs to accelerate the deployment of smart grids across member states.

Asia Pacific, with its rapid urbanization and expanding energy demand, presents immense opportunities for smart grid implementation. China, for instance, has been heavily investing in smart grid projects to modernize its electricity infrastructure and accommodate its growing population's energy needs. Japan, South Korea, and Australia are also actively pursuing smart grid initiatives to improve energy efficiency and grid reliability.

In the Middle East, countries like Saudi Arabia and the United Arab Emirates are investing in smart grid technologies to reduce reliance on fossil fuels and promote sustainability. These countries are leveraging advanced grid management systems, smart meters, and demand response programs to optimize energy usage and integrate RES. Africa and Latin America are also witnessing increased interest in smart grid implementation. Several countries in these regions are deploying smart grid technologies to improve access to electricity, reduce transmission losses, and optimize energy distribution.

Overall, the global outlook for smart grid implementation is positive, with numerous countries recognizing the benefits of these systems in achieving energy security, reducing emissions, and enhancing grid resilience. Continued investments, regulatory support, and collaboration between stakeholders will play crucial roles in driving the widespread adoption of smart grid technologies worldwide.

Condition monitoring is an essential component of a smart grid. The implementation of condition monitoring techniques and systems is crucial to ensure the optimal performance, reliability, and longevity of the smart grid infrastructure. This is because, in a smart grid, numerous electrical assets are integrated, including transformers, circuit breakers, sensors, meters, and communication devices. These assets are critical for the proper functioning of the smart grid and need to be monitored continuously to detect any potential issues or abnormalities.

As shown in Figure 11.1, corrective, preventive, and predictive maintenance (PdM) are three distinct maintenance strategies used to manage the health and reliability of equipment and assets. Each strategy aims to address maintenance needs in different ways, depending on the specific goals and circumstances of the organization. Corrective maintenance, also known as "reactive maintenance" or "breakdown maintenance," involves repairing equipment or assets after they have failed or malfunctioned. In this strategy, maintenance actions are taken in response to an unexpected failure or performance issue. The goal of corrective maintenance is to restore the equipment to its normal operating condition and minimize downtime. Preventive maintenance involves performing maintenance tasks on a scheduled basis, regardless of whether the equipment is showing signs of failure. The goal is to prevent failures before they occur and to keep the equipment in good working condition. Preventive maintenance tasks may include inspections, lubrication, cleaning, component replacements, and other routine activities. PdM is a data-driven approach that uses real-time monitoring and condition-based information to predict when maintenance should be performed. It involves collecting data from sensors, monitoring equipment, and analyzing trends to identify potential issues before they result in failure. Maintenance actions are taken only when the data indicates that they are needed.

Traditionally, condition assessment or maintenance of power system electrical assets typically follows a scheduled or preventive approach. It involves regular inspections, maintenance routines, and equipment replacements based on pre-defined time intervals or usage thresholds.

Here are some common maintenance practices in a traditional power system:

1. **Scheduled inspections**: Utility companies perform routine inspections of electrical equipment such as transformers, circuit breakers, switches, and

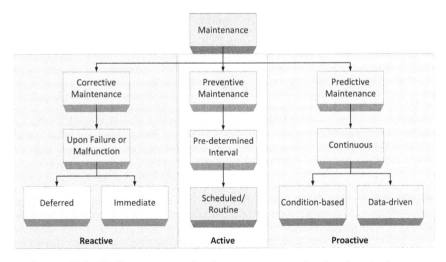

Figure 11.1 Different types of maintenance strategies for electrical assets

cables. These inspections involve visual checks, testing, and measurements to assess the condition and performance of the equipment. Common tests and measurements that are conducted during routine inspections are insulation resistance test, dissolved gas analysis (DGA), circuit breaker timing test, thermal imaging, and partial discharge (PD) testing.

2. **Lubrication and cleaning**: Moving parts of equipment, such as rotating machinery or switches, often require regular lubrication and cleaning to ensure proper operation and reduce friction or corrosion.
3. **Calibration and testing**: Instruments and meters used for monitoring and measurement in the power system are calibrated and tested periodically to maintain their accuracy and reliability.
4. **Preventive replacements**: Certain components in the power system, such as insulators or surge arrestors, may have a predetermined lifespan or recommended replacement interval. These components are replaced on a preventive basis to prevent failures and ensure the system's continued operation.
5. **Overhaul or refurbishment**: Some larger equipment, like generators or transformers, may undergo planned overhauls or refurbishments after a certain number of operating hours or years. These activities involve a comprehensive inspection, repair, and replacement of worn-out or damaged parts.
6. **Emergency maintenance**: In cases of unexpected failures or malfunctions, utilities perform emergency maintenance to restore the system's operation promptly. This may involve troubleshooting, repair, or replacement of faulty equipment.

It is important to note that maintenance practices can vary based on the specific equipment, regulatory requirements, and the utility's maintenance strategy.

The introduction of smart grid technologies and condition monitoring can augment these maintenance practices by providing real-time data and enabling a more proactive and targeted approach to maintenance. This involves the use of sensors, data analytics, and advanced monitoring technologies to gather real-time information about the condition, performance, and health of the different electrical assets. This monitoring allows utilities and grid operators to detect and diagnose potential problems, such as insulation degradation, overheating, excessive vibration, or abnormal behavior, before they lead to equipment failures or grid disruptions.

Monitoring the condition of smart grid electrical assets is crucial for several reasons. First, condition monitoring allows for the early detection of potential failures or malfunctions in electrical assets. By continuously monitoring the condition of transformers, switchgear, cables, solar panels, energy storage devices, and other critical components, utilities can identify signs of degradation or abnormal behavior before they escalate into major issues. This proactive approach helps prevent costly and disruptive equipment failures, minimizing downtime and reducing the risk of power outages.

Second, condition monitoring enables prognostic assessment of asset health condition, thus leading to PdM strategies. By analyzing real-time data collected from smart grid assets, utilities can gain insights into the performance and health of the equipment. This information allows them to schedule maintenance activities based on the actual condition of the assets, rather than relying on fixed time-based schedules. PdM reduces the likelihood of unnecessary maintenance, optimizes resource allocation, and extends the lifespan of electrical assets, resulting in cost savings and improved operational efficiency.

Furthermore, condition monitoring supports optimal asset utilization. By monitoring the performance and condition of electrical assets, utilities can better understand their capacity and capabilities. This knowledge helps utilities make informed decisions about load balancing, asset upgrades, and capacity planning. It allows them to maximize the utilization of existing assets, avoid overloading or underutilization, and optimize the overall grid operation.

Condition monitoring and prognostic health assessment also enhance grid reliability and resilience. By continuously monitoring the condition of electrical assets, utilities can identify potential weaknesses or vulnerabilities in the system. This enables proactive measures to be taken, such as asset replacement or reinforcement, to prevent failures and ensure grid stability. By maintaining a reliable and resilient grid, utilities can minimize disruptions, improve customer satisfaction, and support critical services and industries that depend on uninterrupted power supply.

In summary, condition monitoring of smart grid electrical assets is important because it enables the early detection of potential failures, facilitates PdM, optimizes asset utilization, and enhances grid reliability and resilience. By proactively monitoring and maintaining the health of electrical assets, utilities can improve operational efficiency, reduce downtime, and deliver reliable and high-quality electricity services to consumers.

11.1 Condition monitoring and health prognosis framework

As of 2023, there is no specific universally adopted standard or regulation exclusively focused on condition monitoring in the smart grid. However, several existing standards and guidelines provide recommendations and best practices for condition monitoring within the broader context of the smart grid. For example, ISO 17359 provides general guidelines on the condition monitoring and diagnostics of machines while ISO 13373 focuses on mechanical vibration and shock for vibration condition monitoring of machines. One particular framework to note is ISO 13374 titled "Condition monitoring and diagnostics of machines—Data processing, communication, and presentation—Part 1: General guidelines," which provides general guidelines for data processing, communication, and presentation in the context of condition monitoring and diagnostics of machines [1]. As shown in Figure 11.2, ISO 13374 framework outlines a structured machine condition assessment approach, which comprises seven well-defined processing blocks organized in layers. The first four blocks are technology-specific and involve utilizing specialized signal processing and data analysis functions tailored to the

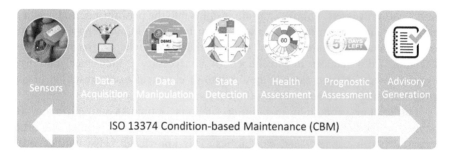

Figure 11.2   ISO 13374 framework for condition-based maintenance

particular technology used. The remaining three blocks focus on the integration of multiple monitoring technologies to evaluate the machine's present health, predict potential future failures, and provide actionable recommendations for operations and maintenance personnel. It's worth noting that regulations and requirements related to condition monitoring in the smart grid can vary across different countries and regions. Regulatory bodies and industry organizations may develop specific guidelines or standards based on local needs and priorities.

The framework for condition-based monitoring (CBM) of asset health typically involves a systematic approach to collecting, analyzing, and utilizing data to assess the condition of assets and make informed maintenance decisions. While specific frameworks can vary depending on the industry and asset type, here is a general outline of the key components of a CBM framework:

1. **Asset selection**: Identify the critical assets within the system that require monitoring based on their impact on operational performance, safety, reliability, or cost implications.
2. **Sensor deployment**: Determine the appropriate sensors and monitoring devices needed to collect relevant data about the asset's condition. This may include sensors for measuring parameters such as temperature, vibration, pressure, current, or fluid levels.
3. **Data acquisition**: Install data acquisition systems to collect real-time or periodic measurements from the deployed sensors. This can involve the use of data loggers, SCADA systems, or other monitoring platforms to capture and store the data.
4. **Data management**: Establish a data management system to organize, store, and process the collected data. This may include data storage infrastructure, databases, and data management software to enable efficient data retrieval and analysis.
5. **Data analysis**: Apply various data analysis techniques to extract meaningful insights from the collected data. This can involve statistical analysis, trend analysis, pattern recognition, machine learning, or other advanced analytics methods to identify anomalies, patterns, or potential failure signatures.
6. **Condition assessment**: Evaluate the asset's condition based on the analysis results. This involves comparing the observed data against predefined

thresholds, historical data, or established models to determine if the asset's health is within acceptable limits or if maintenance actions are required.

7. **Maintenance decision-making**: Utilize the condition assessment results to make informed maintenance decisions. This can include determining the appropriate maintenance actions, such as corrective, preventive, or PdM, based on the identified asset condition and associated risks.

8. **Maintenance execution**: Implement the selected maintenance actions, which can involve repairs, component replacements, or adjustments based on the identified asset condition. Ensure proper documentation and record-keeping of the maintenance activities performed.

9. **Performance evaluation**: Continuously monitor and evaluate the effectiveness of the CBM program by tracking key performance indicators (KPIs) such as asset uptime, maintenance costs, reliability metrics, or other relevant indicators. Use the performance evaluation results to identify areas for improvement and optimize the CBM strategy.

10. **Continuous improvement**: Periodically review and refine the CBM framework based on feedback, lessons learned, and technological advancements. Incorporate new monitoring techniques, data analysis methods, or industry best practices to enhance the effectiveness of the CBM program over time.

This framework provides a structured approach to implementing CBM of asset health. However, it's important to adapt the framework to the specific requirements and characteristics of the assets and industry in question. This includes the types of sensors, data collection intervals, types of assets, and the corresponding anomalies.

11.2 Condition monitoring and health prognosis of smart grid assets

In this chapter, we focus on three crucial assets in a smart grid, namely, substation, solar photovoltaics (PV), and energy storage systems (ESS). These three assets hold paramount significance due to their pivotal roles in ensuring a resilient and efficient electricity network. Substations act as key nodes for power transformation, control, and distribution, making their reliable operation crucial for grid stability. Solar PV systems are integral to the growing adoption of RES, and their efficient functioning is vital for maximizing clean energy generation. ESS play a critical role in grid balancing, enabling the integration of intermittent renewables and enhancing grid resiliency. By implementing robust condition monitoring strategies for these assets, the smart grid can optimize maintenance, detect potential issues early, and bolster overall grid performance, fostering a sustainable and reliable energy ecosystem.

11.2.1 Substation

In a power system, substations play a crucial role in facilitating the transformation, control, and distribution of electricity. In general, a substation is a facility where electricity is transformed from one voltage level to another. It acts as an intermediate point between the power generation sources (power plants) and the distribution network that delivers electricity to consumers. Substations typically consist of transformers, circuit breakers, disconnect switches, busbars, and protective relays. Transformers are used for voltage transformation, while circuit breakers and disconnect switches are employed for control and protection purposes. Substations are usually situated at strategic points in the power grid, such as near power plants, along transmission lines, and at distribution centers. Substations play a critical role in maintaining grid stability, ensuring efficient power transfer, and providing a point of control for grid operators.

In the context of a smart grid, the role and importance of a substation are amplified due to the integration of various RES and distributed generation where the power flow can be highly dynamic and unpredictable. Substations help manage these fluctuations and ensure that the generated power is transformed and transmitted at appropriate voltage levels for efficient and reliable distribution. Smart grid substations are designed for remote control and automation. This allows grid operators to remotely operate switches, circuit breakers, and other devices in the substation, reducing the need for physical intervention and enhancing grid reliability. In addition, smart grid substations facilitate demand response programs by enabling bidirectional communication between consumers and the grid. They can receive signals from the grid operator to curtail or increase electricity consumption during peak or off-peak periods, respectively, contributing to grid stability and energy efficiency.

11.2.1.1   Maintenance and condition monitoring of substations asset

Maintenance of substations involves several tests to ensure their proper functioning and safety. These tests help identify potential issues, prevent failures, and optimize performance. Some common tests performed for maintenance of high voltage substations include:

1. **Visual inspection**: Regular visual inspections are conducted to identify any visible signs of damage, loose connections, or abnormalities in equipment such as circuit breakers, transformers, insulators, and busbars.
2. **Infrared thermography**: This test uses thermal imaging cameras to detect hotspots in electrical components, which may indicate loose connections or excessive electrical resistance. Identifying these hotspots can help prevent equipment failures and potential hazards.
3. **PD testing**: PD testing is carried out to detect PDs within the insulation of high-voltage equipment. It helps to assess the condition of insulation and identify potential insulation weaknesses.
4. **Dielectric withstand test (hipot test)**: The hipot test is performed to check the insulation integrity of electrical equipment by applying a high voltage between the equipment and its enclosure or between different phases.
5. **Circuit breaker testing**: Various tests are conducted on circuit breakers, including timing tests, contact resistance measurements, and insulation resistance tests to ensure their proper functioning during fault conditions.

6. **Transformer testing**: Transformer testing involves assessing the condition of transformers through tests such as turns ratio test, winding resistance measurement, and insulation resistance test.

7. **Oil analysis**: For oil-filled equipment like transformers, regular oil sampling and analysis are performed to check for any degradation or contamination, which could indicate potential problems.

8. **Power factor testing**: This test is carried out to measure the power factor of the equipment, which helps assess the insulation condition and the presence of any defects.

9. **Ground grid integrity testing**: The ground grid of the substation is tested to ensure proper grounding and to detect any issues that may affect the safety or the performance of protective devices.

10. **Control and protection system testing**: Functional tests are conducted on the control and protection systems to verify their proper operation and coordination.

11. **Battery system testing**: The battery systems used for backup power in substations are tested to ensure their capacity and reliability during power outages.

12. **Electromagnetic compatibility (EMC) testing**: EMC testing is performed to assess the electromagnetic interference levels to ensure the proper operation of sensitive equipment and avoid interference with nearby systems.

These tests are typically conducted as part of a comprehensive preventive maintenance program and are tailored to the specific equipment and requirements of the high-voltage and low-voltage substations. Regular maintenance and testing help ensure the substation's reliability, safety, and longevity.

Online condition monitoring systems can measure and monitor these parameters to assess the health and performance of substation components in real time. These parameters provide critical data for PdM, early fault detection, and optimization of substation operations. Some of the different parameters that a substation online condition monitoring system can measure and monitor include:

- **Voltage and current:**
  - Continuous monitoring of voltage and current waveforms helps ensure that the substation is operating within the desired parameters and that the electrical load is balanced correctly.
  - Continuous monitoring of current levels can detect instances of overloading in transformers and switchgear. Overloads can lead to overheating and potential equipment damage, so early detection allows for timely corrective actions.
  - Trends in voltage and current data can provide valuable insights into the condition of insulation and winding health of transformers.
  - Voltage and current measurements provide insights into the efficiency of transformers and switchgear components. Deviations from expected efficiency levels can indicate issues with equipment performance.
  - Unusual voltage and current patterns can be indicative of internal faults or insulation degradation in transformers.

- **Temperature:**
  - Monitoring the temperature of switchgear components and transformer windings helps detect overheating. Overheating is often an early indication of potential issues such as loose connections, high resistance, or insulation degradation.
  - Temperature measurements help assess the thermal stress experienced by switchgear and transformers during normal operation and load changes. Monitoring thermal stress aids in identifying potential hotspots and critical areas prone to failure.
  - For large transformers and some switchgear, temperature measurements are essential for assessing the performance of cooling systems such as fans or radiators. Deviations from normal temperature values may indicate cooling system malfunctions.
  - Continuous temperature monitoring helps identify hotspots in transformers and switchgear, enabling timely corrective actions to prevent damage. Hotspots are localized areas with significantly higher temperatures than the surrounding components.
  - Temperature measurements can help evaluate the performance of on-load tap changers in transformers. Deviations in temperature during tap changer operation may indicate mechanical issues or abnormal load distribution.

- **Oil quality:**
  - The insulating oil in transformers and switchgear provides electrical insulation and helps dissipate heat. Monitoring the oil's quality, including its dielectric strength and breakdown voltage, can indicate the condition of the insulation and detect any degradation.
  - Over time, insulating oil can become contaminated with water, particulates, and other impurities. Oil quality measurements help detect the presence of these contaminants, which can affect the equipment's performance and lead to premature failure.
  - Insulating oil can undergo oxidation due to exposure to high temperatures and oxygen. Oxidation leads to the formation of sludge and acids that can deteriorate the equipment's internal components. Oil quality measurements can detect signs of oxidation and guide timely oil replacement or treatment.
  - Moisture ingress into insulating oil can lead to reduced dielectric strength and insulation performance. Monitoring oil quality for moisture content helps identify potential sources of leakage and prevent insulation breakdown.
  - DGA is a crucial oil quality measurement technique for transformers. DGA helps detect and quantify the concentration of gases dissolved in the oil. The presence of specific gases, such as methane, ethylene, and acetylene, can indicate different types of faults such as arcing, overheating, or PDs.

- **Partial discharge:**
  - Continuous PD monitoring provides valuable information about the condition of insulation in high-voltage transformers and switchgear. It helps

assess the integrity of insulation materials and identifies weak points that may require attention.
- PD patterns can vary depending on the type of fault occurring within the equipment. Analyzing PD data can help diagnose specific fault types such as voids in insulation, deteriorated bushings, or loose connections.
- PD measurement is particularly valuable in detecting corona discharges in high-voltage switchgear. Corona can lead to insulation damage and equipment failure, making early detection crucial.

- **Others:**
  - Monitoring humidity levels is essential to prevent moisture-related issues that could affect the insulation and other equipment components.
  - Monitoring power quality parameters such as harmonics, voltage fluctuations, and frequency variations can help identify disturbances or issues that may affect the performance of connected equipment.
  - Monitoring the condition of bushings in transformers and other equipment is vital to detect any signs of degradation or failure.
  - Monitoring environmental parameters like ambient temperature, humidity, and dust levels can help assess the impact of the external environment on substation equipment.

The continuous monitoring of these parameters allows operators to move from traditional time-based maintenance to condition-based maintenance strategies, maximizing the reliability, efficiency, and lifespan of substation assets. Technical challenges in online condition monitoring of substation assets include:

- **Sensor selection and placement**: Selecting the appropriate sensors for each asset type and determining the optimal sensor placement within the substation can be complex. Different assets may require different types of sensors to monitor various parameters accurately.
- **Data integration and compatibility**: Substations often have a diverse range of equipment from different manufacturers. Integrating data from various sensors and devices into a cohesive monitoring system can be challenging, especially when dealing with different data formats and communication protocols.
- **Data volume and management**: The continuous monitoring of multiple assets generates large volumes of data. Efficiently managing, storing, and analyzing this data requires robust data management systems and advanced data analytics techniques.
- **Real-time monitoring and communication**: Real-time monitoring of substation assets is crucial for proactive fault detection. Ensuring reliable and low-latency communication between sensors, data acquisition systems, and monitoring centers is essential for timely decision-making.
- **Data quality and accuracy**: Ensuring the accuracy and reliability of data collected by sensors is essential for making informed decisions. Calibration, synchronization, and quality control mechanisms are necessary to maintain data integrity.

- **Interference and noise**: Substations are electromagnetically noisy environments, which can introduce interference and noise in the sensor readings. Filtering and signal processing techniques are required to mitigate the impact of such environmental factors.
- **Software and analytics development**: Developing sophisticated algorithms and analytical models to process the collected data and extract meaningful insights requires advanced software development and expertise in data science.
- **Integration with control and automation systems**: Integrating the condition monitoring system with the substation's control and automation systems can be challenging, as it requires seamless communication and coordination between different systems.
- **Cybersecurity and data privacy**: Protecting the online condition monitoring system from cyber-attacks and ensuring data privacy are critical to safeguarding the integrity of the monitoring process and preventing unauthorized access to sensitive data.
- **Environmental considerations**: Substations are often exposed to harsh environmental conditions, such as extreme temperatures, humidity, and corrosive substances. Ensuring the reliability and resilience of monitoring equipment in such conditions is essential.

Addressing these technical challenges requires a multidisciplinary approach, involving expertise in electrical engineering, data science, communication technologies, cybersecurity, and software development. Overcoming these obstacles is crucial to deploying effective online condition monitoring systems that enhance the reliability, safety, and efficiency of substations in the smart grid.

11.2.1.2 Case study – machine learning-based partial discharge waveform identification

PD is a critical phenomenon that occurs in high-voltage switchgear and electrical equipment, often indicating potential insulation defects. Early detection of PD is essential to prevent equipment failure and ensure the safety and reliability of power systems. Three commonly used methods for detecting PD in switchgear are the transient earth voltage (TEV) method, high-frequency current transformer (HFCT) method, and ultrasonic method. These three methods are capable of detecting different types of PDs with different measurement procedures.

(a) Transient earth voltage method:
The TEV method is based on the principle that PD events generate fast transient electromagnetic waves that propagate through the grounding system of the switchgear. The TEV sensors, strategically placed on the equipment's surface, detect the transient voltage signals induced on the enclosure due to PD activity. The TEV method is effective in detecting surface PDs, internal PDs, and corona discharges. Surface discharges occur on the external insulation surface, internal discharges take place within the insulation material, and corona discharges result from the ionization of air around sharp edges or points.

Measurement procedure:
- Install TEV sensors: TEV sensors, also known as antennas or probes, are installed on the outer surface of the switchgear enclosure or nearby grounding connections.
- PD detection: During a PD event, electromagnetic waves are generated, which induce transient voltages on the metal enclosure. The TEV sensors pick up these transient voltage signals.
- Signal analysis: The captured transient voltage signals are analyzed using signal processing techniques to identify the presence and characteristics of PD events.
- PD source correlation: The characteristics of the detected transient voltage signals help determine the location and severity of the PD within the switchgear.

(b) High-frequency current transformer method:

The HFCT method involves the use of HFCTs to measure the high-frequency components of the current flowing through the switchgear. PD events generate characteristic high-frequency signals that can be detected and analyzed using HFCTs. The HFCT method is particularly suitable for detecting internal PDs occurring within the insulation materials of the switchgear.

Measurement procedure:
- Install HFCT sensors: HFCT sensors are placed around the conductors or busbars carrying current within the switchgear.
- Current measurement: HFCT sensors measure the high-frequency components of the current, and any abnormalities in the high-frequency signal are indicative of PD activity.
- Signal analysis: The high-frequency current signals are analyzed to identify and assess the presence of PD events.
- PD source identification: The analysis helps identify the location and severity of the PD within the switchgear.

(c) Ultrasonic method:

The ultrasonic method involves the use of ultrasonic sensors to detect the high-frequency acoustic emissions generated by PD events. The ultrasonic method is effective in detecting surface PDs and corona discharges, which produce characteristic acoustic emissions.

Measurement procedure:
- Install ultrasonic sensors: Ultrasonic sensors are placed near potential PD sources or on the switchgear surface to capture acoustic emissions.
- PD detection: During a PD event, acoustic waves are generated, and the ultrasonic sensors pick up these emissions.
- Signal analysis: The captured ultrasonic signals are analyzed to identify and locate PD sources.
- PD characterization: The analysis helps characterize the type and intensity of the detected PDs.

Early detection and accurate analysis of PD in switchgear are crucial for ensuring the reliability and safety of electrical systems. The TEV, HFCT, and ultrasonic methods are valuable tools for detecting different types of PDs. Each method has its strengths and applicability, and their combined use can provide comprehensive insights into the condition of switchgear insulation. Proper measurement procedures and data analysis play a pivotal role in successful PD detection and subsequent maintenance actions to prevent catastrophic failures. Regular and proactive PD testing can help identify insulation weaknesses, allowing for timely repairs and extending the lifespan of critical electrical assets. However, like any diagnostic method, it has its limitations and may require expertise in data interpretation and correlation to specific PD sources. The TEV method produces the partial discharge phase-resolved (PRPD) plot which is a widely used graphical representation and analysis tool for detecting and diagnosing the presence of PD.

The PRPD plot is a two-dimensional graph that displays the relationship between the phase angle and amplitude of the electrical pulses generated by PD events. Each data point on the PRPD plot represents one PD pulse detected during the monitoring or testing period. On the x-axis of the PRPD plot, the phase angle is typically represented, usually ranging from 0° to 360° (relative to the 360° of an AC cycle). The phase angle indicates the time position of the PD pulse within the electrical cycle, relative to a reference point. On the y-axis, the amplitude of the PD pulse is displayed, often in logarithmic scale (dB) to accommodate a wide range of amplitudes. Figure 11.3 (top) shows the PRPD plot of a cable box with a likely high level of internal discharge. The max amplitude is at 60 dB and with pulses per cycle (ppc) of 2.37. Figure 11.3 (bottom) shows the histogram of the same plot with a high count of PD occurrence with amplitude greater than 30 dB.

Various features observed in the PRPD plot can provide valuable information:

- *Cluster patterns*: Each type of PD source (internal, surface, corona, etc.) generates a characteristic cluster pattern on the PRPD plot. Identifying these clusters helps in differentiating between various PD sources and understanding the severity of each.
- *Center frequency*: The center frequency of the cluster indicates the dominant frequency component of the PD pulse, which can help in understanding the type of insulation defect causing the discharge.
- *Spread and density*: The spread and density of data points within a cluster can provide insights into the distribution and intensity of PD events. A concentrated cluster with low spread may indicate localized PD activity, while a spread-out cluster might suggest widespread PD issues.

However, there can be potential issues with human analysis of PD data. While human experts are skilled at interpreting data and identifying patterns, PD analysis can be challenging and prone to certain limitations when relying solely on human analysis. First, human interpretation of PD data can be subjective and may vary based on the expertise and experience of the analyst. Different analysts might interpret the same data differently, leading to inconsistencies in the diagnosis. Second, PD data can be extensive and complex, especially in large power systems with multiple equipment. Human analysts may struggle to handle and process such vast amounts of data efficiently. Third, manual analysis of PD data can be time-

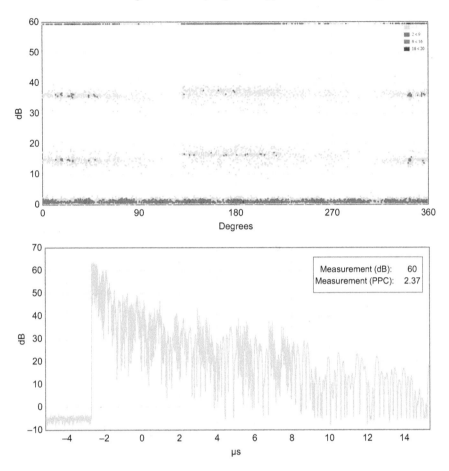

Figure 11.3 PRPD plot (top) and PD pulse waveform (bottom) of a cable box in a switchgear

consuming, particularly when dealing with historical data or continuous monitoring datasets. Timely decision-making becomes difficult, especially in real-time monitoring scenarios. Fourth, human analysts may not always be able to identify all relevant features in PD data due to the vast array of potential variables affecting PD events. Some crucial features might go unnoticed, affecting the accuracy of the analysis. Finally, prolonged human analysis can lead to fatigue, which may impact the accuracy and consistency of the interpretation. Human errors can occur during long hours of analysis. In addition to all these challenges in the human analysis of PD data, human analysis may not be scalable to handle large-scale data from widespread monitoring systems, leading to limited applications in extensive power grids.

To address these challenges and enhance the efficiency and accuracy of PD analysis, machine learning methods have gained popularity. Machine learning algorithms can process vast amounts of data, identify complex patterns, and provide consistent and automated analysis, thereby overcoming many of the limitations associated with human analysis. However, it is essential to note that machine learning models should be developed and validated using reliable and high-quality labeled data to ensure their accuracy and effectiveness in PD analysis. Additionally, the combination of human expertise and machine learning technology can lead to a more comprehensive and reliable approach to PD detection and diagnosis, leveraging the strengths of both human analysts and automated algorithms.

A machine learning-based PRPD waveform analysis software was developed by the Singapore Institute of Technology (SIT) [2]. The proposed technique was tested using 452 PRPD plots gathered from distribution substations throughout a six-month duration. The proposed method unsupervised utilizes unsupervised learning to first extract possible PD clusters. The features listed in Table 11.1 were then extracted from these individual PD clusters instead of the entire PRPD plot.

The reason for selecting these four features from the PRPD plots is their ability to differentiate between authentic PD clusters and noise or interference. To demonstrate this, an ablation study was conducted, and the findings were provided in [2]. Subsequently, these four features were utilized as inputs to different machine learning algorithms, and their respective accuracy rates were compared. After evaluating a limited test dataset comprising approximately 122 plots, it was observed that the tree-based techniques outperformed both neural networks and support vector machines (SVM) techniques marginally. Specifically, the decision tree and random forest algorithms exhibited the most favorable results, achieving zero false negatives. In addition to the PRPD plot, a PD waveform analysis algorithm was developed to determine the number of pulses (including rising and falling edges) and the pulse width of the PD data. This is needed to determine the presence of PD in a more accurate manner. A software was then established as shown in Figure 11.4 to package the machine learning algorithm for PRPD pattern recognition and the PD pulse waveform. The software will be able to provide a first-cut decision on the PD occurrence.

Table 11.1   Feature and description of PD cluster

S/N	Feature	Description
F1	Length of the cluster	Within a cluster, the rightmost x-value deducted from the leftmost x-value.
F2	Height of the cluster	Within a cluster, the top y-value deducted from the bottom y-value.
F3	Gradient from top right to bottom left of the cluster	This gradient is calculated from the point with the largest y-value and rightmost x-value to the point with the lowest y-value and leftmost x-value.
F4	Gradient from top left to bottom right of the cluster	This gradient is calculated from the point with the largest y-value and leftmost x-value to the point with the lowest y-value and rightmost x-value.

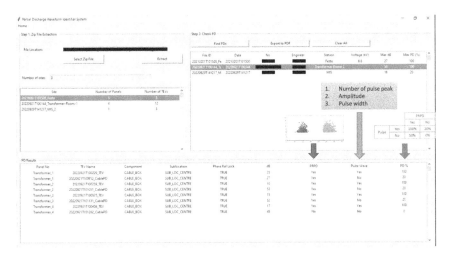

Figure 11.4 PRPD plot (top) and PD pulse waveform (bottom) of a cable box in a switchgear

11.2.2 Solar photovoltaics

In recent times, solar energy has risen to prominence as one of the most significant RES globally, finding widespread deployment in numerous countries. This surge in popularity can be attributed to its abundant availability, decreasing costs, and enhanced safety measures. According to the World Energy Outlook 2021, as estimated by the National Renewable Energy Laboratory (NREL), the global solar capacity is projected to experience an exponential growth of almost 20 times between 2020 and 2050, potentially contributing to 12%–32% of the world's electricity generation by 2050 [3].

Integrating solar power generation into the existing energy infrastructure provides numerous benefits. The scalability and versatility of solar PV technology allow it to be deployed on various scales, from individual residential systems to large-scale solar farms. This adaptability makes solar PV suitable for a wide range of applications, including rooftop installations, solar parks, and even portable solar panels for off-grid use or in remote areas. The declining costs of solar PV systems, coupled with technological advancements, have contributed to the rapid growth of the industry. Improved efficiency, increased durability, and streamlined manufacturing processes have made solar PV more affordable and accessible to a broader range of consumers and businesses. Solar PV energy also offers economic benefits. It reduces dependence on traditional energy sources, decreases electricity bills, and creates job opportunities in manufacturing, installation, and maintenance. Moreover, solar PV systems can provide energy resilience by generating electricity at the point of use, reducing vulnerability to power outages or disruptions in the grid.

Despite its advantages, solar PV energy faces challenges, including intermittency (reduced generation during cloudy periods or at night) and the need for effective energy storage and grid integration solutions. PV systems' output is highly variable due to their high dependence on environmental conditions. Solar intermittency due to sudden changes in the weather conditions even in the time-scales of several minutes could significantly impact the PV output. Furthermore, PV systems are subject to various environmental constraints, for example, irradiance, ambient temperature, humidity, dust, and other droppings, dependent on the specific type of solar PV system. Copper *et al.* indicate that condition monitoring of a solar PV system requires a large number of sensors to measure environmental conditions and suitable tools [4]. Designing a conditional monitoring (CM) system is considered an important approach to keep PV panels healthy against the frequently changing working environment while attempting to find the optimal control sequences [5].

11.2.2.1 Maintenance and condition monitoring of solar PV system

Condition monitoring involves the continuous monitoring and analysis of various measured parameters and performance indicators of a PV system. By actively assessing environmental information, such as irradiance, temperature, voltage, and current, condition monitoring systems enable operators and maintenance personnel to detect, diagnose, and predict faults or performance deviations in the system.

One of the primary objectives of condition monitoring in PV systems is to identify issues early on, preventing their progression into severe problems. By detecting and diagnosing faults promptly, operators can take proactive measures to mitigate risks, minimize downtime, and optimize energy production. This approach not only enhances system reliability and efficiency but also contributes to maximizing the return on investment (ROI) of PV installations.

Figure 11.5 shows the overall architecture of solar PV condition monitoring system which includes sensor data acquisition systems, fault detection

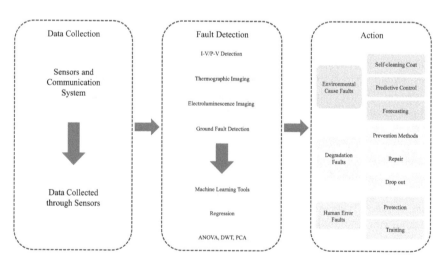

Figure 11.5   Overall architecture of solar PV condition monitoring system

algorithms, and recommended actions. These enable the collection, analysis, and interpretation of data from different components of the PV system such as solar panels, inverters, batteries, and monitoring systems. The data obtained through condition monitoring is typically processed and analyzed using advanced analytics and machine learning algorithms [6]. This enables the detection of anomalies, fault patterns, or deviations from expected performance levels. By leveraging historical data and predictive modeling, condition monitoring systems can also forecast potential failures or performance degradation, allowing for proactive maintenance and intervention.

The benefits of condition monitoring in PV systems extend beyond identifying faults and optimizing maintenance. It also facilitates the identification of underperforming components, suboptimal system designs, or incorrect installation practices. By addressing these issues, operators can enhance the overall performance and energy yield of the PV system, leading to improved financial viability and sustainability.

Modeling simulation of PV generator panels is considered a popular method for PV fault detection [7–9]. Some other research focuses on using electrical signal handling such as irradiance, ambient temperature, wind speed, voltage and current, power, communication condition, and so on. There are many efforts have been made in looking into the various ways of condition monitoring technologies, with new diagnostic techniques and algorithms proposed. We can classify conditions into the following categories according to the types of causes:

1. **Environmental cause failures**: Environmental-related failures include many weather-related temporal faults and long-term ones. It is generally hard to prevent temporal faults as they are considered unavoidable and likely to emerge frequently. Typical temporal weather faults include cloud presence, partial shadowing, rain, temperature and humidity change, and insect presence. Cloud presence is considered when a set of clouds passes over the PV panels, resulting in a temporal loss of power [10–12]. Partial shadowing is brought up and analyzed [13,14], which would cause the amount of light that reaches the PV panel surface to become non-uniform due to shadows generated by external elements like trees or buildings. Such conditions not only yield a decrease in power generation but lead to temporal hot spots on the PV panels. For these aforementioned temporal weather-related faults, most people invite introduce panel and ambient temperature sensing, irradiance forecasting, and satellite image analysis to detect the trend of irradiance changing. Predictive control or AI-related algorithms might be applied to analyze the specific scenario for forecasting or image processing. Furthermore, there are also long-term or even permanent environmental faults. Dust accumulation is an important and common issue to be delved into [15–17]. The dust suspended in the environment is accumulated on the surface of the PV panel, blocking the light to be received by the PV cells. In addition to the power loss and PV hot spot issues, dust accumulation may bring about corrosion of the PV panel. Light, heat, lightning, storm, and water infiltration would also result in the deterioration of PV

cells' functioning, leading to destruction, rust, corrosion, and related performance loss. Such issues are relatively more severe than temporal faults; also, there are many preventive or early-stage fault detection strategies. Regression algorithms and artificial neural network (ANN) tools are widely used to estimate PV conversion efficiency decline due to natural dust or long-term weather influence. Self-cleaning coats, electrodynamic cleaners, and robotic arms are also involved as useful strategies in PV condition monitoring systems.

2. **Degradation failures**: PV panels are mostly designed to serve electric power generation for a lifespan of 25–30 years. As the degradation of PV system elements deteriorates, the performance of the PV system would significantly drop. Poor isolation between the module and inverter would result in short circuits, module destruction, and even fire. After nature weathering and storms, the pedestal of the PV panel may become weak and destructed over time. Another significant cause of performance degradation is the so-called "potential-induced degradation" (PID), where voltage differences between the cells and the grounded frame lead to efficiency losses. PV system inverters can experience electronic component degradation, such as capacitors or semiconductors, which can lead to reduced efficiency and output. Furthermore, excessive heat or inadequate cooling can accelerate inverter degradation. Lastly, the connections between PV modules and inverters, as well as other electrical components, are crucial for proper system functioning. Loose or corroded connections can cause power losses, arcing, and even electrical fires. Proper installation and regular maintenance are essential to prevent connection-related failures. Generally, we would classify these faults again into short circuit faults, open circuit faults, line-to-line faults, inverter faults, and ground faults. $I–V$ (current–voltage) and $P–V$ (power–voltage) curve testing are commonly used in degradation diagnosis to detect abnormality in PV panels; other detection techniques like electroluminescence image analysis and thermographic images from UAVs are also drawing more attention in recent years.

3. **Human error failures**: Human behaviors also have an impact on PV systems. For example, the incorrect module inclination would create shading and may obviously diminish the output power of PV panels. Inappropriate installation and maintenance of PV panels, and wires also lead to bad performance of PV panels; inadequate installation practices, such as improper mounting, faulty wiring, or incorrect positioning of PV modules, can result in reduced energy capture and potential safety risks. Similarly, neglecting regular maintenance and inspection can lead to the accumulation of dust, dirt, and debris on the panels, lowering their efficiency and output. Improper installation also leads to overloading a PV system beyond its capacity, for instance, connecting additional electrical loads can lead to equipment damage and safety hazards. Additionally, using PV systems for unintended purposes or improper applications may void warranties and compromise the system's efficiency.

There are many condition monitoring methods and state-of-the-art detection techniques related to the aforementioned failures in PV systems. By continuously assessing the health and status of system components, condition monitoring helps identify potential issues, degradation, and faults early on, enabling timely interventions and improved system reliability. Various detection methods are employed to monitor the condition of PV systems effectively. Here are some common detection methods:

1. **Current–voltage (*I–V*) curve testing**: *I–V* curve testing, also known as current–voltage characterization, measures the electrical behavior of PV modules under different conditions. It helps assess the module's efficiency and potential degradation, and identify mismatched or faulty modules within an array.
2. **Power–voltage (*P–V*) curve testing**: Like *I–V* curve testing, *P–V* curve testing measures the power output of PV modules at different voltage levels. By determining the maximum power point (MPP), *P–V* curve testing helps optimize system performance and identify issues related to module mismatch or degradation.
3. **Thermographic imaging**: Infrared thermography is a non-contact technique used to detect temperature variations in PV modules. Hotspots or temperature anomalies can indicate potential defects or malfunctions such as cell cracks or poor connections. Regular infrared inspections can help identify and address issues before they escalate, minimizing power losses and enhancing safety.
4. **Electroluminescence imaging**: Electroluminescence imaging is used to inspect the internal condition of PV cells by capturing the light emitted during a reverse-biased operation. EL images reveal cell cracks, inactive areas, and other cell-level defects, helping diagnose module degradation and performance issues.
5. **Performance ratio monitoring**: Performance ratio monitoring involves comparing the actual energy output of the PV system with the expected output based on weather conditions and other parameters. Deviations from the expected performance can signal possible issues such as soiling, shading, or module degradation.
6. **Ground fault detection**: Ground fault detection involves monitoring the PV system's electrical grounding for potential faults or disruptions. Detecting ground faults promptly is essential for safety and preventing damage to the system components.
7. **Data analytics and machine learning**: Advanced data analytics and machine learning techniques are used to process large volumes of data from PV system sensors and components. These methods can help identify patterns, predict failures, and optimize system operation based on historical data and real-time performance.

In conclusion, condition monitoring plays a pivotal role in maintaining the reliability, efficiency, and longevity of PV systems. By continuously monitoring critical parameters and analyzing performance data, operators can identify faults, predict failures, optimize maintenance schedules, and maximize energy production. With the growing importance of renewable energy, condition monitoring has become an essential tool for ensuring the effective operation and optimization of PV systems.

11.2.2.2   Case study: detecting shading and over-heating conditions

An IoT-based solar PV condition monitoring system was designed and developed to effectively monitor PV systems at the SIT building in Nanyang Polytechnic (NYP) campus (Figure 11.6). The data is obtained from the system which comprises 12 PV modules, arranged as four parallel strings. Five distinct types of data from the PV panels are captured, namely, 12 PV panel temperature, 12 PV panel voltage, four string's current, irradiance, and combined ambient temperature and humidity. These comprehensive data measurements will be transmitted to a LoRa terminal and further stored in a SQL database after decoding.

Let's look at the panel temperature monitoring and how the analysis of variance (ANOVA) technique contributes to the overall solar PV system functioning and fault detection. ANOVA is a statistical technique used to compare the means of two or more groups to determine whether there are statistically significant differences among them. It is commonly employed when dealing with categorical independent variables and continuous dependent variables. ANOVA assesses whether the variations between group means are greater than the variations within the groups. ANOVA compares the variance between the group means (explained variance) to the variance within the groups (unexplained variance). If the explained variance is significantly larger than the unexplained variance, it suggests that the group means are different from each other and not just due to chance.

In our case, panel temperature data for panels 2, 3, 4, and 5 will be used; a one-way ANOVA test will be performed. The ANOVA test will provide a test statistic (F-statistic) and a p-value. The p-value indicates the probability of obtaining the

Figure 11.6   PV panels at the Singapore Institute of Technology building in Nanyang Polytechnic, Singapore

observed results by chance alone. If the *p*-value is below a predetermined significance level (commonly 0.05), the null hypothesis will be rejected, and it can be concluded that there are significant differences in the mean temperatures of the panels.

If the *p*-value is above the significance level, it means the result fails to reject the null hypothesis, implying that there is no strong evidence of significant differences in the temperatures among the panels.

ANOVA is based on sound statistical principles and mathematical proofs. ANOVA is a valid and widely used statistical method to compare means among two or more groups and assess whether there are statistically significant differences between the means. ANOVA relies on assumptions that the data is normally distributed within each group and has equal variances across the groups. When these assumptions are met, ANOVA produces valid and reliable results. However, if the assumptions are violated, the results may not be accurate, and alternative methods may be required. Furthermore, ANOVA is based on partitioning the total variance in the data into different components: variance between groups (explained variance) and variance within groups (unexplained variance). The *F*-statistic in ANOVA represents the ratio of the explained variance to the unexplained variance. When there are genuine differences in the group means, the explained variance will be relatively large compared to the unexplained variance, leading to a significant *F*-statistic.

If ANOVA indicates that there are significant differences among the groups, post hoc tests (e.g., Tukey's HSD, Bonferroni) can be performed to identify which specific groups' means differ significantly. These post hoc tests help to make more detailed and accurate comparisons when there are multiple groups.

In the experiment, a total of three different scenarios were compared in order to provide a clear and comprehensive overview of the CM during different environmental, degradation or communication faults. For the result comparison, we look into the *p*-value given by the ANOVA method to judge if there is an unexpected environmental condition or degradation failure on the PV panels.

1. **Scenario 1**: No special condition is discovered in the testing week.
2. **Scenario 2**: From day 5, there is unexpected shading on panel 4.
3. **Scenario 3**: From day 5 to day 6, there is overheating on panel 3 and hot spot.

Figure 11.7 shows a normal scenario in a week, where there is no significant difference in temperature among all four panels.

Shading is a very common environmental cause of faults in PV systems. The influence of shading on a PV system can have significant impacts on its performance and efficiency. Shading occurs when certain parts of the PV system receive reduced or obstructed sunlight compared to other parts, leading to partial or complete shading of the solar panels. Shading on a solar panel reduces the amount of sunlight reaching the shaded cells, resulting in a decrease in electricity generation. Even a small amount of shading on a single solar cell can significantly affect the overall output of the entire panel and the entire PV system. Figure 11.8 shows the decline of temperature on PV panel 4, with a potential decrease in power output.

Due to environmental factors or partial shading, overheating is another common challenge in PV systems. The influence of overheating on PV panels can

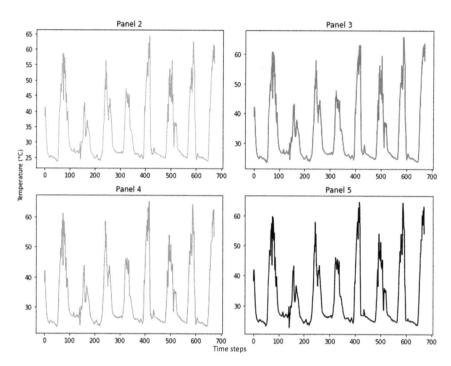

Figure 11.7   Scenario 1: temperature of PV panels 2, 3, 4, and 5 for a week, without any conditions

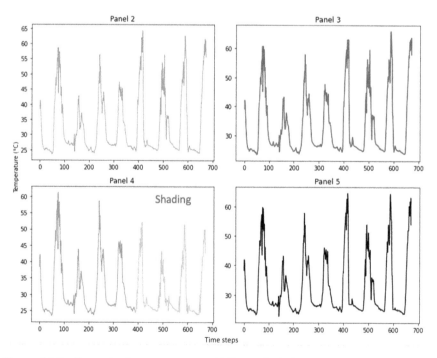

Figure 11.8   Scenario 2: temperature of PV panels 2, 3, 4, and 5 for a week, with shading on panel 4

have significant negative impacts on the performance, efficiency, and reliability of the solar power system. Overheating occurs when the temperature of the PV panels exceeds their optimal operating range. Overheating leads to a decrease in the efficiency of the PV panels. The output power of solar cells decreases as their temperature increases. When the temperature exceeds the manufacturer's recommended operating temperature, the conversion of sunlight into electricity becomes less efficient. High temperatures can accelerate the aging process of the PV panels. Elevated temperatures cause increased degradation of the materials used in solar cells such as the semiconductor materials and encapsulation layers. This can lead to a decrease in the panel's performance over time and shorten its lifespan. Overheating can also create localized areas of high temperature known as "hot spots." Hot spots occur when a portion of the PV panel is shaded or when there is an electrical mismatch in the cells. The shaded or mismatched cells act as resistive loads, converting sunlight into heat instead of electricity, leading to potential damage and performance degradation. Figure 11.9 shows overheating on panel 3.

As it can been seen from Table 11.2, we conclude that only in Scenario 1, the null hypothesis is accepted as the p-value is larger than the significance level of 0.05. So only in Scenario 1, the ANOVA method provides sufficient proof to

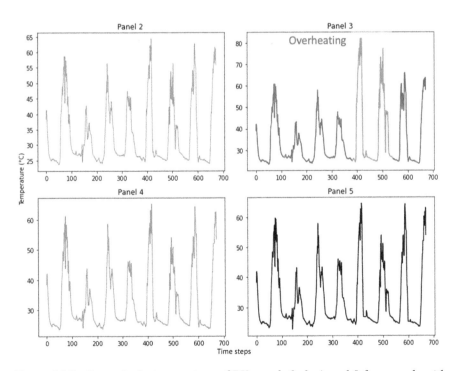

Figure 11.9   Scenario 3: temperature of PV panels 2, 3, 4, and 5 for a week, with overheating on panel 3

Table 11.2   ANOVA test results for different scenarios

ANOVA test	*F*-statistic	*p*-Value
Scenario 1: No condition	0.5198	0.6687
Scenario 2: Shading	5.9101	0.0005
Scenario 3: Overheating	5.3754	0.0011

confirm that there does not exist a significant difference among four testing PV panels. As for Scenarios 2 and 3, the *p*-value is too small, and the null hypothesis would be rejected. A warning would be triggered in the condition monitoring system and further action would be required to make sure the long-term healthiness and performance of PV panels.

11.2.3   Energy storage

The continuous use of non-RES such as fossil energy in the form of coal, oil, and gas has caused severe environmental crises and emission of high amounts of greenhouse gases (GHG) such as carbon dioxide. The ever-increasing world population further causes higher fossil energy depletion rates than they can be replaced. Hence, RES such as sunlight, wind, geothermal, and hydro emerge as clean and pollution-free solutions to the energy exhaustion limitation. However, the intermittent nature of RES due to the dependency on climate is considered its major drawback [18]. Also, in situations where the generated energy surpasses the power load, the potential to extend the utilization of this surplus energy would go to waste in the absence of efficient energy storage mechanisms.

   Therefore, an ESS plays a vital role in addressing the intermittent nature of RES and optimizing grid operations for the reliable and continuous operation of a system [19]. The asset within an ESS refers to the physical infrastructure responsible for storing and delivering energy when required. ESS are often used for energy arbitrage, load shifting, and ancillary services provision, which not only increases the power system reliability but also benefits the grid operators and end-users by addressing power generation shortages, mitigating load spikes through short-term ride-through capabilities, minimizing network losses, and enhancing system protection through their contribution to fault currents.

   ESS have demonstrated their unique capabilities in addressing various critical aspects of electricity [20]. First, ESS play a vital role in reducing power costs by efficiently storing electricity during off-peak hours when prices are lower, thereby avoiding the need to purchase electricity at higher costs during peak hours. Second, ESS contribute to enhancing power supply stability, particularly during disruptions caused by natural disasters or other network interruptions. They serve as valuable backup power sources in such situations. Additionally, ESS help maintain and improve power quality, frequency, and voltage levels. The concept of a smart grid involves modernizing power grid infrastructure. By integrating smart grid technologies, the grid becomes more adaptable and responsive, enabling real-time data sharing between electricity producers and consumers. This integration fosters a more sustainable and efficient power supply. In this context, ESS emerge as a key component in the development of a smart grid, providing the necessary energy storage capabilities to support its advanced functionalities.

ESS play a crucial role in three main contexts: power generation, grid operations, and end-user applications. On the power generation side, ESS serve as a demand endpoint for power plants [21]. Given the diverse impacts of various power sources on the grid and the inherent challenge of predicting load fluctuations, ESS encounter multiple demand scenarios in power generation. These scenarios include energy time shifting, capacity augmentation, load tracking, system frequency regulation, standby capacity provision, and integration of renewable energy into the grid. On the grid side, energy storage applications primarily focus on alleviating transmission and distribution bottlenecks, deferring the need for infrastructure expansion, and providing reactive power support. Although the number of applications on the grid side may be fewer compared to the power generation side, their impact is characterized more by substituting conventional practices. On the customer side, energy storage pertains to electricity consumption at the terminal level, where customers serve as consumers and users of electricity. The cost and benefit derived from power generation, transmission, and distribution are reflected in the electricity price, which directly affects customer demand.

ESS have been utilized to address the inherent variability of RES, ensuring a consistent power supply even during intermittent generation periods. By effectively mitigating the effects of intermittence, ESS not only enhance the reliability of renewable energy but also offer cost-saving benefits. Furthermore, these systems optimize the sizing of system components, leading to reduced operational costs. Ultimately, the integration of ESS in grids facilitates access to affordable, reliable, and sustainable energy solutions [22].

As shown in Figure 11.10, energy storage techniques can be classified into five main categories, namely, electromechanical, electromagnetic, chemical, thermal, and electrochemical. Each of these can be further expanded into various types, which will be elaborated below [23].

(a) Electromechanical: Electromechanical ESSs employ sophisticated mechanisms involving heat, water, or air to store mechanical energy. These systems offer reliable alternatives to electrochemical battery storage, utilizing compressors, turbines, and other machinery. Pumped hydroelectric systems and compressed air energy storage are prominent examples of electromechanical energy storage [24].
(b) Electromagnetic: In electromagnetic-based ESS, energy is stored by magnetic fields generated by coils carrying electric current that have the capacity to store electromagnetic energy [25]. Superconducting magnetic energy storage is one example of electromagnetic-based ESS.
(c) Chemical: Hydrogen is a prominent example of chemical storage, which involves the conversion of electrical energy into hydrogen through a chemical process [26]. The stored hydrogen can be utilized as fuel in a fuel cell or a combustion engine to regenerate electricity. Electrolysis is a straightforward process for converting water into hydrogen, provided that affordable power is available.

Figure 11.10   ESS types and challenges

(d) Thermal: Thermal ESS involves storing energy by heating or cooling a medium [27]. A simple example is a water tank used as a heat backup system, where excess energy is used to heat the water for later use. Different methods exist for thermal energy storage, including thermochemical storage, sensible heat storage, and latent heat storage.

(e) Electrochemical: Batteries are one of the earliest methods of energy storage, converting electrical energy into chemical energy. Electrochemical ESSs are divided into two main branches, namely, secondary batteries and flow batteries. Sodium-sulfur, vanadium redox flow, lithium-ion batteries, and supercapacitors are some examples of electrochemical-based ESS [28].

11.2.3.1   Maintenance and condition monitoring of energy storage system

There are various faults or hazards that can be associated with ESS such as a battery, which subsequently leads to premature aging. Battery cell faults encompass two main categories: mild faults, such as aging-related cell issues, and incipient faults, exemplified by internally shorted cells. Of particular significance is the early detection of internal short circuits that can avert thermal runaway, fires, and explosions, thus ensuring battery safety. Thermal runaway refers to the uncontrolled self-heating of a battery cell, which occurs when the internal heat generated surpasses the amount of heat dissipated to the surroundings. Consequently, the initially overheated cell produces flammable and toxic gases, and the heat level can become high enough to ignite these gases. This critical situation can rapidly propagate to neighboring cells and spread throughout the entire ESS, hence earning the term "runaway." Understanding and mitigating the risks associated with thermal runaway are crucial for ensuring the safety and reliability of ESS applications, particularly in scenarios where multiple cells are interconnected in the system [29].

Internal faults occur due to inadequate design, low-quality materials, or manufacturing deficiencies can lead to internal faults in the battery [30]. Environmental factors can also contribute to ESS failure. Extreme ambient temperatures, seismic activity, floods, debris ingress, corrosive mists such as dust or salt fog, or rodent damage to wiring can lead to system malfunction. Additionally, locations with rapid temperature variations, such as mountains, may experience dewing, potentially causing damage to outdoor ESS if not properly controlled.

Physical or mechanical abuse occurs when the battery undergoes physical compromise such as crushing, dropping, penetration, vibration, shock, or distortion due to external mechanical forces, resulting in failure [31]. Any ESS is designed to operate within a specific temperature range set by the manufacturer. Operating outside this range can cause the ESS to malfunction, prematurely age the battery, and even lead to complete failure, resulting in fire and explosions. These can be categorized as thermal abuse, which can occur due to contact with overheated adjacent cells, exposure to elevated temperatures, or external heat sources in the storage environment. Electrical abuse on the other hand arises from overcharging, rapid charging, or external short-circuiting, for example on a battery [32]. Rapid discharging or over-discharging the battery below its specified end voltage can also cause electrical abuse. These situations can render the ESS inoperable, leading to overheating, fire, and potential explosions.

Hence, condition monitoring emerges as a vital solution as it allows for early fault detection, PdM, and improved system reliability [33]. PdM or CBM represents an advanced diagnostic technique geared towards the early detection of machinery faults to avoid unexpected breakdowns. By closely analyzing sensor signals, this approach enables timely and proactive maintenance actions to be undertaken, ensuring optimal machinery performance and longevity. By continuously monitoring the health and performance of ESS components, potential issues can be identified and addressed proactively, minimizing downtime and maintenance costs. Safety is enhanced through the detection of potential hazards, and optimal performance is achieved by maintaining efficiency. Additionally, condition monitoring enables data-driven decision-making and cost optimization, leading to better overall effectiveness and utilization of energy storage technology in various applications. In contemporary industry, maintenance quality holds a direct impact on the operational uptime and efficiency of the equipment. Consequently, through the utilization of machinery condition monitoring, PdM techniques can effectively mitigate machine downtime and potential losses, thereby optimizing overall performance.

The EMS plays a vital role in overseeing, coordinating, and efficiently managing the various tasks assigned to each subsystem within the system to facilitate diverse system operations and functions [34]. When proper supervision is ensured throughout the entire system, it becomes achievable to attain high levels of operational efficiency and functionality. In hybrid systems, a significant challenge lies in the condition monitoring and effective management of power distribution.

In an electrochemical-based ESS, the battery management system (BMS) is used to continuously monitor the health of batteries through a set of algorithms, for condition monitoring, fault diagnosis, and fault prognosis. By employing these algorithms, the BMS aims to enhance the operational performance, safety, reliability, and overall lifespan of batteries. Condition monitoring for batteries involves tracking changes in critical parameters and operational states such as state of charge (SOC) and state of health (SOH). Unfortunately, these parameters and states often require estimation based on measurable data obtained from battery cells, such as voltage, current, and surface temperature, since they are not directly measured by sensors in BMS.

Despite being an ideal energy storage solution with high energy density, low pollution, and prolonged service life, lithium-ion batteries frequently function as ESS in electric vehicles (EVs) and RES. Nevertheless, continuous charging and discharging, coupled with exposure to harsh environments like vibration and high temperature, can lead to irreversible battery damage and performance degradation, posing safety risks to regular operations.

Once these batteries are retired after a certain period, they are still capable of being used in other applications as supplementary ESS such as in microgrids or EV charging stations [35]. For example, a retired battery from an EV usually preserves 70%–80% of its original capacities. These retired, second-life batteries still have the potential to be utilized in other less demanding applications such as microgrids and uninterruptible power supplies (UPS). Compared to purchasing a new battery, secondary use of batteries can contribute to significant cost reduction but are subject to their capacity retention rate (CRR). A battery's CRR is defined as the ratio of the battery's current capacity to its rated capacity. For retired batteries with a lower value than the CRR threshold, their secondary use value is limited and may not be used in the future. Those with a higher CRR value can be used in accommodating RES, storing energy purchased from the grid, and feeding power back into the grid.

Battery SOH can be effectively assessed by evaluating the extent of performance degradation, mainly quantified through battery capacity and internal resistance. To ensure system safety and stability, battery failure is typically considered when the capacity declines to 70%–80% of its nominal capacity. Remaining useful life (RUL) denotes the number of cycles left for the battery to reach a predefined failure threshold based on its output capacity [36]. The assessment of the SOH and RUL of lithium-ion batteries play a pivotal role in determining their current aging status and ensuring safe usage [37].

The condition prognosis and diagnosis techniques allow for identifying and categorizing faults through the implementation of data-driven ML models. The ideal procedure for evaluating a machine's condition encompasses five key stages [38]. Initially, the processed data is harnessed to train the intelligent algorithm for efficient fault detection. Subsequently, during the fault classification phase, further information is extracted, shedding light on the nature, extent, and specific type of failure encountered. The subsequent step involves the precise localization of the fault within the equipment, followed by the fault quantification stage, which accurately estimates and quantifies the magnitude of the detected fault. Finally, this approach enables us to predict the RUL of the monitored machinery, ensuring timely maintenance actions. However, it is noteworthy that the exact implementation of the procedure may vary, tailored to specific requirements and the unique characteristics of the system under consideration. Numerous studies may focus on exploring one or two of the aforementioned features, making it essential to comprehensively investigate the potential of this approach for reliable condition monitoring and prognosis.

Fault diagnosis for battery cells is a critical technique that aims to detect faulty cells and identify the specific fault types [39]. These failures in lithium batteries can stem from inherent defects, improper usage, or harsh environmental conditions. While redesigning battery structures and materials for improved safety presents long-term challenges, accurate prediction and diagnosis of power battery faults remain crucial to ensuring the safe and reliable operation of EVs. To this end, various fault diagnosis methods have been proposed and generally classified into several categories, namely, knowledge-based, model-based, data-driven, and hybrid methods [40].

Knowledge-based methods rely on subjective analysis techniques such as manual analysis, inferential analysis, and logical judgment to qualitatively diagnose faults [41]. Model-based methods rely on battery models, including electrochemical, electrical circuits, and thermal models, to estimate parameters and evaluate the residuals by comparing the observed signal with the model output signal to obtain the residual value, which is then analyzed to determine the presence of faults which serve as robust indicators of battery faults. For model-based condition monitoring, advanced algorithms like sliding mode observer (SMO), adaptive unscented Kalman filter (UKF), and structural analysis with sequential residual generators have been applied for model-based fault diagnosis [42].

On the other hand, model-free or data-driven methods do not require precise analytical mathematical models or expert knowledge of the battery system. Instead, they focus on analyzing and processing the operating machinery lifecycle data of the system to detect and isolate faults. These methods often utilize real-time data from battery systems and involve machine learning to extract fault symptoms from signals using signal processing techniques like wavelet transform and Shannon entropy, along with artificial intelligence approaches such as fuzzy logic and ANNs. Hybrid methods combine any two of the above-mentioned techniques [22].

11.2.3.2 Case study: battery state of health prediction

In the quest to develop sustainable and reliable energy storage solutions, the analysis of battery performance has emerged as a crucial area of investigation when used in ESS. Understanding the impact of environmental factors on battery lifespan is paramount to optimizing their usage and ensuring safe and efficient operation.

SOH estimation and RUL prediction are vital aspects in assessing battery health and anticipating its operational lifespan. Accurate and efficient battery health estimation models are essential for ensuring the safe and optimal utilization of batteries in various applications. For this study, we utilized the NASA Prognostics Center of Excellence Data Repository battery dataset as a source of information [43]. By leveraging the NASA battery dataset and employing advanced data analytics and machine learning techniques, we aim to shed light on the intricate relationship between temperature and battery performance, thereby contributing to the development of more resilient and sustainable energy storage solutions. The primary objective of our analysis is to estimate the SOH and subsequently predict the RUL of the batteries, focusing specifically on the influence of temperature.

Correlation of battery features
The dataset offers a comprehensive collection of multiple battery performance data, including crucial parameters such as temperature, voltage, and capacity, which were run through three different operating profiles of charging, discharging, and impedance at different operating temperatures of a colder 4 °C, a room temperature of 24 °C, and a higher temperature of 43 °C.

Generally, a battery's capacity will decrease as the number of cycles increases. As shown in Figure 11.11, the temperature at which a lithium-ion battery is operating has a significant impact on its life cycle. The batteries that were operating at room temperature of 24 °C have the longest life cycle, while the batteries that were operating at a higher temperature of 43 °C had the shortest life cycle of around 40. However, batteries that were operating at a lower temperature of 4 °C also had a

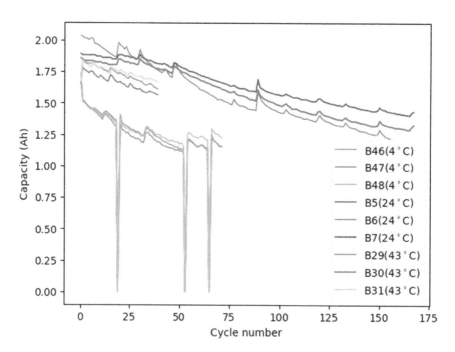

Figure 11.11   Li-ion battery capacity degradation at varying temperatures

reduced life cycle of around 72 compared to batteries operating at room temperature. This could be due to several factors such as temperature sensitivity and other physical and chemical stress endured by the battery which affected the internal processes. Lithium-ion batteries are temperature-sensitive, and their performance and lifespan can be significantly affected by operating outside of their optimal temperature range, which subsequently impacts the other parts of the battery such as the SEI layer and electrolytes.

Three Li-ion battery datasets numbered B0005, B0006, and B0007 provided by NASA were used for the following analysis. All the charging and discharging processes of the batteries were carried out at room temperature. Charging was carried out in constant current (CC) mode at 1.5 A until the battery voltage reached 4.2 V and then continued in a constant voltage (CV) mode until the charge current dropped to 20 mA. The discharge was carried out at a CC level of 2 A until the battery voltage fell to 2.7 V, 2.2 V, and 2.5 V for the batteries B0005, B0006, and B0007, respectively. The experiments were stopped when the batteries reached their end-of-life criteria, which was a 30% fade in rated capacity (from 2 Ah to 1.4 Ah). These features are extracted and plotted to study their respective lifecycle trend as shown in Figure 11.12.

SOH estimation
Data normalization is commonly applied to in-depth modeling algorithms where it is appropriate to improve the convergence of the model and the accuracy of the

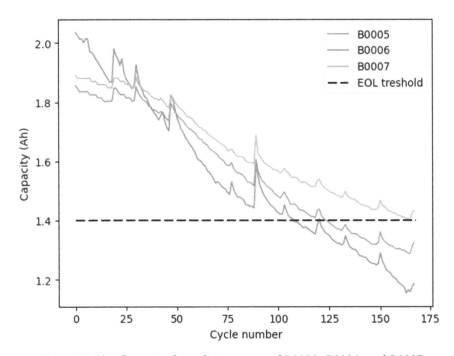

Figure 11.12   Capacity degradation curve of B0005, B0006, and B0007

prediction. The data will be scaled between 0 and 1 using the minimum-maximum approach of normalization. Then, the datasets are divided into training and testing sets, which serve to train the model and test, respectively, for assessing the performance of the model. The training and testing set cycles for each of the chosen dataset vary from 30%, 50%, and 70%, as illustrated in Table 11.3.

Three ML models are used in this study for SOH estimation and subsequently to evaluate the most suitable algorithm for this dataset, namely, multi-layer perceptron (MLP), convolutional neural network (CNN), and long short-term memory (LSTM). The MLP contains two hidden layers of computational units and a single output layer as shown in Table 11.4. A one-dimensional (1D) CNN architecture is adopted for the battery SOH prediction as 1D CNNs are well-suited for processing temporal data. CNN normally consists of the convolutional layer and the pooling layer. The convolutional layer is responsible for extracting features from the input data. The pooling layer is responsible for reducing the size of the output from the convolutional layer while preserving the most important features. The amount of battery cycle is relatively small. If a pooling layer were to be added to a CNN, it would lose a large amount of information. Therefore, the structure of the CNN model consists of only a convolutional layer followed by a single hidden layer and an output layer.

The LSTM network is employed due to its ability to effectively capture sequential patterns in time-series data. The input to the LSTM would be a sequence

Table 11.3   Dataset segmentation

Dataset	Training set	Testing set
B0005	(30%) 50 cycles	(70%) 118 cycles
	(50%) 84 cycles	(50%) 84 cycles
	(70%) 118 cycles	(30%) 50 cycles
B0006	(30%) 50 cycles	(70%) 118 cycles
	(50%) 84 cycle	(50%) 84 cycle
	(70%) 118 cycle	(30%) 50 cycle
B0007	(30%) 50 cycle	(70%) 118 cycle
	(50%) 84 cycle	(50%) 84 cycle
	(70%) 118 cycle	(30%) 50 cycle

Table 11.4   Algorithm's parameters

Parameters	MLP	CNN	LSTM
Learning rate α	0.01	0.01	0.01
Conv1D layer filter number	–	14	–
Conv1D layer filter size	–	5	–
LSTM layers	–	–	20
Hidden layer	[6,70]	50	80
Mini-batch size	25	25	25
Dropout rate	0.1	0.2	0.2

of the last ten cycles of historical HI data. The LSTM layer learns from this sequence and passes its output forward to the hidden layer. Ultimately, the output of the LSTM would be the predicted SOH capacity value.

The capacity measurement requires a full charge and discharge cycle, which can be time-consuming. Therefore, the goal is to develop a new HI that can replace the capacity and predict the battery's RUL or SOH.

Figures 11.13–11.15 show the timing curve characteristics variations of the voltage, current, and temperature of the B0005 battery at different SOH levels during the charging and discharging processes. These figures show that the time taken for voltage, current, and temperature to reach the maximum or minimum point changes as the battery ages.

As seen in Figure 11.13, the time required for the charging voltage of different charging cycles to reach 4.2 V is different. Similarly, the time required for the discharging voltage of different discharging cycles to reach 2.7 V is also different. Additionally, the time required for the charging temperature of different charging and discharging cycles to reach a maximum value is different.

From the correlation analysis, the time taken for the temperature to reach the peak during the discharge, the time at which the discharge voltage reaches its minimum point, and the time required for the charging voltage to reach 4.2V are 0.999813, 0.999947, and 0.914449, respectively, as shown in Table 11.5. It can be

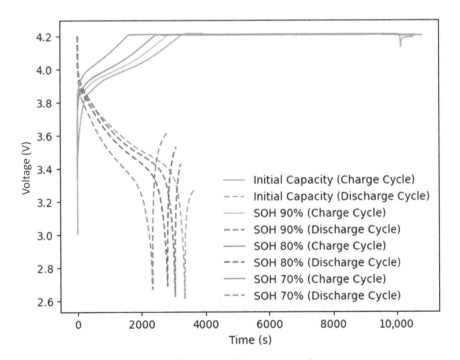

Figure 11.13   Charge and discharge voltage curve

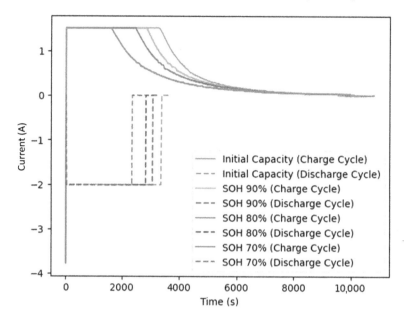

Figure 11.14   Charge and discharge current curve the battery's RUL or SOH

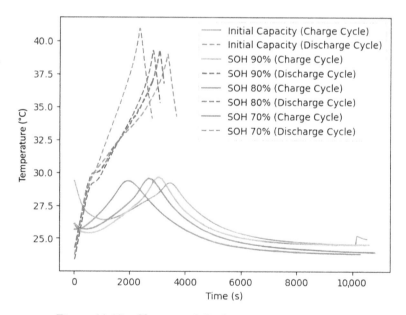

Figure 11.15   Charge and discharge temperature curve

Table 11.5   Correlation analysis of HI and capacity of B0005

HI	Pearson
Time taken for the temperature to reach the peak value during the discharge process	0.999813
Time at which the discharge voltage reaches its minimum point	0.999947
The time required for the charging voltage to reach 4.2 V	0.914449

Table 11.6   SOH estimation performance comparison of different ML algorithms

Dataset	Start point	Algorithm	MAE	RMSE	R2	MAPE
B0005	50	LSTM	0.049	0.057	0.81	0.035
		CNN	0.075	0.077	0.66	0.05
		MLP	0.052	0.06	0.80	0.03
	84	LSTM	0.01	0.015	0.96	0.007
		CNN	0.015	0.019	0.94	0.011
		MLP	0.052	0.061	0.41	0.037
	118	LSTM	0.011	0.013	0.77	0.008
		CNN	0.009	0.014	0.76	0.006
		MLP	0.05	0.062	−1.55	0.038
B0006	50	LSTM	0.058	0.077	0.70	0.04
		CNN	0.04	0.052	0.86	0.03
		MLP	0.08	0.098	0.51	0.06
	84	LSTM	0.036	0.053	0.71	0.02
		CNN	0.048	0.068	0.52	0.037
		MLP	0.068	0.085	0.24	0.054
	118	LSTM	0.054	0.067	−0.04	0.044
		CNN	0.058	0.072	−0.19	0.047
		MLP	0.052	0.065	0.012	0.042
B0007	50	LSTM	0.044	0.051	0.76	0.029
		CNN	0.069	0.07	0.55	0.044
		MLP	0.043	0.051	0.76	0.028
	84	LSTM	0.041	0.048	0.43	0.02
		CNN	0.025	0.028	0.8	0.016
		MLP	0.04	0.05	0.37	0.027
	118	LSTM	0.056	0.059	0.29	0.018
		CNN	0.027	0.029	0.31	0.018
		MLP	0.023	0.027	0.41	0.015

observed that the constructed indirect HI is very similar to Li-ion battery capacity, and the Pearson correlation coefficient between indirect HI series and Li-ion battery capacity series is equal to 0.999.

Results and analysis

To assess the precision and robustness of the model more thoroughly in SOH prediction, SOH monitoring results at various starting points were examined. Table 11.6 analyzes the predictions performance made by the three different models for the three different batteries at the 50th, 84th, and 118th cycles. From the results, the LSTM and CNN models performed similarly, especially for the B0005 and B0007 datasets.

Figures 11.16–11.18 show the prediction performances for battery #7. It can be observed that the LSTM and CNN models better fit the slope of the degradation trend, but they do not fit the capacity recharge phenomenon as well as the MLP model. The graphs show that the prediction curve starting at 30% of the cycles is not as accurate as the prediction curves for starting at 50% and 70% of the cycles. This means that the model is not as accurate when it is trained with data from the beginning of the battery's lifespan.

A battery's RUL can be accurately determined by looking at its SOH. A battery's SOH declines with time. This means that ultimately the battery's capacity will decline to the point where it is insufficient for the purpose for which it was designed. A battery's SOH can be used as a starting point to calculate the RUL of the battery. SOH prediction is a useful technique for managing batteries overall. It is practical to take action to prolong the life of batteries and avoid unanticipated downtime by predicting the SOH of batteries.

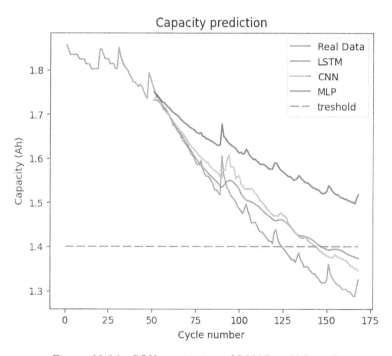

Figure 11.16   SOH monitoring of B0007 at 50th cycle

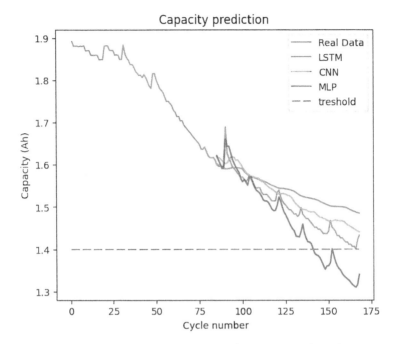

Figure 11.17   SOH monitoring of B0006 at 84th cycle

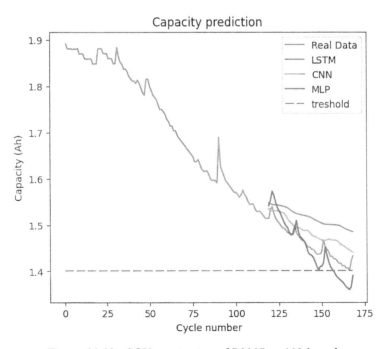

Figure 11.18   SOH monitoring of B0007 at 118th cycle

11.3 Challenges and future directions

11.3.1 Technical challenges

The implementation of condition monitoring and prognosis in smart grids presents a range of intricate technical challenges that necessitate careful consideration for successful deployment and operation. Notable among these challenges is the efficient acquisition, storage, and processing of vast amounts of data generated by diverse sensors and devices within the smart grid. Ensuring data quality and consistency across multiple sources, as well as integrating data from disparate legacy systems, poses a significant hurdle in the pursuit of accurate condition monitoring.

Furthermore, the real-time analysis of data emerges as a crucial aspect of promptly detecting anomalies and potential faults within the smart grid. Accomplishing this requires the development and deployment of sophisticated algorithms and techniques that can process data in real time while maintaining low latency. The accurate identification and diagnosis of faults within large-scale and interconnected smart grid systems present another formidable technical challenge, demanding advanced fault detection methodologies.

Incorporating machine learning and artificial intelligence algorithms for predictive prognostics adds complexity to the implementation, requiring expertise in model selection, training, and validation. Additionally, achieving seamless interoperability between diverse monitoring systems and devices from multiple vendors demands meticulous planning and adherence to standardization protocols.

Cybersecurity and privacy considerations are critical aspects in the implementation of condition monitoring and prognosis in smart grids. As smart grids rely heavily on digital communication, data sharing, and remote access, they become potential targets for cyberattacks. Moreover, the vast amount of sensitive operational data exchanged within the smart grid raises concerns about data privacy and unauthorized access. Addressing these cybersecurity and privacy considerations requires a comprehensive and proactive approach, involving collaboration between grid operators, cybersecurity experts, regulators, and policymakers. Implementing industry best practices, conducting thorough risk assessments, and staying abreast of evolving cybersecurity threats are fundamental to building a resilient and secure smart grid ecosystem.

The scalability of the implemented systems is paramount, ensuring their capacity to handle expanding data volume and increasing system complexity as the smart grid evolves. Cost-effectiveness is also a critical consideration to strike a balance between financial constraints and the attainment of high-performance and reliable monitoring.

The integration of condition monitoring and prognosis with grid operations in smart grids is accompanied by a series of intricate challenges that warrant careful attention for successful implementation. Foremost among these challenges is the management and analysis of the vast and diverse data generated by an extensive array of sensors, devices, and applications within the smart grid. Coping with the volume and complexity of such big data necessitates efficient data handling and processing techniques. In addition, integrating multiple condition monitoring systems and prognostic models from different vendors presents a formidable interoperability challenge. Standardization and data-sharing protocols are required to achieve seamless integration, facilitating the exchange of information among diverse systems.

To address these multifarious technical challenges effectively, a collaborative and interdisciplinary approach is imperative. Expertise from diverse fields such as electrical engineering, data analytics, cybersecurity, and policy-making must converge to drive research and innovation, yielding robust methodologies and technologies that underpin accurate condition monitoring and prognostics within smart grid systems.

11.3.2   Future direction

The field of smart grid health prognosis is witnessing rapid advancements, driven by ongoing research and technological innovations. Several emerging trends and future directions are shaping the landscape of smart grid health prognosis. In this section, we explore some of these trends and discuss potential avenues for future research:

Data-driven prognostics: Data-driven approaches are gaining prominence in smart grid health prognosis. Advanced machine learning and artificial intelligence techniques are being employed to analyze large volumes of real-time data from smart grid sensors and devices. These data-driven models can provide more accurate and adaptive prognostic assessments, enabling better predictions of equipment failures and degradation.

Predictive maintenance with prescriptive insights: The focus is shifting from mere fault prediction to prescriptive prognostics. Instead of solely indicating when equipment might fail, prognostic models are now offering actionable insights, and recommending optimal maintenance strategies to mitigate potential failures. Prescriptive prognostics enable grid operators to optimize maintenance schedules, reduce downtime, and extend the lifespan of grid assets.

Integrated prognostics and control: The integration of health prognosis with control strategies is an emerging trend in smart grids. Prognostic models are increasingly being embedded within the control systems, enabling real-time adaptation of grid operations based on equipment health status. This integration enhances grid resilience, improves load balancing, and facilitates dynamic grid reconfiguration in response to equipment degradation.

Federated and collaborative prognostics: Collaborative and federated prognostic approaches are gaining traction, especially in large-scale smart grid systems involving multiple stakeholders. These models leverage collective intelligence by aggregating data and insights from various sources, promoting more accurate and comprehensive prognostic assessments.

Prognostics for emerging grid components: As smart grids evolve, new and emerging components such as RES, EES, and EV infrastructure are becoming integral to grid operations. The future direction of health prognosis involves developing specialized prognostic models tailored to these innovative grid components.

Explainable AI and uncertainty quantification: Ensuring the transparency and interpretability of prognostic models is critical for their acceptance and practical application. Research is focusing on developing explainable AI techniques that provide insights into the reasoning behind prognostic predictions. Additionally, quantifying and managing uncertainties in prognostic assessments are receiving increased attention.

Cybersecurity in prognostics: With the integration of prognostics into smart grid control systems, the cybersecurity of prognostic models becomes a significant concern. Research is exploring methods to safeguard prognostic models from cyber threats, ensuring the integrity and confidentiality of prognostic insights.

Edge computing for real-time prognostics: Edge computing, where data processing occurs closer to the data source, is emerging as a potential solution for real-time prognostics. Edge computing can reduce latency, enhance data privacy, and enable faster response times in critical prognostic applications.

Self-healing grids: Prognostics is becoming a key component in self-healing grid concepts. Self-healing grids leverage prognostic insights to autonomously detect, diagnose, and mitigate faults, promoting grid resilience and minimizing human intervention.

Long-term prognostics and asset life-cycle management: Research is delving into long-term prognostics that predict the RUL of grid assets. Long-term prognostics enable better asset life-cycle management, aiding in asset replacement planning and optimizing capital expenditures.

11.4   Summary

This chapter highlights the importance of condition monitoring and health prognosis in smart grids. It explains the significance of real-time data-driven prognostics, enabling PdM, improving reliability, and enhancing resilience. The chapter underscores the potential of these advanced techniques in revolutionizing grid operations, ensuring reliable and efficient energy delivery, and paving the way for a sustainable and intelligent future for smart grids. However, the implementation of condition monitoring and prognosis in smart grids presents multifaceted challenges. Managing vast and diverse data volumes, ensuring data quality and interoperability, and achieving real-time processing are among the critical technical hurdles. Successfully navigating these obstacles holds the key to unlocking the full potential of condition monitoring and prognosis, empowering smart grids to enhance reliability, optimize performance, and contribute to a sustainable energy future. Furthermore, leveraging data-driven approaches, integrated control strategies, and collaborative models can revolutionize condition monitoring, assessment, and maintenance practices. The incorporation of explainable AI, cybersecurity measures, and edge computing offers potential solutions to critical challenges. Prognostics for emerging grid components and long-term asset life-cycle management ensure grid adaptability and efficiency.

Embracing these trends through interdisciplinary research and innovation will empower smart grids to optimize performance, enhance resilience, and pave the way for a sustainable energy future.

References

[1] International Organization for Standardization, "ISO 13374-1:2003: Condition monitoring and diagnostics of machines—Data processing, communication, and presentation—Part 1: General guidelines," Geneva, ISO, 2003.

[2] D. Soh, B. Sivaneasan, J. Abraham, K.X. Lai, and J.F. Yongyi, "Partial discharge diagnostics: Data cleaning and feature extraction," *Energies*, vol. 15, no. 2, p. 508, 2022.

[3] National Renewable Energy Laboratory. 2021 Standard Scenarios Report: A U.S. Electricity Sector Outlook, Technical Report No. NREL/TP-6A40-80641, 2021. Retrieved from www.nrel.gov/publications (accessed July 20, 2023).

[4] J. Copper, A. Bruce, T. Spooner, M. Calais, T. Pryor, and M. Watt, "Australian technical guidelines for monitoring and analysing photovoltaic systems," *The Australian Photovoltaic Institute (APVI), (Version 1)*, vol. 7, 2013.

[5] H. Zitouni, A. Azouzoute, C. Hajjaj, *et al.*, "Experimental investigation and modeling of photovoltaic soiling loss as a function of environmental variables: A case study of semi-arid climate," *Solar Energy Materials and Solar Cells*, vol. 221, p. 110874, 2021.

[6] T. Berghout, M. Benbouzid, T. Bentrcia, X. Ma, S. Djurović, and L.-H. Mouss, "Machine learning-based condition monitoring for PV systems: State of the art and future prospects," *Energies*, vol. 14, no. 19, pp. 6316, 2021.

[7] K.-H. Chao, S.-H. Ho, and M.-H. Wang, "Modeling and fault diagnosis of a photovoltaic system," *Electric Power Systems Research*, vol. 78, no. 1, pp. 97–105, 2008.

[8] D. Guasch, S. Silvestre, and R. Calatayud, "Automatic failure detection in photovoltaic systems," in *Proceedings of the 3rd World Conference on Photovoltaic Energy Conversion*, vol. 3, IEEE, 2003, pp. 2269–2271.

[9] M. Hamdaoui, A. Rabhi, A. El Hajjaji, M. Rahmoun, and M. Azizi, "Monitoring and control of the performances for photovoltaic systems," in *Proceedings of the International Renewable Energy Congress*, Tunisia, 2009, pp. 69–71.

[10] C. Y. Lau, C. K. Gan, K. A. Baharin, and M. F. Sulaima, "A review on the impacts of passing-clouds on distribution network connected with solar photovoltaic system," *International Review of Electrical Engineering (IREE)*, vol. 10, no. 3, pp. 449–457, 2015.

[11] K. Lappalainen and S. Valkealahti, "Photovoltaic mismatch losses caused by moving clouds," *Solar Energy*, vol. 158, pp. 455–461, 2017.

[12]   K. Lappalainen and S. Valkealahti, "Number of maximum power points in photovoltaic arrays during partial shading events by clouds," *Renewable Energy*, vol. 152, pp. 812–822, 2020.

[13]   G. Mostafaee and R. Ghandehari, "Power enhancement of photovoltaic arrays under partial shading conditions by a new dynamic reconfiguration method," *Journal of Energy Management and Technology*, vol. 4, no. 1, pp. 46–51, 2020.

[14]   S. Bodkhe, P. Sawarkar, M. Bopche, P. Kumbhare, and V. Deshpande, "Partial shading, effects and solution for photovoltaic string: A review," *Helix—The Scientific Explorer—Peer Reviewed Bimonthly International Journal*, vol. 10, no. 2, pp. 58–62, 2020.

[15]   M. Jaszczur, A. Koshti, W. Nawrot, and P. Sedor, "An investigation of the dust accumulation on photovoltaic panels," *Environmental Science and Pollution Research*, vol. 27, pp. 2001–2014, 2020.

[16]   H. A. Kazem, M. T. Chaichan, A. H. Al-Waeli, and K. Sopian, "A review of dust accumulation and cleaning methods for solar photovoltaic systems," *Journal of Cleaner Production*, vol. 276, p. 123187, 2020.

[17]   S. Fan, Y. Wang, S. Cao, T. Sun, and P. Liu, "A novel method for analyzing the effect of dust accumulation on energy efficiency loss in photovoltaic (PV) system," *Energy*, vol. 234, p. 121112, 2021.

[18]   B. S. Pali and S. Vadhera, "Renewable energy systems for generating electric power: A review," in *IEEE 1st International Conference on Power Electronics, Intelligent Control and Energy Systems (ICPEICES)*, 2016, pp. 1–6.

[19]   K. Kpoto, A. M. Sharma, and A. Sharma, "Effect of energy storage system (ess) in low inertia power system with high renewable energy sources," in *5th International Conference on Electrical Energy Systems (ICEES)*, 2019, pp. 1–7.

[20]   T. S. Babu, K. R. Vasudevan, V. K. Ramachandaramurthy, S. B. Sani, S. Chemud, and R. M. Lajim, "A comprehensive review of hybrid energy storage systems: Converter topologies, control strategies and future prospects," *IEEE Access*, vol. 8, pp. 148702–148721, 2020.

[21]   C. K. Das, O. Bass, G. Kothapalli, T. S. Mahmoud, and D. Habibi, "Overview of energy storage systems in distribution networks: Placement, sizing, operation, and power quality," *Renewable and Sustainable Energy Reviews*, vol. 91, pp. 1205–1230, 2018. [Online]. Available: https://www.sciencedirect.com/science/article/pii/S1364032118301606.

[22]   E. I. Come Zebra, H. J. van der Windt, G. Nhumaio, and A. P. Faaij, "A review of hybrid renewable energy systems in mini-grids for off-grid electrification in developing countries," *Renewable and Sustainable Energy Reviews*, vol. 144, p. 111036, 2021. [Online]. Available: https://www.sciencedirect.com/science/article/pii/S1364032121003269.

[23]   J. Mitali, S. Dhinakaran, and A. Mohamad, "Energy storage systems: a review," *Energy Storage and Saving*, vol. 1, no. 3, pp. 166–216, 2022. [Online]. Available: https://www.sciencedirect.com/science/article/pii/S277268352200022X.

[24] E. O. Ogunniyi and H. Pienaar, "Overview of battery energy storage system advancement for renewable (photovoltaic) energy applications," in *International Conference on the Domestic Use of Energy (DUE)*, 2017, pp. 233–239.

[25] X. Chen, Q. Xie, X. Bian, and B. Shen, "Energy-saving superconducting magnetic energy storage (smes) based interline dc dynamic voltage restorer," *CSEE Journal of Power and Energy Systems*, vol. 8, no. 1, pp. 238–248, 2022.

[26] L. Fan, Z. Tu, and S. H. Chan, "Recent development of hydrogen and fuel cell technologies: A review," *Energy Reports*, vol. 7, pp. 8421–8446, 2021. [Online]. Available: https://www.sciencedirect.com/science/article/pii/S2352484721006053.

[27] S. Chavan, R. Rudrapati, and S. Manickam, "A comprehensive review on current advances of thermal energy storage and its applications," *Alexandria Engineering Journal*, vol. 61, no. 7, pp. 5455–5463, 2022. [Online]. Available: https://www.sciencedirect.com/science/article/pii/S1110016821007328.

[28] T. S. Costa, M. de Fátima Rosolem, J. L. d. S. Silva, and M. G. Villalva, "An overview of electrochemical batteries for ESS applied to PV systems connected to the grid," in *14th IEEE International Conference on Industry Applications (INDUSCON)*, 2021, pp. 1392–1399.

[29] X. Feng, M. Ouyang, X. Liu, L. Lu, Y. Xia, and X. He, "Thermal runaway mechanism of lithium ion battery for electric vehicles: A review," *Energy Storage Materials*, vol. 10, pp. 246–267, 2018. [Online]. Available: https://www.sciencedirect.com/science/article/pii/S2405829716303464.

[30] R. Fajardo, E. Mallari, P. Vivo, M. Pacis, and J. Martinez, "Impact study of a microgrid with battery energy storage system (BESS) and hybrid distributed energy resources using MATLAB simulink and t-test analysis," in *IEEE 12th International Conference on Humanoid, Nanotechnology, Information Technology, Communication and Control, Environment, and Management (HNICEM)*, 2020, pp. 1–6.

[31] J. A. Jeevarajan, T. Joshi, M. Parhizi, T. Rauhala, and D. Juarez-Robles, "Battery hazards for large energy storage systems," *ACS Energy Letters*, vol. 7, no. 8, pp. 2725–2733, 2022. [Online]. Available: https://doi.org/10.1021/acsenergylett.2c01400.

[32] Y. Huang, J. Lu, Y. Lu, and B. Liu, "Investigation into the effects of emergency spray on thermal runaway propagation within lithium-ion batteries," *Journal of Energy Storage*, vol. 66, p. 107505, 2023.[Online]. Available: https://www.sciencedirect.com/science/article/pii/S2352152X23009027.

[33] S. K. Pradhan and B. Chakraborty, "Battery management strategies: An essential review for battery state of health monitoring techniques," *Journal of Energy Storage*, vol. 51, p. 104427, 2022. [Online]. Available: https://www.sciencedirect.com/science/article/pii/S2352152X22004509.

[34] T. Kim, D. Makwana, A. Adhikaree, J. S. Vagdoda, and Y. Lee, "Cloud-based battery condition monitoring and fault diagnosis platform for large-scale lithium-ion battery energy storage systems," *Energies*, vol. 11, no. 1, 2018. [Online]. Available: https://www.mdpi.com/1996-1073/11/1/125.

[35]   M. Muhammad, M. Ahmeid, P. S. Attidekou, Z. Milojevic, S. Lambert, and P. Das, "Assessment of spent EV batteries for second-life application," in *IEEE 4th International Future Energy Electronics Conference (IFEEC)*, 2019, pp. 1–5.

[36]   Q. Xu, M. Wu, E. Khoo, Z. Chen, and X. Li, "A hybrid ensemble deep learning approach for early prediction of battery remaining useful life," *IEEE/CAA Journal of Automatica Sinica*, vol. 10, no. 1, pp. 177–187, 2023.

[37]   J. Qu, F. Liu, Y. Ma, and J. Fan, "A neural-network-based method for rul prediction and soh monitoring of lithium-ion battery," *IEEE Access*, vol. 7, pp. 87178–87191, 2019.

[38]   O. Surucu, S. A. Gadsden, and J. Yawney, "Condition monitoring using machine learning: A review of theory, applications, and recent advances," *Expert Systems with Applications*, vol. 221, p. 119738, 2023. [Online]. Available: https://www.sciencedirect.com/science/article/pii/S0957417423002397.

[39]   Y. Qiu, T. Dong, D. Lin, B. Zhao, W. Cao, and F. Jiang, "Fault diagnosis for lithium-ion battery energy storage systems based on local outlier factor," *Journal of Energy Storage*, vol. 55, p. 105470, 2022. [Online]. Available: https://www.sciencedirect.com/science/article/pii/S2352152X22014621.

[40]   M. Z. Khaneghah, M. Alzayed, and H. Chaoui, "Fault detection and diagnosis of the electric motor drive and battery system of electric vehicles," *Machines*, vol. 11, no. 7, 2023. [Online]. Available: https://www.mdpi.com/2075-1702/11/7/713.

[41]   B. Zou, L. Zhang, X. Xue, *et al.*, "A review on the fault and defect diagnosis of lithium-ion battery for electric vehicles," *Energies*, vol. 16, no. 14, 2023. [Online]. Available: https://www.mdpi.com/1996-1073/16/14/5507.

[42]   Y. Song, D. Liu, and Y. Peng, "Lithium-ion battery pack on-line health diagnosis based on multi uncertainty model fusion," in *IEEE International Instrumentation and Measurement Technology Conference (I2MTC)*, 2023, pp. 1–6.

[43]   B. Saha and K. Goebel, "Battery data set," NASA Prognostics Data Repository, NASA Ames Research Center, Moffett Field, CA, 2007. [Online]. Available: https://ti.arc.nasa.gov/tech/dash/groups/pcoe/prognostic-data-repository/ [Accessed July 10, 2023].

Chapter 12

Power electronics in a PV-integrated grid-connected electric vehicle charging system for V2G/G2V operation

Rajendra Prasad Upputuri[1] and Bidyadhar Subudhi[1]

Vehicle-to-grid (V2G) technology holds great promise as it enables electric vehicles (EVs) to serve as distributed energy resources (DERs), contributing to peak load management, reactive power support, load balancing, and the stability of the power grid. In order for V2G technology to function, EV batteries and the AC power grid must be able to exchange energy in both directions. This chapter presents a review of the most widely used dc–dc converter and bidirectional power factor correction (PFC) topologies in V2G and G2V applications. Each converter topology's efficacy is assessed based on the results of the experimental investigation. The total harmonic distortion (THD) of the grid current is used to evaluate the performance of PFC topologies. Moreover, the efficiency and THD characteristics of each converter topology are compared and analysed. Furthermore, this chapter sheds light on the challenges and future directions in the development of bidirectional converter topologies. It highlights the areas that require further attention and research to enhance the performance and adoption of these chargers in V2G applications.

12.1 Introduction

V2G technology has gained significant popularity due to its ability to decrease dependence on imported fuels, minimise pollution, promote decarbonisation in transportation, and encourage the adoption of innovative urban mobility approaches [1]. This technology facilitates the transfer of power from electric vehicle batteries (EVBs) to the grid [2]. By leveraging EVBs as energy storage systems (ESS) directly linked to the grid, V2G technology represents a cutting-edge approach that moves beyond considering EVs solely as grid consumers. Additionally, V2G technology allows EVs to serve as valuable DERs by utilising their EVBs for charging purposes. The DER, in this context, serves as a temporary

[1]School of Electrical Sciences, Indian Institute Technology Goa, India

ESS that can supply power to the utility grid when needed [3]. This capability improves the overall reliability, stability, and efficacy of the grid, while simultaneously expanding its power generation capacity [4]. Grid-integrated EVs can operate in V2G mode, grid-to-vehicle (G2V) mode, or standby mode. EVs equipped with V2G functionality provide valuable assistance to utility operators in tackling various challenges, including voltage fluctuations, active power control, reactive power support, load management, current harmonic filtering, peak demand management, and reserve capacity needs [5]. To facilitate V2G and G2V operations, a high-performance bidirectional medium is essential between the grid and the EVB. This interface is responsible for charging the EVB and efficiently transferring power sent back to the utility grid while maintaining standards. Figure 12.1 illustrates the generation, transmission, distribution, and EV charging infrastructure. The charging infrastructure offers three primary options for conductively recharging the EVB: ac slow charging, ac fast charging (FC), and dc FC. In general, ac slow charging involves charging an EVB using the onboard charger (OBC) at home/workplace with a single-phase power supply. AC FC, on the other hand, uses a three-phase power supply to charge the EVB at higher power levels, enabling faster charging. In the dc FC method, an additional offboard charger can be used to charge an EVB. In practice, the charger's performance significantly depends on various factors, including the type of converter topology, power rating, and charging and discharging techniques [6]. Therefore, charger design, development, and control play a part in the development of an efficient charging system. Desired features of a bidirectional charger involve high power density and efficiency, low weight, low cost, reliability, and scalability.

In recent studies, state-of-the-art converter topologies were comprehensively reviewed in [7–15]. A typical ac-connected charging station comprises the ac–dc stage, also known as the PFC stage, and the dc–dc stage. The harmonics generated

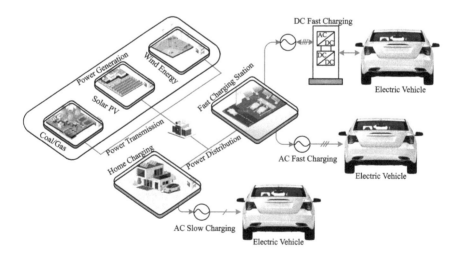

Figure 12.1   Structure of EV charging infrastructure

by the chargers can be compensated by the PFC stage [16]. In the literature, several PFC converter topologies for charging an EVB were presented. One such topology is the single-phase totem pole PFC, which operates in triangular current mode (TCM). It offers high power density, high efficiency, low current ripple, and less circuit complexity [17]. Additionally, it minimises THD and enhances current sharing in parallel circuits, even though it necessitates more intricate control. Furthermore, for high-power applications, the three-phase PFC topologies are used in chargers. These topologies provide bidirectional power transfer, possess a simple configuration, experience low device current stress, are simple in control, offer high efficiency, and exhibit low THD [18]. Among these topologies, the boost rectifier with six switches is commonly used and enables high-power operation with a modular design [19]. To minimise circulating currents, a zero-sequence-based current control strategy is employed. However, in such systems, an oversized dc-link capacitor is required to address the issue of harmonics caused by an unbalanced system at the dc-link [20]. To address these issues, an active dual current control was proposed in [21]. The PFC with the proposed control [21] offers a low input current THD and unity power factor (UPF) from the grid. But the primary concern with PFC topologies is the presence of the third harmonic component in the grid current under light load conditions [22]. Several compensation methods were reported in the literature for reducing the third harmonic component [23,24]. In one approach described in [25], duty compensation was used to reduce the harmonics produced by the switch node of a bidirectional H-bridge-based converter topology. In [23], another technique based on average mode current control was implemented to tackle the frequency instability issue [26]. The researchers in [24] investigated the low-pass filter's stability and a feed-forward compensation technique to reduce the third-harmonic current. However, this approach involves intricate calculations, making it computationally intensive.

The dc–dc converter topology is cascaded to the PFC stage as another stage of an OBC. It is responsible for converting one value of voltage to another. Bidirectional power transfer, high efficiency, high power density, a wide range of voltage gain, and low device voltage stress are the key features of a bidirectional dc–dc converter topologies. These converter topologies are widely categorised as non-isolated and isolated. In [27], a non-isolated-based interleaved buck converter (IBC) was reported with multiple inductors. This design offers advantages such as reduced ripple in current, lower inductor volume, modularity, enhanced power capabilities, and improved thermal management. But the switches experience voltage stress because of the applied voltage in conventional IBCs, leading to high switching and recovery losses in high voltage-rated EVB charging applications. To address this issue, Esteki *et al.* [28] introduced a distinct architecture of a modified IBC. This modified design reduces switching losses, eases voltage stress, minimises current ripple, and boosts the step-down conversion ratio.

The work presented in [29] revealed that when the active switches in the modified IBC are turned off or on, they are subjected to a voltage stress equal to 50% of the input voltage. Importantly, the IBC stabilises the current without any extra circuitry. Due to the intrinsic charge balance in the blocking capacitors, the current is shared between interleaved phases in converters [28] in the absence of extra circuitry. Furthermore, Drobnic *et al.* [30] proposed a modular-based interleaved converter featuring three inductors and six identical switches. This design aims to improve performance through increased interleaving. In [31], the authors introduced an eight-phase synchronous IBC to improve voltage regulation, frequency control, and current equalisation of phases. The current equalisation can be achieved by employing the sliding mode control technique. In [32], the authors proposed a modular-based three-level converter for high-power operation. In addition, bipolar dc-link structure-based neutral point clamped (NPC) PFC eliminates the necessity of balancing circuitry. This can be achieved by employing a power management strategy.

To ensure a safe operation, it is essential to incorporate galvanic isolation between the utility grid and the EVB. One way of doing it is by employing isolation in the dc–dc stage, which offers benefits such as high efficiency, high power density, and smaller size, making it popular in EVB charging applications. Among the isolated converter options, the inductor-inductor-capacitor (LLC) converter is popular due to its compact size, high efficiency, and ability to achieve zero current switching (ZCS) and zero voltage switching (ZVS) over a wide range [33]. In [34], the LLC converter was reported for the purpose of EVB charging, which demonstrated unity as a maximum voltage gain during V2G mode. However, traditional LLC converters are not suitable for grid-tied operations in reverse mode. It is because reduced EVB voltage results in a dc-link voltage that is insufficient for PFC buck-type converters, which require a higher voltage at the dc-link than the maximum grid voltage [35]. To address this limitation, Park and Choi [36] proposed an LLC converter as a dc–dc transformer, where a bidirectional buck/boost converter topology helps in controlling the EVB voltage. But this approach makes chargers more bulky and costly and increases power loss. This is due to the need for an additional converter for power conversion. Therefore, the bidirectional operation of LLC converters requires additional control methods. To address the aforementioned issue and improve performance, the CLLLC converter topologies, which are derived from LLC converters, employ an additional resonant pair of LC on the secondary side of the converter. Such configuration results in high efficiency due to the resonant frequency operation at the cost of an extra LC resonant pair [37]. Nonetheless, CLLC converters suffer from difficulty maintaining high efficiency over a wide range of battery voltage [38]. In [39], another CLLC converter was reported for an uninterrupted power supply (UPS) application. It possesses ZCS and ZVS features to reduce switching losses, enhancing overall efficiency.

In [40], a control strategy employed in a CLLC converter for soft-start was presented with an efficiency of 97.8% at 4 kW power output. For bidirectional power conversion in a battery charger, Xue *et al.* [41] employed a dual active bridge (DAB) dc–dc converter topology. However, DAB converters often experience high currents during turn-off, which results in reduced efficiency due to turn-off losses. To improve the efficiency of dc–dc converters, synchronous rectification can be employed. In [42], various current-sensing synchronous rectification techniques were discussed. However, the use of current sensing requires additional components, which can result in significant power losses. In [43], the authors developed a DAB

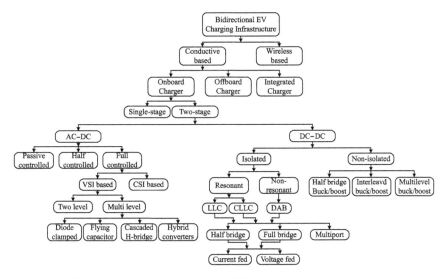

Figure 12.2   Classification of EV charging infrastructure

converter for UPS applications with an efficiency of 96%. The selection of a converter topology for V2G/G2V operation depends on various factors such as power rating, efficiency, component size, THD, and reliability. Thus, this chapter aims to review, analyse, and compare the performance, efficiency, and applicability of various bidirectional ac–dc and dc–dc topologies for conductive-based two stage EV charging infrastructure. The scope of this chapter specifically excludes wireless charging and single-stage chargers, as depicted in Figure 12.2. The effectiveness of the considered ac–dc and dc–dc converters is evaluated experimentally, considering parameters such as THDs, battery current, battery voltage, grid current, grid voltage, and loss analysis. In addition, a performance comparison and analysis of the presented PFCs and dc–dc converters is carried out by considering various performance factors such as device stress, current THD, efficiency, and device count.

The remaining chapters are structured as follows: The review of popular ac–dc and dc–dc topologies for V2G and G2V applications is presented in Section 12.2. Bidirectional converter topologies' performance assessment is presented in Section 12.3, and their comparison analysis is provided in Section 12.4. Section 12.5 presents the challenges and future trends that EV chargers will encounter in the future. Finally, Section 12.6 provides conclusions.

12.2   Converter topologies

EV chargers can be categorised into two types: conductive-based chargers and wireless-based chargers, as depicted in Figure 12.2. While wireless/inductive charging infrastructure is in the progressing stage [15], the prime focus of EV chargers (i.e., onboard and offboard) growth lies in conductive charging, which uses wire connections to sustain power flow to EVs. In this chapter, we specifically emphasise conductive-based EV chargers. These chargers allow bidirectional power transfer and can be implemented in either a single-stage [44] and two-stage architecture. The scope of this chapter targets two-stage EV charging infrastructure, offering advantages such as simple control strategies for various stages and galvanic isolation in the dc–dc stage [15]. The two-stage power conversion of a bidirectional charger comprises an ac–dc converter stage cascading to the dc–dc converter stage.

12.2.1   AC–DC converter topologies

The front-end converter's (FEC) design is crucial for bidirectional power flow, PFC, and harmonic reduction, according to IEEE 519-2014. In G2V and V2G modes, the FEC performs the functions of a rectifier and an inverter. Table 12.1 provides a summary of the most recent bidirectional ac–dc converter topologies. According to Table 12.1, the most commonly used PFC architecture is the full-bridge (FB) PFC because of its UPF operation and simple control. In addition, most FB PFCs reported in the literature rely on a large electrolytic dc bus capacitor to function as a secondary energy storage device. Therefore, the system's cost and footprint increase while its reliability and capability for high power density decrease. Single-stage conversion topologies, such as matrix converters, eliminate the need for electrolytic dc-bus capacitors. However, additional stabilising circuits are required to maintain UPF and guarantee a minimal THD. The multilevel PFC architecture requires more devices than the FB topology while being scalable and giving decreased THD without filters.

Numerous single-phase bidirectional ac–dc converter topologies have been extensively researched in the literature. These include FB-based PFCs [57], T-type converters [58], six-level flying capacitor converter (FCC) [59], and three-level NPC [60]. These PFCs provide advantages like low device voltage stress and

Table 12.1   State-of-the-art ac–dc converter topologies

Reference	Power (kW)	Battery voltage (V)	Switches	Filters	PF	THD	Observation
[45]	3.6	270–360	4	L	0.99	<3%	Low THD
[46]	3	120	4	NA	1	4.5%	High THD
[47]	3.3	400–450	4	LCL	Variable	NA	Better transient performance
[48]	30	177–201	8	LC	1	2.72%	Low THD
[49]	3.5	300–340	4	RLC	1	NA	No galvanic isolation
[50]	18	345.6	6	NA	1	2.35%	More devices
[51]	0.5	60–120	4	LC	1	NA	High switching loss
[52]	0.4	120	4	LC	Variable	6.15%	VAR compensation, high THD
[53]	3.3	280–430	8	2 LC	0.985	<5%	97.8% High efficiency and complex control
[54]	30	500	16	3-ϕ LC, LC	1	< 5%	Better transient performance
[55]	4	200	8	LC	~1	<4%	Cost-effective
[56]	20	800	6	LCL	NA	~3.3%	99% High efficiency

increased power level. Furthermore, three-phase PFCs, such as FB [61], T-type [62], and NPC [63], have also been studied and analysed in the literature. However, single-phase PFCs have yet to find widespread use in the market because of their high price and relatively small size.

12.2.1.1 Totem-pole PFC

To overcome the conduction losses of traditional boost PFCs, bridge-less totem pole interleaved PFCs, as depicted in Figure 12.3(a), have emerged as a viable alternative [17,64]. It has two interleaved boost phases and an extra leg for synchronous rectification. Introducing a 180-degree phase shift in one of the interleaved phases leads to double the switching frequency.

Compared to traditional PFCs, totem pole PFCs offer advantages such as reduced components, higher efficiency, higher power density, and lower THD. However, they do exhibit higher THD compared to other topologies. The drawbacks of totem pole PFCs include the limited voltage range of the input and the high complexity of the control. A 3.3 kW silicon carbide (SiC)-based PFC was reported with 99.2% efficiency and a power factor of 0.99 for V2G/G2V applications [17].

12.2.1.2 Swiss rectifier

The desired features of an ac–dc rectifier include PFC, low THD, high efficiency, and high-power density. The three-phase buck-type rectifier (TPBR) is well-suited for the ac–dc stage due to its ability to meet these criteria [65]. Also, TPBR provides phase leg shoot-through protection, inherent over-current protection during short circuits, a larger output voltage control range, and inrush current-free startup over boost-type three-phase rectifiers [66]. The TPBR known as the Swiss rectifier (SR) was presented with an eight-switch unidirectional configuration [67]. The SR

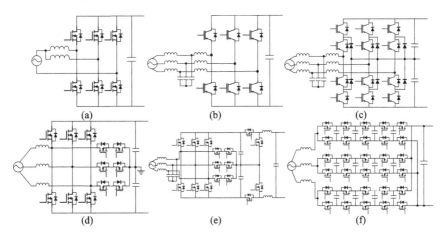

Figure 12.3 *PFC topologies: (a) totem pole PFC, (b) two-level PFC, (c) three-level NPC PFC, (d) T-type PFC, (e) Swiss rectifier type PFC, and (f) six-level flying capacitor multilevel PFC*

exhibits higher efficiency, reduced switching losses, and higher conduction losses in power devices over six-switch buck rectifier. Its circuit design allows for the implementation of dc–dc converter control techniques. For V2G/G2V applications, a bidirectional 7.5 kW SR with low input current THD was reported in [68], as depicted in Figure 12.3(e). Further, introducing interleaving in SRs minimises the current and voltage ripple, which reduces the need for filters, increases power delivery, and enhances reliability. [67]. In [69], the SR is connected to an inter-leaved dc–dc topology, which achieved 99.3% efficiency at 8 kW.

12.2.1.3   Full-bridge PFC

To expedite EV charging times, three-phase PFCs are advantageous due to their high power transfer capability [61]. The three-phase FB PFC is one such topology, as depicted in Figure 12.3(b). BYD has commercially implemented this PFC as the FEC in their bidirectional OBC. In this configuration, the PFC functions as inverter to supply power to different loads at a rate of 3.3 kW. ON Semiconductors has developed an 11 kW three-phase FB PFC using SiC technology, achieving an efficiency of 98.3% [70]. Various NPC topologies were reported in the literature to minimise THD and voltage rate-of-change (dv/dt), which are common issues in traditional three-phase FB PFCs [63]. An NPC-based boost converter is shown in Figure 12.3(c), offering advantages such as low switching losses, less device vol-tage stress, and improved THD and dv/dt performance. In [71], an NPC topology with lower input current THD is presented, featuring a distinct bipolar dc bus configuration. Furthermore, a study [72] introduces an interleaved six-level flying capacitor multilevel converter (FCMLC) that achieves high efficiency and power density. In [73], a three-phase FCMLC was presented, connecting each input phase to a six-level FCMLC as depicted in Figure 12.3(f). Another benefit of FCMLC is its reduced reliance on pulse width modulation (PWM) due to the higher switching node frequency, which enhances PWM resolution [74]. To address the issue of double line frequency pulsation, an active buffer is employed as a power decou-pling stage [72]. However, utilising smaller inductors in the FCMLC for size reduction makes controlling the PFC a challenge.

12.2.1.4   T-type PFC

The three-phase T-type PFC topology offers advantages over the conventional NPC configuration, including less number of devices and increased reliability. Also, it provides minimal device stress, reduced device's conduction and switching losses, and simple working principle. Figure 12.3(d) illustrates the three-level T-type PFC, which is built with six active devices. However, it commonly faces issues such as unbalance of neutral point voltage (NPV) and variations in common-mode voltage (CMV). To address these challenges, a medium vector PWM technique was pre-sented in [75]. This technique ensures the NPV balance and completely prevents variations in CMV simultaneously. In [76], a T-type PFC was presented with V2G capability, which was designed for off-board chargers. The experimental study has shown that the PFC achieved a THD of 4.92% even under abnormal operating conditions such as voltage unbalance and distortions of the grid. Another

10 kW rated SiC-based T-type PFC was introduced in [77]. To reduce switching losses while still allowing all switches to function in ZVS operation, this design implements a new ZVS PWM technique. The T-type PFCs based on wind band gap (WBG) devices such as SiC or gallium nitride (GaN) offer minimal on-state resistance, leading to reduced volume and ratings of passive components without compromising current capability [78]. Lower resistance values result in decreased conductance losses. Also, WBG devices enable the utilisation of increased switching frequencies, which results in high power density capability. Therefore, WBG-based T-type PFCs are superior choices for bidirectional PFC applications.

12.2.2   DC–DC converter topologies

Bidirectional dc–dc converters are used to step up or down the input or output voltage, which can be achieved by controlling the switches accordingly. During G2V or charging operation, the power can be transferred to the EVB from dc-bus. Conversely, in the discharging process, the power can be transferred from the EVB to the dc-bus in the V2G or discharging process. Operating these converter topologies at high frequencies results in high power density. However, this can lead to increased system noise. This is because of rapid ON/OFF of devices, leading to electromagnetic interference (EMI) in the grid. As a result, the performance of other grid-connected equipment may be compromised. Consequently, it becomes crucial to adopt an efficient converter topology that can address these challenges effectively.

Table 12.2 provides a summary of dc–dc converters reported in the literature. As per Table 12.2, the popular topologies are buck-boost and FB. Non-isolated topologies have the advantages of being smaller in size, low in cost, and simple to

Table 12.2   State-of-the-art dc–dc converter topologies

Reference	Power (kW)	Battery voltage (V)	Switches	Filter	Isolation	Switching	f_s (kHz)
[45]	3.5	270–360	2	LC	No	Hard	20
[46]	1.2	106–136	2	LC	No	Hard	50
[79]	3.5	250–450	8	C	Yes	Soft	145
[47]	3.3	235–430	8	CLC	Yes	Soft	250
[80]	0.25	150–300	8	C	Yes	Soft	100
[81]	1	250–450	6	C	Yes	Soft	100
[48]	30	177–201	4	C	No	Hard	20
[51]	0.5	60–120	2	C	No	Hard	20
[52]	0.4	120	4	C	No	Hard	20
[82]	6.6	250–415	8	LC	Yes	Soft	50
[83]	3.3	360	4	C	Yes	Soft	180–200
[84]	2.5	50	8	C	Yes	Soft	50
[85]	3.3	250–420	4	C	Yes	Soft	100–200
[86]	3.3	250–410	4	LC	Yes	Soft	Not known
[87]	10	400	4	LC	Yes	Soft	90–150

control since they don't require an isolation transformer. These topologies max-imise component utilisation by using fewer switching devices. However, the high current through the switching devices results in significant conduction loss, limited power density capability, and low efficiency. Interleaving is commonly employed in non-isolated topologies to address these limitations, particularly for high-power FC applications. Isolated topologies provide more protection against system fail-ures but have a larger converter footprint because of the need for a high-frequency transformer (HFT). International standards often mandate isolation, especially for high-power operations. Non-resonant topologies, with fewer passive components, tend to be more reliable. However, they are less efficient and incur more conduc-tion loss due to the slightly larger current flowing through the switches. Resonant converter topologies can mitigate these issues, providing higher efficiency. However, they exhibit a slower transient response.

12.2.2.1   Non-isolated converter topologies

Non-isolated converters are characterised by their simple architecture and requires less number of switches for power conversion. These topologies do not require an HFT, resulting in reduced cost and size compared to isolated topologies [48]. While they lack galvanic isolation, non-isolated topologies offer advantages such as wide range of voltage gain, soft-switching capability, and flexible control, leading to greater efficiency compared to isolated converter topologies [88]. One widely used non-isolated topology is the buck-boost type, [49], as shown in Figure 12.4(a). It

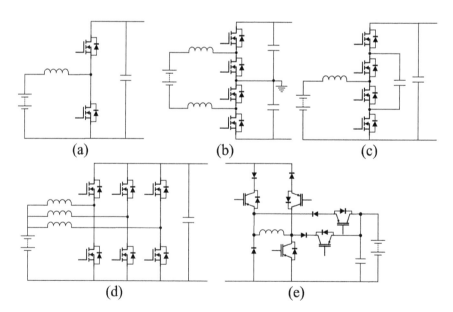

*Figure 12.4   Non-isolated converter topologies: (a) half-bridge boost converter,
(b) interleaved boost converter, (c) three-level boost converter, (d)
three-level FCC, and (e) noninverted buck boost converter*

operates in both buck and boost modes for charging and discharging the EVB. Figure 12.4(e) depicts the non-isolated converter that utilises five switches [50]. Interleaving the legs in the non-isolated topology leads to an increment in current-carrying capability and a decrement in current ripple.

A three-phase interleaved boost converter topology, as illustrated in Figure 12.4(d), was reported in [89]. It has gained significant attention for V2G applications as it offers a simple structure, enhanced power transfer, and improved efficiency. An optimal discharging algorithm is employed to maintain a C-rate of 1C. To generate 30 kW, six-phase legs are interconnected in parallel and inter-leaved to maximise efficiency of the converter in [89]. In discontinuous conduction mode (DCM) of operation, it allows for the use of smaller inductors and achieves ZVS across all devices, resulting in better efficiency [90].

Three-level boost dc–dc converter topologies, as illustrated in Figure 12.4(b), were reported with lower THD compared to boost converters [91]. They require a smaller inductor size (i.e., one-fourth of the size). A comparative analysis of boost, three-level boost and interleaved boost topologies was reported in [92], high-lighting the high efficiency and compact magnetics of the three-level boost topol-ogy. However, it is noted that this topology produces significant EMI, leading to EVB degradation. A three-level FCC was introduced (Figure 12.4(c)), although its switching commutation loop causes unexpected spikes (over-voltage) during switching instances.

12.2.2.2 Isolated converter topologies

According to safety standards, it is necessary to isolate the EVB from the grid during ac and dc charging processes. The isolation can be established in two ways: The first method involves a system comprising a line frequency transformer (LFT) (50 Hz or 60 Hz) connected to an ac–dc converter stage and a dc–dc converter stage connected to an EVB. The second method replaces the LFT with a HFT and incorporates it in the dc–dc stage, i.e., an isolated dc–dc converter, resulting in a more compact and flexible overall system [93]. Several isolated dc–dc converters were studied in the literature. Among them, the voltage-fed DAB dc–dc converter is a better solution for isolated converters, offering a wide voltage range for V2G operation, as shown in Figure 12.5(a) [94]. The DAB configuration consists of two H-bridges separated by a HFT. Power transfer in both V2G and G2V modes is obtained using the phase-shift modulation (PSM) strategy. It provides advantages such as galvanic isolation, simple control, high efficiency, high power density, soft-switching capability and reduced device peak voltage stress [95]. Furthermore, they are scalable, symmetrical in structure, modular, and have high power transfer capability [96]. However, it should be noted that the DAB topology with single-phase-shift (SPS) modulation exhibits significant backflow power through the switches and passive elements.

In [47], an alternative voltage-fed DAB bidirectional dc–dc converter was introduced, incorporating snubber capacitors across the switches. On the contrary, the current-fed DAB dc–dc converter topology offers advantages such as a wide voltage range, high voltage gain [97], minor diode ringing, a low turns ratio of the

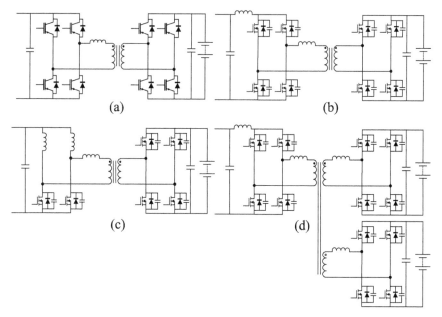

Figure 12.5   Isolated converter topologies: (a) dual voltage-fed DAB, (b) snubber-less current-voltage-fed half-full-bridge DAB, (c) snubber-less current–voltage-fed DAB, and (d) multi-port DAB

transformer, easy current control, and zero duty cycle loss [98]. Also, it has a minimal input current and a large turn-off loss. Snubbers are additionally utilised to decrease turn-off losses, surges in voltage, diode reverse recovery, and EMI. But the addition of snubbers increases the cost of the converter [81]. Proper selection of inductors is crucial to enhance the range of soft-switching and optimise the circulating currents in such converters. A current-voltage-fed DAB converter, as displayed in Figure 12.5(b), was demonstrated in [80] to obtain an enhanced soft-switching range. This converter exhibits ZVS for secondary devices and ZCS for primary devices due to the existence of a large magnetising inductance and naturally clamped soft-switching. It is noteworthy that the converter achieves ZCS and ZVS regardless of the load conditions. Additionally, Xuewei and Rathore [80] present a current-voltage-fed based DAB employing PSM as depicted in Figure 12.5(c).

For personal computer (PC) applications, Morrison and Egan [99] introduced an isolated DAB converter utilising MOSFETs. The converter comprises a voltage-fed half-bridge on the primary side and a voltage-fed FB on the transformer's secondary side. For low-power applications, a half-bridge-based DAB converter was presented in [100]. Authors in [101] presented an interleaved-based dual half-bridge converter. It lowers the devices' current stress and turns ratio while increasing their voltage boost capability. To interface the EVs with various energy sources, such as renewable energy (RE) for RE system applications [102], a multi-source technology is employed, necessitating a multi-port converter topology. An isolated bidirectional multiple-input based DAB dc–dc converter topology was reported in [103], as illustrated in Figure 12.5(d). It utilises a multiple-winding transformer and provides a lower footprint, low cost, and fewer power conversion stages.

12.2.2.3 Resonant converter topologies

During light-load operation, the DAB converters fail to possess the ZVS feature. Therefore, to further enhance soft-switching capability over a wide range of input and output voltages, resonant converter topologies are viable solutions. They provide notable advantages such as high efficiency, operation at very high frequencies, reduced power device requirements by eliminating the need for clamp circuits or snubbers, and negligible EMI [79]. Also, resonant topologies exhibit comparable performance and high efficiency to DAB topologies, particularly in high-power applications [38]. However, they face challenges related to severe start-up current, leading to minimal input impedance. A resonant-based DAB dc–dc converter topology, specifically an LLC-based configuration with four switches, was presented in [104]. Another resonant-based DAB topology was presented in [82], as depicted in Figure 12.6(a). For V2G operation, a CLLLC-based DAB was reported in [79], as illustrated in Figure 12.6(b). For power conversion, this topology employs the PSM technique, with a variable load switching frequency method to assure ZVS over a wide range of operation. Figure 12.6(c) depicts a three-winding transformer-based dual-output converter presented in [83] for high-voltage to low-voltage EV charging applications. Incorporating three transformer windings into one single transformer reduces

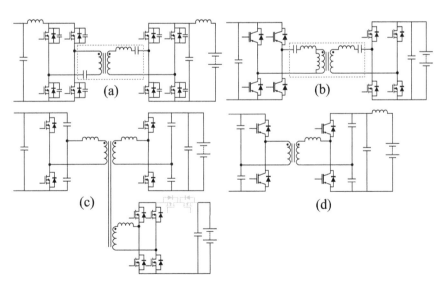

Figure 12.6 *Resonant converter topologies: (a) full-bridge PWM-based DAB, (b) dual output-based DAB, (c) dual voltage-fed CLLLC-based DAB, and (d) half-bridge series-based DAB*

costs, minimises saturation effects, occupies less space, and reduces magnetic losses. Furthermore, the ZVS turn-on feature enables high efficiency and high power density. But the design of a huge power transfer inductor is challenging and costly. In [86], a resonant DAB converter with a sinusoidal charging method was presented as illustrated in Figure 12.6(d). It explicitly eliminates the need for an large electrolytic capacitor. Also, this technique achieves ZCS turn-off and ZVS turn-on, resulting in enhanced efficiency over the entire power transmission range. But it does suffer from significant conduction losses.

12.3   Results and discussion

To show their effectiveness, the presented PFC and dc–dc topologies are validated using a hardware-in-loop (HIL) experimental setup. A maximum power rating of 3.3 kW is chosen for assessment, which is chosen based on the power ratings of most chargers listed in Table 12.1.

12.3.1   AC–DC converter topologies

The grid voltage and frequency are set to 415 V_{rms} and 50 Hz, respectively. Additionally, the dc bus voltage is established at 600 V. Figure 12.7 displays the experimental performance results of grid voltages and grid currents for considered PFCs. To regulate the grid current, a direct-quadrature (dq) control strategy is utilised with a reference current of 30 A. It can be seen that all of the converters' grid voltages and currents are in phase while operating in G2V mode and out of phase when operating in V2G mode.

12.3.2   DC–DC converter topologies

The dc link voltage is set at 110 V. The non-isolated topologies are evaluated considering 72 V and 26 Ah as the ratings of EVB. Figure 12.8 displays the performance results of the voltages and currents of the EV battery for the considered non-isolated converter topologies. It is observed that all non-isolated converters exhibit a step change in battery current of 30 A, which depicts the mode transition between the G2V and V2G operations.

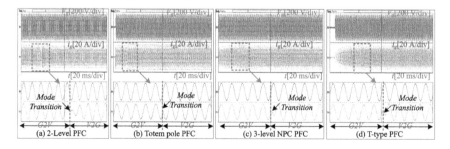

Figure 12.7   Results of bidirectional PFC converter topologies

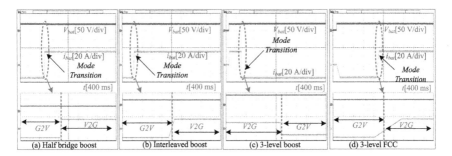

Figure 12.8   *Results of bidirectional non-isolated dc–dc converter topologies*

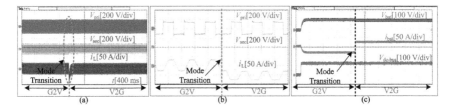

Figure 12.9   *Results of non-resonant bidirectional DAB dc–dc converter topology:*
(a) dynamic performance waveforms of DAB converter, (b) zoomed
version of the dynamic performance waveforms of DAB converter,
and (c) dynamic performance waveforms of the DAB converter at the
battery side and dc-link side

For the isolated dc–dc converters, 100 V and 26 Ah are chosen as the ratings of
an EV battery to validate each topology. Figures 12.9–12.11 depict the results of
the DAB-, LLC-, and CLLC-based dc–dc converter topologies, respectively. A
PSM technique is employed for the transmission of power in both V2G and G2V
modes of operation. As illustrated in Figure 12.9, the voltage at the primary side of
the transformer leads to the voltage at the secondary side in G2V mode, whereas
the voltage at the secondary side leads to the voltage at the primary side in V2G
mode. Figure 12.10 displays the results of the LLC bidirectional converter topology. It is noticed that the primary side current is in phase with the primary side
voltages in G2V mode and out-of-phase in V2G mode. Figure 12.11 depicts the
results of the CLLC resonant topology under V2G and G2V operation. The driving
pulses for the primary and secondary bridges are produced using pulse frequency
modulation (PFM) strategy. The CLLC converter's primary side current is shown
to be out of phase when operating in V2G mode and in phase with the voltages
when operating in G2V mode. Figures 12.9(c), 12.10(c), and 12.11(c) show that a
step change of 30 A in battery current is seen in the DAB, LLC, and CLLC converters during mode transition from G2V to V2G operation, respectively.

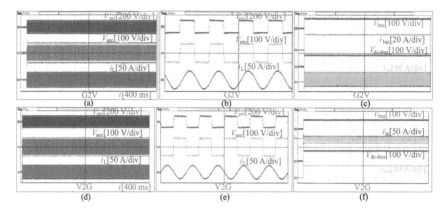

Figure 12.10   Results of resonant-based bidirectional LLC converter topology: (a) steady-state performance waveforms of LLC converter under G2V mode, (b) zoomed version of the Steady-state performance waveforms of LLC converter under G2V mode, (c) steady-state performance waveforms of LLC converter under G2V mode at both dc-link side and battery side, (d) steady-state performance waveforms of LLC converter under V2G mode, (e) zoomed version of the Steady-state performance waveforms of LLC converter under V2G mode, and (f) steady-state performance waveforms of LLC converter under V2G mode at both dc-link side and battery side

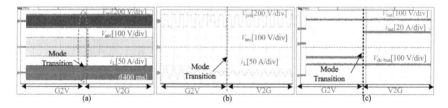

Figure 12.11   Results of resonant-based bidirectional CLLC converter topology: (a) dynamic performance waveforms of CLLC converter, (b) zoomed version of the dynamic performance waveforms of CLLC converter, and (c) dynamic performance waveforms of the CLLC converter at both the battery side and dc-link side

12.4   Comparative analysis and performance assessment

The effectiveness of the PFCs is assessed by measuring the current THD of the grid under low power (300 W) and high power (3.3 kW) operating conditions in both V2G and G2V modes, following the IEC 61000-3-2 class-A standard. The measurements are conducted using the Tektronix Mixed Domain Oscilloscope (MDO) 3024. Figure 12.12(a) and (b) represents the comparison of the THDs of each PFC,

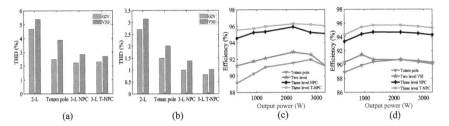

Figure 12.12 *THDs comparison of PFC:. (a) low power operation and (b) high power operation. Efficiency's comparison of PFCs. (c) G2V mode, and (d) V2G mode*

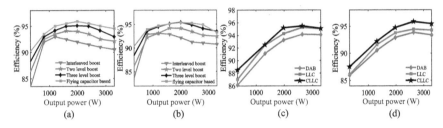

Figure 12.13 *Comparison of efficiencies of dc–dc converter topologies. Non-isolated topologies: (a) G2V operation, and (b) V2G operation, Isolated topologies: (c) G2V operation, and (d) V2G operation*

and it is observed that the PFCs operate nearer to their maximum efficiency point during high power, leading to a lower THD compared to PFCs operating at low power. Also, it can be seen that the THD of PFCs in G2V mode is lower than in V2G mode of operation for both low- and high-power operations. This difference can be attributed to the fact that the PFC gets power from the EV and sends it to the grid, while the grid itself contains THD because of non-linear loads used by consumers, leading to increased harmonic distortion. Among the evaluated PFCs, the NPC topologies exhibit lower THD values. This can be attributed to their utilisation of a NPV, which leads to input current balancing and reduces THD. Figures 12.12 (c) and (d) and 12.13 display the efficencies of PFCs and dc–dc topologies, respectively, with respect to the output power. In general, the efficiency of a converter shows an initial increase as the operating power increases, primarily because the switching and conduction losses of the devices are minimal. But the efficiency begins to decline during high-power operation. This is attributed to increased losses in passive components and the voltage drop across semiconductor switches and diodes. Additionally, the proportion of power losses contributed by switching and conduction losses becomes more significant at higher power levels, further contributing to the decrease in efficiency.

12.4.1   Observations

A performance comparison of the considered PFC topologies is carried out and presented in Table 12.3. It is noticed that the efficiency of the three-level NPC and T-NPC PFCs is higher than the other PFCs. But the trade-off is that the power density of the T-NPC PFC is greater than the NPC PFC. This is attributed to the additional switching devices employed in the three-level NPC, resulting in higher conduction losses and a reduced power density than the T-NPC PFC. Further, the use of the inverter leg's neutral point reduces the voltage stress on the semiconductor devices and the THD in the case of NPC PFCs.

A comparison of bidirectional non-isolated dc–dc converter topologies is shown in Table 12.4. In non-isolated converters, the higher drain-source voltage that builds up across the diodes when the switch is turned off is the main reason why reverse recovery losses are higher. However, power dissipation is minimised in the case of the three-level FCC converter since the voltage developed across the diodes is only 50% of the rated output voltage. As a result, at the specified switching frequency, gate losses and voltage stress on the devices become insignificant. Half-bridge-based boost dc–dc converters use fewer switches than traditional converters, which results

Table 12.3   Observations of bidirectional PFC topologies

Parameter	Two-level	Totem pole	Three-level NPC	Three-level TNPC
Switching loss	High	Medium	Low	Medium
Switches/diodes	6/0	6/0	12/6	12/0
Conduction loss	Low	Medium	High	Medium
Grid current THD	High	Low	Very low	Very low
Efficiency	Low	Low	High	High
Power density	Low	Low	Medium	High
Peak voltage stress	High	Medium	Low	Low
Control	Easy	Medium	Medium	Medium

Table 12.4   Observations of non-isolated bidirectional dc–dc topologies

Parameter	Half-bridge boost	Interleaved boost	Three-level boost	Three-level FCC
Efficiency	Medium	Low	High	High
Power rating	Low	High	High	High
Peak devices voltage stress	High	Medium	Medium	Low
Switches/diodes	2/0	6/0	4/0	4/0
Switching loss	High	High	Low	Low
Conduction losses	Medium	Medium	High	High
Control complexity	Easy	Medium	Medium	Medium
Current ripple	Low	Medium	Very low	Very low
Switching Frequency	Low	Medium	Medium	High

Table 12.5   Observations of isolated bidirectional dc–dc topologies

Parameter	DAB	LLC	CLLC
Efficiency	Medium	High	Very high
Power density	High	High	Very high
Switches/diodes	8/0	8/0	8/0
Conduction losses	Lowest	High	Medium
Turn ON switching loss	ZVS	ZVS	ZVS
Total losses	Medium	Low	Low
Turn OFF switching loss	High	Low	Low
Paralleling Modules	Easy	Intensive	Intensive
Control complexity	Simple to complex	Medium	Medium
Switching frequency	High	High	Very high

in lower conduction losses. However, the use of a larger number of diodes leads to increased diode losses. The efficiency of each converter is illustrated in Figure 12.13. Despite having higher conduction losses, the three-level FCC-based non-isolated topology exhibits high efficiency among the compared non-isolated converters. As a half-bridge-based boost topology has half as many devices, a carrier signal is enough to generate two PWM switching signals. Both the three-level boost and FCC converters exhibit better performance in terms of efficiency than other considered non-isolated converters. But the trade-off is that the three-level FCC topology offers less device voltage stress and lower conduction losses compared to the three-level boost-based non-isolated dc–dc converter, indicating better performance.

Table 12.5 presents insights and a performance comparison of all considered isolated dc–dc converters. It is noticed that the efficacy of the CLLC-based isolated topology is better than that of the DAB converter, as the switches in the DAB converter are not able to maintain ZVS during light-load power operations. Moreover, employing the SPS modulation strategy in DAB topology generates significant circulating currents, resulting in low converter efficiency. Additionally, the presence of harmonics in the AC waveform increases transformer power losses. Hence, conventional DAB topology requires extra filtering circuits to mitigate the harmonic distortions. Due to this reason, the CLLC-based DAB topology's power density is better than the conventional non-resonant DAB topology. In terms of control complexity, the DAB converter exhibits lower complexity than the CLLC-based DAB topology. This is due to the fact that the voltage gains in a non-resonant DAB topology are linearly proportional to the phase angle and loads. Conversely, the gains in the CLLC converter have a nonlinear relationship with the loads. However, careful consideration must be given to the selection of the Q value in the CLLC-based DAB topology to narrow the range of working frequencies, particularly at lower Q values. Hence, the trade-off is that the CLLC DAB's power density is better than conventional DAB topology. As a result, resonant-based DAB topologies demonstrate superior performance over non-resonant-based DAB topologies for V2G and G2V operations. In addition, Figures 12.12(c) and (d) and 12.13 provide a performance comparison for PFCs, non-isolated, and isolated topologies in terms of efficiencies, respectively.

12.5   Challenges, recommendations, and future directions

This section focuses on the key factors that play an important role in improving the performance and widespread adoption of bidirectional chargers. These factors include achieving high power density, attaining high efficiency, minimising costs, and enhancing power transmission capabilities. This section provides a comprehensive discussion on the challenges associated with converter topologies, the utilisation of WBG devices, optimising thermal design, mitigating harmonics, and advancements in inductive charging systems.

12.5.1   Challenges

The power electronics industry and academic research community have made significant developments towards achieving low cost, enhanced efficiency, improved performance, and higher power density. Extensive efforts have been dedicated to the development of PFCs, non-isolated and isolated dc–dc converter topologies. However, a comprehensive analysis is essential for the advancement of V2G technology. While bidirectional converters in V2G applications already exhibit high efficiency, there are ongoing challenges such as compact design, optimal thermal management, increased power density, reduced cost, and high reliability. A modular-based architecture has emerged as a viable solution for high-power operations in bidirectional chargers. However, ensuring the reliability of such chargers poses a challenge. In an effort to enhance reliability, a fault-tolerant approach was presented in [105]. However, this approach requires additional circuitry, leading to an increase in the cost and size of the entire system. Another approach for reliable operation is the use of a modular-based power converter, as suggested in [106]. Additionally, peak load shaving and frequency regulation are crucial parameters to be considered in V2G infrastructure. But implementing these functions significantly impacts the power quality (PQ) of the utility grid and contributes to the degradation of EVBs. This is attributed to the frequent charging and discharging cycles of EVBs. Therefore, maintaining satisfactory PQ levels while preventing battery degradation presents another challenging task.

12.5.2   Recommendations and future directions

1. It is expected that bidirectional chargers will continue to adopt a two-stage design in the coming decade. The development of T-NPC PFCs in ac–dc stages is currently underway for high-power applications. This is due to their simplicity in design, high efficiency, and ease of operation. For the dc–dc stages of the charger, DAB topologies are commonly chosen due to their high efficiency, allowing for efficient power transfer. However, these converters may involve additional switches and more complex control regulations. In the case of EVB charging, the T-NPC PFC topology, despite its additional switches and intricate control requirements, is recommended because of its high efficiency. Alternatively, single-stage-based converter topologies offer high power density

and lower footprint. However, they face challenges related to hardware design and achieving high efficiency [107].

2. In the EV sector, WBG-based switches are strategically utilised to enhance efficiency and achieve a lower footprint. The SiC MOSFETs, with their high blocking voltage of 3.3 kV, can handle 800 V of battery voltage, making them suitable for most EV applications [108]. On the other hand, GaN devices offer reduced on-state resistance, allowing for lower ratings and sizes of passive components while maintaining current capabilities. GaN also exhibits better performance than SiC at high temperatures, making it preferable for high-frequency applications. Commercially available GaN MOSFETs typically have a voltage tolerance of 650 V, making them more suitable for EVB charging applications. Additionally, companies like Texas Instruments contribute to reducing the costs of WBG devices and improving overall performance. However, the adoption of these devices also presents challenges related to gate drivers, printed circuit board layout, reduction of EMI, and dead/blanking time that require specific attention in design and implementation.

3. The compact nature of designs and the presence of passive component losses such as skin effects and eddy currents pose challenges for thermal management in integrated power electronics at high switching frequencies. In particular, the eddy currents can lead to reduced charger efficiency and reliability. To address these challenges, the bidirectional PFCs and dc–dc converter topologies of the future should accommodate advanced cooling techniques like high-end liquid cooling plates and establish effective connections with the cooling system that has been fixed in the corresponding vehicle [109]. Experts must develop the vehicle's thermal management system precisely so that the EV charger and EVB pack are kept at an optimal temperature. Innovative approaches like additive printing can be employed to create new heat transfer or exchange systems with improved surface densities. Vehicle cooling systems can additionally employ the use of heat pipes due to their huge heat transfer or exchange capacity, inexpensive price, and low thermal resistance.

4. Fast chargers often experience deterioration of voltage profiles and high emissions of harmonics, which can pose challenges for their widespread adoption. The presence of power PQ issues in FCSs highlights the need for better understanding of PQ standards and the implementation of mitigation measures. Consequently, there is ample scope for further research in the field of PQ. It is essential to assess FCS using PQ criteria outlined in relevant standards. However, existing standards primarily focus on addressing the harmonic emissions of fast chargers. To mitigate the impact on the grid's PQ, a potential solution is to employ a methodology that involves coordinating many FCs in parallel. Such an approach improves the performance of FCs by optimising the challenges associated with PQ.

5. Conductive EV charging is associated with several challenges, including high charging times, limited charging infrastructure availability, range anxiety, and the long waiting line. To address these challenges while enhancing user convenience and comfort, inductive power transfer (IPT) technology would be a

viable solution [110]. However, IPT also faces its own set of challenges, such as high cost, lower power density compared to conductive charging, relatively lower efficiency, and high design and development complexity. Moreover, enabling distant charging poses additional difficulties in high-power applications. In spite of these obstacles, several original equipment manufacturers (OEMs) are investigating IPT technology for EVs. For instance, IPT technology has been developed for the BMW 530e plug-in EV in California, and Honda has developed the WiTricity wireless V2G concept, which utilises IPT for bidirectional power transmission.

12.6   Conclusion

This chapter presents a comprehensive review of the bidirectional ac–dc and dc–dc topologies for V2G and G2V applications. Experimental testing was carried out to assess the performance of these converter topologies in V2G and G2V modes. The effectiveness of the PFCs was assessed by analysing the THD in both V2G and G2V modes. A comparative analysis was carried out to show the effectiveness of the considered bidirectional ac–dc and dc–dc topologies, and the observations are presented. The results indicate that among the PFC topologies, the three-level T-NPC PFC provides the highest efficiency and exhibits the lowest THD. Among the isolated converter topologies, the CLLC converter demonstrates superior performance compared to others. In the non-isolated bidirectional dc–dc topologies, both the three-level boost and FCC-based topologies exhibit high efficiency. Furthermore, the challenges, recommendations, and future directions in the development of bidirectional chargers for V2G technology were discussed.

References

[1] Ghazanfari A, and Perreault C The path to a vehicle-to-grid future: Powering electric mobility forward. *IEEE Ind Electron Mag.* 2021;16(3):4–13.
[2] Hannan M, Mollik M, Al-Shetwi AQ, *et al.* Vehicle to grid connected technologies and charging strategies: Operation, control, issues and recommendations. *J Cleaner Prod.* 2022;339:130587.
[3] Kempton W, Tomic J, Letendre S, Brooks A, and Lipman T. *Vehicle-to-grid power: battery, hybrid, and fuel cell vehicles as resources for distributed electric power in California.* 2001. Available at: https://escholarship.org/uc/item/5cc9g0jp.
[4] Tomić J, and Kempton W Using fleets of electric-drive vehicles for grid support. *J Power Sources.* 2007;168(2):459–468.
[5] Rahman MS, Rafi FH, Hossain MJ, and Lu J. Power control and monitoring of the smart grid with EVs. In Lu J and Hossain J (eds), *Vehicle-to-Grid: Linking Electric Vehicles to the Smart Grid* (pp. 107–156). Stevenage: Institution of Engineering and Technology; 2015

[6]   Rachid A, El Fadil H, Gaouzi K, *et al.* Electric vehicle charging systems: Comprehensive review. *Energies.* 2022;16(1):255.

[7]   Vadi S, Bayindir R, Colak AM, *et al.* A review on communication standards and charging topologies of V2G and V2H operation strategies. *Energies.* 2019;12(19):3748.

[8]   Upputuri RP, and Subudhi B A comprehensive review and performance evaluation of bidirectional charger topologies for V2G/G2V operations in EV applications. *IEEE Trans Transp Electrif.* 202310(1):583–595.

[9]   Khaligh A, and D'Antonio M Global trends in high-power on-board chargers for electric vehicles. *IEEE Trans Veh Technol.* 2019;68(4):3306–3324.

[10] Brenna M, Foiadelli F, Leone C, *et al.* Electric vehicles charging technology review and optimal size estimation. *J Electr Eng Technol.* 2020;*15* (6):2539–2552.

[11] Ginart A, and Sharifipour B High penetration of electric vehicles could change the residential power system: Public dc fast chargers will not be enough. *IEEE Electrific Mag.* 2021;9(2):34–42.

[12] Khalid MR, Khan IA, Hameed S, *et al.* A comprehensive review on structural topologies, power levels, energy storage systems, and standards for electric vehicle charging stations and their impacts on grid. *IEEE Access.* 2021; 9:128069–128094.

[13] Yuan J, Dorn-Gomba L, Callegaro AD, *et al.* A review of bidirectional on-board chargers for electric vehicles. *IEEE Access.* 2021;9:51501–51518.

[14] Sayed SS, and Massoud AM Review on state-of-the-art unidirectional non-isolated power factor correction converters for short-/long-distance electric vehicles. *IEEE Access.* 2022;10:11308–11340.

[15] Safayatullah M, Elrais MT, Ghosh S, *et al.* A comprehensive review of power converter topologies and control methods for electric vehicle fast charging applications. *IEEE Access.* 2022;10:40753–40793.

[16] Singh B, Singh BN, Chandra A, *et al.* A review of three-phase improved power quality AC–DC converters. *IEEE Trans Ind Electron* 2004;51(3):641–660.

[17] Tang Y, Ding W, and Khaligh A A bridgeless totem-pole interleaved PFC converter for plug-in electric vehicles. In: *Proc. IEEE Applied Power Electronics Conference and Exposition (APEC)*. Piscataway, NJ: IEEE; 2016, pp. 440–445.

[18] Mallik A, Ding W, Shi C, *et al.* Input voltage sensorless duty compensation control for a three-phase boost PFC converter. *IEEE Trans Ind Appl.* 2017; 53(2):1527–1537.

[19] Ye Z, Boroyevich D, Choi JY, *et al.* Control of circulating current in two parallel three-phase boost rectifiers. *IEEE Trans Power Electron.* 2002;17(5):609–615.

[20] Enjeti PN, and Choudhury SA A new control strategy to improve the performance of a PWM AC to DC converter under unbalanced operating conditions. *IEEE Trans Power Electron.* 1993;8(4):493–500.

[21] Malinowski M, Jasinski M, and Kazmierkowski MP Simple direct power control of three-phase PWM rectifier using space-vector modulation (DPC-SVM). *IEEE Trans Ind Electron.* 2004;51(2):447–454.

[22]   Tyagi S, Mallik A, Sankar A, *et al.* Third harmonic compensation in a single-phase H-bridge PFC. In: *European Conference on Power Electronics and Applications (EPE'17 ECCE Europe)*; 2017, pp. P.1–P.10.

[23]   Lu W, Han J, Li S, *et al.* Mitigating line frequency instability of boost PFC converter under proportional outer-voltage loop with additional third current-harmonic feedforward compensation. *IEEE Trans Circuits Syst I Reg Papers*. 2019;66(11):4528–4541.

[24]   IEEE Recommended Practice and Requirements for Harmonic Control in Electric Power Systems. IEEE Std 519-2014 (Revision of IEEE Std 519-1992). 2014, pp. 1–29.

[25]   Sankar UA, Mallik A, and Khaligh A Duty compensated reduced harmonic control for a single-phase H-bridge PFC converter. In: *IEEE Applied Power Electronics Conference and Exposition (APEC)*; 2018, pp. 1996–2000.

[26]   Shin JW, Cho BH, and Lee JH Average current mode control in digitally controlled discontinuous-conduction-mode PFC rectifier for improved line current distortion. In: IEEE Applied Power Electronics Conference and Exposition (APEC); 2011, pp. 71–77.

[27]   Garcia O, Zumel P, de Castro A, *et al.* Automotive DC–DC bidirectional converter made with many interleaved buck stages. *IEEE Trans Power Electron*. 2006;21(3):578–586.

[28]   Esteki M, Poorali B, Adib E, *et al.* Interleaved buck converter with continuous input current, extremely low output current ripple, low switching losses, and improved step-down conversion ratio. *IEEE Trans Ind Electron*. 2015;62(8):4769–4776.

[29]   Lee IO, Cho SY, and Moon GW. Interleaved buck converter having low switching losses and improved step-down conversion ratio. *IEEE Trans Power Electron*. 2012;27(8):3664–3675.

[30]   Drobnic K, Grandi G, Hammami M, *et al.* An output ripple-free fast charger for electric vehicles based on grid-tied modular three-phase interleaved converters. *IEEE Trans Ind Appl*. 2019;55(6):6102–6114.

[31]   Repecho V, Biel D, Ramos-Lara R, *et al.* Fixed-switching frequency interleaved sliding mode eight-phase synchronous buck converter. *IEEE Trans Power Electron*. 2018;33(1):676–688.

[32]   Tan L, Wu B, Rivera S, *et al.* Comprehensive DC power balance management in high-power three-level DC–DC converter for electric vehicle fast charging. *IEEE Trans Power Electron*. 2016;31(1):89–100.

[33]   Zhang Z, Wu YQ, Gu DJ, *et al.* Current ripple mechanism with quantization in digital LLC converters for battery charging applications. *IEEE Trans Power Electron*. 2018;33(2):1303–1312.

[34]   Musavi F, Craciun M, Gautam DS, *et al.* Control strategies for wide output voltage range LLC resonant DC–DC converters in battery chargers. *IEEE Trans Veh Technol*. 2014;63(3):1117–1125.

[35]   Krismer F, and Kolar JW Accurate power loss model derivation of a high-current dual active bridge converter for an automotive application. *IEEE Trans Ind Electron*. 2010;57(3):881–891.

[36] Park J, and Choi S Design and control of a bidirectional resonant DC–DC converter for automotive engine/battery hybrid power generators. *IEEE Trans Power Electron*. 2014;29(7):3748–3757.

[37] Li B, Lee FC, Li Q, *et al*. Bi-directional on-board charger architecture and control for achieving ultra-high efficiency with wide battery voltage range. In: *2017 IEEE Applied Power Electronics Conference and Exposition (APEC)*; 2017, pp. 3688–3694.

[38] Li H, Zhang Z, Wang S, *et al*. A 300-kHz 6.6-kW SiC bidirectional LLC onboard charger. *IEEE Trans Ind Electron*. 2019;67(2): 1435–1445.

[39] Chen W, Rong P, and Lu Z Snubberless bidirectional DC–DC converter with new CLLC resonant tank featuring minimized switching loss. *IEEE Trans Ind Electron*. 2010;57(9):3075–3086.

[40] Jung JH, Kim HS, Ryu MH, *et al*. Design methodology of bidirectional CLLC resonant converter for high-frequency isolation of DC distribution systems. *IEEE Trans Power Electron*. 2013;28(4):1741–1755.

[41] Xue L, Shen Z, Boroyevich D, *et al*. Dual active bridge-based battery charger for plug-in hybrid electric vehicle with charging current containing low frequency ripple. *IEEE Trans Power Electron*. 2015;30(12):7299–7307.

[42] Wu X, Hua G, Zhang J, *et al*. A new current-driven synchronous rectifier for series–parallel resonant (*LLC*) DC–DC converter. *IEEE Trans Ind Electron*. 2011;58(1):289–297.

[43] Jeong DK, Ryu MH, Kim HG, *et al*. Optimized design of bi-directional dual active bridge converter for low-voltage battery charger. *J Power Electron*. 2014;14(3):468–477.

[44] Sayed MA, Suzuki K, Takeshita T, *et al*. PWM switching technique for three-phase bidirectional grid-tie DC–AC–AC converter with high-frequency isolation. *IEEE Trans Power Electron*. 2017;33(1):845–858.

[45] Pinto J, Monteiro V, Gonçalves H, *et al*. Onboard reconfigurable battery charger for electric vehicles with traction-to-auxiliary mode. *IEEE Trans Veh Technol*. 2013;63(3):1104–1116.

[46] Verma AK, Singh B, and Shahani D Grid to vehicle and vehicle to grid energy transfer using single-phase bidirectional AC–DC converter and bidirectional DC–DC converter. In: *Proceedings of the International Conference on Energy, Automation and Signal*. Piscataway, NJ: IEEE, 2011, pp. 1–5.

[47] Pahlevani M, and Jain P A fast DC-bus voltage controller for bidirectional single-phase AC/DC converters. *IEEE Trans Power Electron*. 2014;30(8): 4536–4547.

[48] Hegazy O, Van Mierlo J, and Lataire P Control and analysis of an integrated bidirectional DC/AC and DC/DC converters for plug-in hybrid electric vehicle applications. *J Power Electron*. 2011;11(4):408–417.

[49] Pinto J, Monteiro V, Gonçalves H, *et al*. Bidirectional battery charger with grid-to-vehicle, vehicle-to-grid and vehicle-to-home technologies. In:

Proceedings of the IEEE Industrial Electronics Society. Piscataway, NJ: IEEE; 2013, pp. 5934–5939.

[50]   Onar OC, Kobayashi J, Erb DC, *et al.* A bidirectional high-power-quality grid interface with a novel bidirectional noninverted buck–boost converter for PHEVs. *IEEE Trans Veh Technol.* 2012;61(5):2018–2032.

[51]   Han H, Liu Y, Sun Y, *et al.* A single-phase current-source bidirectional converter for V2G applications. *J Power Electron.* 2014;14(3):458–467.

[52]   Peng T, Yang P, Dan H, *et al.* A single-phase bidirectional AC/DC converter for V2G applications. *Energies.* 2017;10(7):881.

[53]   Jauch F, and Biela J Single-phase single-stage bidirectional isolated ZVS AC–DC converter with PFC. In: *Proceedings of the International Power Electronics and Motion Control Conference (EPE/PEMC).* Piscataway, NJ: IEEE; 2012, p. LS5d–1.

[54]   Sandoval JJ, Essakiappan S, and Enjeti P A bidirectional series resonant matrix converter topology for electric vehicle DC fast charging. In: *Proceedings of the IEEE Applied Power Electronics Conference and Exposition (APEC).* Piscataway, NJ: IEEE; 2015, pp. 3109–3116.

[55]   Thrimawithana DJ, Madawala UK, Twiname R, *et al.* A novel matrix converter based resonant dual active bridge for V2G applications. In: *Proceedings of the International Power Energy Conference (IPEC).* Piscataway, NJ: IEEE; 2012, pp. 503–508.

[56]   Choi W, Han D, Morris CT, *et al.* Achieving high efficiency using SiC MOSFETs and reduced output filter for grid-connected V2G inverter. In: *Proceedings of the IEEE Industrial Electronics Society*; 2015, pp. 003052–003057.

[57]   Liu B, Qiu M, Jing L, *et al.* Design of ac/dc converter for bidirectional on-board battery charger with minimizing the amount of SiC MOSFET. In: *IEEE Transportation Electrification Conference and Expo, Asia-Pacific (ITEC Asia-Pacific)*; 2017, pp. 1–6.

[58]   Chou D, Fernandez K, and Pilawa-Podgurski RCN An interleaved 6-level GaN bidirectional converter for level II electric vehicle charging. In: *IEEE Applied Power Electronics Conference and Exposition (APEC)*; 2019, pp. 594–600.

[59]   Sfakianakis GE, Everts J, and Lomonova EA Overview of the requirements and implementations of bidirectional isolated AC–DC converters for automotive battery charging applications. In: *10th International Conference on Ecological Vehicles and Renewable Energies (EVER)*; 2015, pp. 1–12.

[60]   Lin B, and Yang T Single-phase half-bridge rectifier with power factor correction. *IEE Proceedings—Electric Power Applications.* 2004;151(4):443–450.

[61]   Moia J, Lago J, Perin AJ, *et al.* Comparison of three-phase PWM rectifiers to interface AC grids and bipolar DC active distribution networks. In: *Proceedings of the IEEE International Symposium on Power Electronics for Distributed Generation Systems (PEDG).* Piscataway, NJ: IEEE; 2012, pp. 221–228.

[62]   Schweizer M, and Kolar JW Design and implementation of a highly efficient three-level T-type converter for low-voltage applications. *IEEE Trans Power Electron.* 2012;28(2):899–907.

[63] Tan L, Wu B, Yaramasu V, *et al.* Effective voltage balance control for bipolar-DC-bus-fed EV charging station with three-level DC–DC fast charger. *IEEE Trans Ind Electron.* 2016;63(7):4031–4041.

[64] Marxgut C, Krismer F, Bortis D, *et al.* Ultraflat interleaved triangular current mode (TCM) single-phase PFC rectifier. *IEEE Trans Power Electron.* 2013;29(2):873–882.

[65] Kolar JW, and Friedli T The essence of three-phase PFC rectifier systems—Part I. *IEEE Trans Power Electron.* 2013;28(1):176–198.

[66] Ancuti MC, Sorandaru C, Musuroi S, *et al.* High efficiency three-phase interleaved buck-type PFC rectifier concepts. In: *Proceedings of the IEEE Industrial Electronics Society*; 2015, pp. 004990–004995.

[67] Soeiro TB, Friedli T, and Kolar JW Swiss rectifier – a novel three-phase buck-type PFC topology for electric vehicle battery charging. In: *IEEE Applied Power Electronics Conference and Exposition (APEC)*; 2012, pp. 2617–2624.

[68] Schrittwieser L, Kolar JW, and Soeiro TB Novel SWISS rectifier modulation scheme preventing input current distortions at sector boundaries. *IEEE Trans Power Electron.* 2017;32(7):5771–5785.

[69] Schrittwieser L, Leibl M, Haider M, *et al.* 99.3% efficient three-phase buck-type all-SiC SWISS rectifier for DC distribution systems. *IEEE Trans Power Electron.* 2018;34(1):126–140.

[70] ON Semiconductor Corporation *On board charger (OBC) three-phase PFC converter*; March 2020.

[71] Celanovic N, and Boroyevich D A comprehensive study of neutral-point voltage balancing problem in three-level neutral-point-clamped voltage source PWM inverters. *IEEE Trans Power Electron.* 2000;15(2):242–249.

[72] Liao Z, Chou D, Fernandez K, *et al.* Architecture and control of an interleaved 6-level bidirectional converter with an active energy buffer for level-II electric vehicle charging. In: *2020 IEEE Energy Conversion Congress and Exposition (ECCE)*; 2020, pp. 4137–4142.

[73] Syu YL, Liao Z, Fu NT, *et al.* Design and control of a high power density three-phase flying capacitor multilevel power factor correction rectifier. In: *2021 IEEE Applied Power Electronics Conference and Exposition (APEC)*; 2021, pp. 613–618.

[74] Syed A, Ahmed E, Maksimovic D, *et al.* Digital pulse width modulator architectures. In: *2004 IEEE 35th Annual Power Electronics Specialists Conference (IEEE Cat. No.04CH37551)*, vol. 6; 2004, pp. 4689–4695.

[75] Lak M, Chuang BR, and Lee TL A common-mode voltage elimination method with active neutral point voltage balancing control for three-level T-type inverter. *IEEE Trans Ind Appl.* 2022;58(6):7499–7514.

[76] Thanakam T, and Kumsuwan Y A high-performance grid-tied three-level T-type converter for off-board battery chargers under AC-side voltage distortions and unbalances. In: *19th International Conference on Electrical Engineering/Electronics, Computer, Telecommunications and Information Technology (ECTI-CON)*; 2022, pp. 1–4.

[77] Deng J, Hu C, Shi K, *et al.* A ZVS-PWM scheme for three-phase active-clamping T-type inverters. *IEEE Trans Power Electron.* 2023;38(3):3951–3964.

[78] Mitova R, Ghosh R, Mhaskar U, *et al.* Investigations of 600-V GaN HEMT and GaN diode for power converter applications. *IEEE Trans Power Electron.* 2014;29(5):2441–2452.

[79] Zahid ZU, Dalala ZM, Chen R, *et al.* Design of bidirectional DC–DC resonant converter for vehicle-to-grid (V2G) applications. *IEEE Trans Transp. Electrif.* 2015;1(3):232–244.

[80] Xuewei P, and Rathore AK Novel bidirectional snubberless naturally commutated soft-switching current-fed full-bridge isolated DC/DC converter for fuel cell vehicles. *IEEE Trans Ind Electron.* 2013;61(5):2307–2315.

[81] Rathore AK, and Prasanna U Analysis, design, and experimental results of novel snubberless bidirectional naturally clamped ZCS/ZVS current-fed half-bridge DC/DC converter for fuel cell vehicles. *IEEE Trans Ind Electron.* 2012;60(10):4482–4491.

[82] Lee BK, Kim JP, Kim SG, *et al.* An isolated/bidirectional PWM resonant converter for V2G (H) EV on-board charger. *IEEE Trans Veh Technol.* 2017;66(9):7741–7750.

[83] Tang Y, Lu J, Wu B, *et al.* An integrated dual-output isolated converter for plug-in electric vehicles. *IEEE Trans Veh Technol.* 2017;67(2): 966–976.

[84] Twiname RP, Thrimawithana DJ, Madawala UK, *et al.* A new resonant bidirectional DC–DC converter topology. *IEEE Trans Power Electron.* 2013;29(9):4733–4740.

[85] Zou S, Lu J, Mallik A, *et al.* Bi-directional CLLC converter with synchronous rectification for plug-in electric vehicles. *IEEE Trans Ind Appl.* 2017;54(2):998–1005.

[86] Kwon M, and Choi S. An electrolytic capacitorless bidirectional EV charger for V2G and V2H applications. *IEEE Trans Power Electron.* 2016;32 (9):6792–6799.

[87] Wang X, Jiang C, Lei B, *et al.* Power-loss analysis and efficiency maximization of a silicon-carbide MOSFET-based three-phase 10-kW bidirectional EV charger using variable-DC-bus control. *IEEE J Emerg Sel Top Power Electron.* 2016;4(3):880–892.

[88] Koç Y, Birbir Y, and Bodur H Non-isolated high step-up DC/DC converters—an overview. *Alex Eng J.* 2022;61(2):1091–1132.

[89] Kang T, Kim C, Suh Y, *et al.* A design and control of bi-directional non-isolated DC–DC converter for rapid electric vehicle charging system. In: *Proceedings of the IEEE Applied Power Electronics Conference and Exposition (APEC).* Piscataway, NJ: IEEE; 2012. pp. 14–21.

[90] Zhang J, Lai JS, Kim RY, *et al.* High-power density design of a soft-switching high-power bidirectional dc–dc converter. *IEEE Trans Power Electron.* 2007;22(4):1145–1153.

[91] Grbović PJ, Delarue P, Le Moigne P, *et al.* A bidirectional three-level DC–DC converter for the ultracapacitor applications. *IEEE Trans Ind Electron.* 2009;57(10):3415–3430.

[92] Dusmez S, Hasanzadeh A, and Khaligh A Comparative analysis of bidirectional three-level DC–DC converter for automotive applications. *IEEE Trans Ind Electron.* 2014;62(5):3305–3315.

[93] Rituraj G, Mouli GRC, and Bauer P A comprehensive review on off-grid and hybrid charging systems for electric vehicles. *IEEE Open J Ind Electron Soc.* 2022;3:203–222.

[94] Xu B, Wang H, Sun H, *et al.* Design of a bidirectional power converter for charging pile based on V2G. In: *Proceedings of the IEEE International Conference on Industrial Technology (ICIT)*. Piscataway, NJ: IEEE; 2017, pp. 527–531.

[95] Du Y, Lukic S, Jacobson B, *et al.* Review of high power isolated bidirectional DC–DC converters for PHEV/EV DC charging infrastructure. In: *Proceedings of the IEEE Energy Conversion Congress and Exposition*. Piscataway, NJ: IEEE; 2011, pp.553–560.

[96] Zhao B, Song Q, Liu W, *et al.* Overview of dual-active-bridge isolated bidirectional DC–DC converter for high-frequency-link power-conversion system. *IEEE Trans Power Electron.* 2013;29(8):4091–4106.

[97] Dobakhshari SS, Milimonfared J, Taheri M, *et al.* A quasi-resonant current-fed converter with minimum switching losses. *IEEE Trans Power Electron.* 2016;32(1):353–362.

[98] Shi Y, Li R, Xue Y, *et al.* Optimized operation of current-fed dual active bridge DC–DC converter for PV applications. *IEEE Trans Ind Electron.* 2015;62(11):6986–6995.

[99] Morrison R, and Egan MG A new power-factor-corrected single-transformer UPS design. *IEEE Trans Ind Appl.* 2000;36(1):171–179.

[100] He P, and Khaligh A Comprehensive analyses and comparison of 1 kW isolated DC–DC converters for bidirectional EV charging systems. *IEEE Trans Transp Electrif.* 2016;3(1):147–156.

[101] Park S, and Song Y An interleaved half-bridge bidirectional DC–DC converter for energy storage system applications. In: *Proceedings of the International Conference on Power Electronics—ECCE Asia*. Piscataway, NJ: IEEE; 2011, pp. 2029–2034.

[102] Gorji SA, Ektesabi M, and Zheng J Double-input boost/Y-source DC–DC converter for renewable energy sources. In: *Proceedings of the IEEE 2nd Annual Southern Power Electronics Conference (SPEC)*. Piscataway, NJ: IEEE; 2016, pp. 1–6.

[103] Gorji SA, Sahebi HG, Ektesabi M, *et al.* Topologies and control schemes of bidirectional DC–DC power converters: An overview. *IEEE Access.* 2019; 7:117997–118019.

[104] Cesiel D, and Zhu C A closer look at the on-board charger: The development of the second-generation module for the Chevrolet Volt. *IEEE Electr Mag.* 2017;5(1):36–42.

[105]   Alharbi M, Bhattacharya S, and Yousefpoor N Reliability comparison of fault-tolerant HVDC based modular multilevel converters. In: *IEEE Power Energy Society General Meeting*; 2017, pp. 1–5.

[106]   Raveendran V, Andresen M, and Liserre M Improving onboard converter reliability for more electric aircraft with lifetime-based control. *IEEE Trans Ind Electron*. 2019;66(7):5787–5796.

[107]   Varajao D, Miranda LM, Araújo RE, *et al.* Power transformer for a single-stage bidirectional and isolated ac–dc matrix converter for energy storage systems. In: *Proceedings of the IEEE Industrial Electronics Society*. Piscataway, NJ: IEEE; 2016, pp. 1149–1155.

[108]   Kicin S, Burkart R, Loisy JY, *et al.* Ultra-fast switching 3.3 kV SiC high-power module. In: *International Exhibition and Conference for Power Electronics, Intelligent Motion, Renewable Energy and Energy Management*; 2020, pp. 1–8.

[109]   Abramushkina E, Zhaksylyk A, Geury T, *et al.* A thorough review of cooling concepts and thermal management techniques for automotive WBG inverters: Topology, technology and integration level. *Energies*. 2021;14(16):4981.

[110]   Liu J, Xu F, Sun C, *et al.* A soft-switched power-factor-corrected single-phase bidirectional AC–DC wireless power transfer converter with an integrated power stage. *IEEE Trans Power Electron*. 2022;37(8):10029–10044.

Chapter 13

Smart grid scenarios of different countries

Mousa Marzband[1], Shobhit Nandkeolyar[2],
Bidyadhar Subudhi[3] and Pravat Kumar Ray[2]

13.1 Background

Smart grid technology presents a promising solution for enhancing the generation, transmission, and distribution of electric power. In comparison to conventional grids, smart grids offer several advantages such as increased flexibility in installation and reduced spatial requirements. The fundamental concept behind smart grid design revolves around achieving improved power system performance along with enhanced security. Furthermore, it seeks to optimize the economic aspects of operations, maintenance, and planning. It is noteworthy that smart grid technology can be scaled down to the microgrid level. These microgrids can subsequently be interconnected to create an expansive network of smart grids. This interconnected system holds significant potential and can address the reliability challenges in power transmission and distribution, particularly in developing countries with inadequate infrastructure. In the United States, a substantial portion of carbon dioxide (CO_2) emissions, approximately 40%, stems from electricity generation, overshadowing the 20% contribution from the transportation sector. This disparity is attributed to the escalating demand for electricity. Smart grids are poised to play a pivotal role in mitigating this issue by facilitating the efficient distribution of electric power, thereby reducing greenhouse gas emissions and pollutants such as NO_x and SO_x. Additionally, smart grid technology empowers consumers to forecast their energy demands and optimize energy utilization in a cost-effective manner [1].

The inception of the smart grid concept can be traced back to the evolution of electrical distribution systems. Over time, the growing demands for improved control, monitoring, pricing mechanisms, and service quality in the transmission and distribution of electrical power have driven the development of smart grid technology. Smart grid implementation is often associated with the deployment of smart meters. In the 1970s and 1980s, these devices were utilized to relay consumer

[1]Electrical Power and Control Systems Research Group, Northumbria University, UK
[2]Department of Electrical Engineering, National Institute of Technology Rourkela, India
[3]School of Electrical Sciences, Indian Institute of Technology Goa, India

information back to the grid. However, it is essential to understand that, even with the latest advancements, the primary objective is to enhance the reliability and efficiency of energy transmission and distribution through the electric power grid [2]. Notably, ongoing research in the field of advanced grid systems extends beyond the conventional boundaries of transmission and distribution. It now encompasses a critical role in the generation of clean and sustainable energy, with the overarching aim of reducing greenhouse gas emissions and mitigating our carbon footprint [3–6].

Figure 13.1 lists out some of the key application areas of smart grid. The role of smart grid in enabling these applications is as follows:

- **Improving grid's resiliency and flexibility:** Smart grids allow for the integration of distributed energy resources (DERs) and distributed storage systems. These resources can supply power independently in case of grid disruptions, enhancing grid resiliency. The smart grid enhances grid flexibility by incorporating demand-side management (DSM) for specific functions, thus reducing the need for capacity expansion to a certain extent.
- **Minimize carbon emissions:** Smart grids use advanced algorithms and real-time data to optimize the dispatch of power generation sources. This minimizes the use of fossil fuels and reduces carbon emissions by favoring cleaner energy sources.
- **Smart net-zero energy buildings:** Smart grids can communicate with smart buildings, adjusting their energy usage based on real-time pricing and demand signals. By incorporating energy storage systems (ESS), smart grids can store excess energy during off-peak times and supply it to buildings during peak demand, thus supporting "net-zero" energy consumption.
- **Renewable energy integration:** Smart grids possess distributed storage, which helps to mitigate the intermittency issue of the renewable energy sources

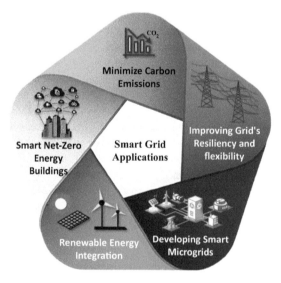

Figure 13.1   Core application areas of the smart grid landscape

(RES). Advanced communication systems enable real-time data exchange between the grid and RES. This facilitates efficient integration by allowing the grid to adapt to the fluctuations in renewable energy (RE) generation.

• **Developing smart microgrids:** Microgrids within the smart grid framework enable localized energy generation and distribution. In the event of grid failures, these microgrids can operate independently, supplying power to critical infrastructure. They also have islanding capabilities, which ensure continuity of power supply to specific areas during emergencies or outages.

When a traditional grid incorporates advanced automation and control systems, it qualifies as a smart grid. From a technical perspective, the smart grid represents an evolution of traditional grids, enhanced by modern, automated features that significantly improve their reliability and sustainability. Historically, conventional grids primarily served the purpose of transmitting and distributing electric power. In contrast, the contemporary concept of the smart grid is characterized by its ability to not only transmit power but also engage in communication, data storage, and decision-making based on prevailing conditions. In accordance with the strategic deployment document (SDD) for Europe's Electricity Networks of the future, a smart grid is defined as an intelligent electricity network that harmonizes the activities of all stakeholders, including power generators, consumers, and those who assume both roles [7]. The ultimate objective is to deliver electricity with enhanced efficiency, sustainability, cost-effectiveness, and securely. It's important to note that the smart grid is not a solitary technology to be implemented; instead, its scope and effectiveness expand in direct correlation with the involvement of various stakeholders. The smart grid offers stakeholders a unique opportunity to optimize the efficiency, reliability, economic viability, and security of their electrical networks as stated in the definition. An architectural overview of the smart grid is provided in Figure 13.2. The key constituents of a smart grid include bulk power generating stations, DERs, transmission and distribution utilities, grid operators, aggregators, and prosumers. The roles of each of them have been thoroughly discussed in the previous chapters. This figure demonstrates how complex the architecture of a smart grid is, along with advanced communication channels among its various constituents.

Development of smart grids may get hindered due to several reasons. First, there is a substantial capital requirement for planning and constructing the smart grid architecture, necessitating the replacement of outdated technologies with newer ones. Second, there is a lack of cooperation between the private and public sectors. The government should incentivize private sector investments, considering the significant environmental benefits associated with smart grids. Lastly, jurisdictional ambiguity arises. Central jurisdictions often oversee high-voltage networks, while distribution networks fall under local jurisdiction. This can discourage local companies from making substantial investments in regional network upgrades, resulting in a focus on profit maximization rather than comprehensive grid development. To address this, a coordinated investment strategy is needed to encourage local utility companies to invest in smart grid research and development.

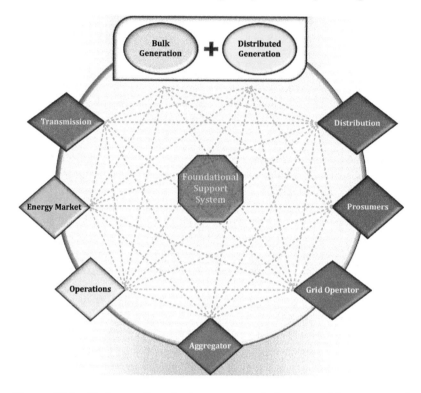

Figure 13.2 Understanding the interconnected elements of the smart grid

In the last couple of decades, various countries around the world have started recognizing the significance and potential of smart grids. Governments in several countries, including Australia, the United States, India, China, the United Kingdom, South Korea, and Japan, have proactively explored smart grid options as a means to reduce carbon emissions and enhance energy security. They have been actively engaging in smart grid pilot projects and have initiated endeavors to explore this concept for rigorous testing and research. Some of these initiatives related to the smart grid development will be discussed in the subsequent sections.

13.2 The global landscape of smart grids

Until a few years ago, fossil fuels predominantly served as the primary energy source in various critical energy consumption sectors, including transportation, power generation, building infrastructure, and industrial operations. The limited environmental awareness within mainstream society, along with a dearth of regulatory policies addressing the environmental impact, economic competitiveness, and affordability of fossil fuels, fostered a heavy dependence on these non-renewable resources for sustaining human activities.

13.2.1 Key drivers for smart grid technology adoption

A transformative shift has occurred over the years, driven by significant developments and global initiatives. Landmark events such as the 1997 Kyoto Protocol and the 2015 Paris Agreement have played pivotal roles in reshaping the energy–planet relationship, emphasizing sustainability as the cornerstone [8]. Presently, substantial technological advancements and investments have been channeled into each energy consumption sector which includes power generation, transportation, industry, and building infrastructure. These endeavors have yielded a spectrum of innovative solutions, giving rise to both fresh challenges and opportunities for increased adoption of eco-friendly energy sources. These alternatives have the potential to either fully or partially take the place of fossil fuels. This is due to the fact that they are comprehensively and effectively integrated into our energy systems.

In accordance with the findings of the U.S. Energy Information Administration's research, as presented in the International Energy Outlook for 2016, it is anticipated that the largest increments in CO_2 emissions during the forthcoming quarter-century will originate from developing countries that do not belong to the Organisation for Economic Co-operation and Development (OECD) [9]. The outlook forecasts a significant rise, amounting to one-third, in global energy-related CO_2 emissions between the years 2012 and 2040. This sustained upward trajectory in overall emissions is expected to persist in the upcoming years as well. Figure 13.3 presents a visual representation of energy-related CO_2 emissions in the years 2012–2025. It distinguishes between emissions originating from OECD and non-OECD countries.

RE electrification is a strategy involving the electrification of end-use sectors combined with the integration of renewable sources in power generation. A report of International Renewable Energy Agency (IRENA) titled "Global Energy Transformation: A Roadmap to 2050" projects that the adoption of RE electrification has the potential to achieve a substantial 44% reduction in CO_2-eq (carbon dioxide equivalent of greenhouse gas) emissions associated with total energy consumption [10]. This fact has been illustrated in Table 13.1 with the help

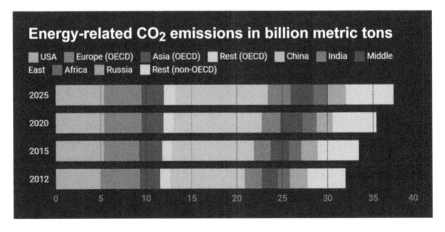

Figure 13.3 Comparing past and projected CO₂ emissions worldwide

Table 13.1   Renewable energy impact on carbon dioxide equivalent emissions

Sector wise CO_2-eq emissions (in Mt)	2050 (raw projection)	2050 (with RE electrification)
Power	1,12,120	14,339
Buildings	13,139	12,361
Transportation	1,10,215	14,730
Industries	16,818	15,745
Heating	12,167	12,167

of estimated data. The table provides a breakdown of these reductions, showcasing a significant 64% decrease in greenhouse gas emissions within the power generation sector, a noteworthy 25% reduction in energy requirements for buildings, a substantial 54% reduction in energy demands for transportation, and a commendable 16% reduction in end-use energy requirements. But the process of electrification, particularly in the context of RE integration, is poised to bring about significant transformation within the existing power system. This exerts additional pressure on its already strained capacities. Consequently, it becomes imperative to initiate innovation and enhancement within the prevailing power infrastructure to effectively accommodate the expanding realm of renewable electrification. In this context, the remedy for addressing the inefficiencies prevalent within the system lies in the form of a technologically advanced and innovative electric power system commonly referred to as a "smart grid."

A universally accepted single definition for the smart grid seems unobtainable. Its description can range from simplistic explanations to more intricate and complex depictions. First, let's examine the various definitions of smart grids across different countries. As per the National Institute of Standards and Technology (NIST) of the United States Department of Commerce, the smart grid can be defined as a grid system that seamlessly integrates a diverse array of digital computing, communication technologies, and services into the existing power system infrastructure. It extends beyond the realm of smart meters for both residential and commercial premises, as it encompasses the bidirectional energy flow and the bidirectional communication and control capabilities that usher in novel functionalities. According to the SDD for Europe's Electricity Networks of the Future, a smart grid can be characterized as an electricity network that demonstrates the capacity to intelligently incorporate the activities of all entities linked to it. These entities encompass generators, consumers, and those fulfilling both roles, all working in tandem to facilitate the efficient delivery of sustainable, economically viable, and secure electricity supplies. According to the National Smart Grid Mission (NSGM) of the Government of India, a smart grid is an electrical grid that uses automation, communication, and IT systems to monitor and manage the flow of electricity starting from power generation sites to points of consumption. It possesses the capability to regulate power flow and curtail electrical loads to maintain real-time equilibrium between power generation and demand. The Korean Smart Grid Roadmap 2030 provides a definition of the smart grid, characterizing it as a next-generation network that incorporates information technology within the preexisting power grid. This integration aims to enhance energy efficiency by enabling a bidirectional exchange of electricity-related information in real time between suppliers and consumers.

The NIST, U.S. has introduced a conceptual model designed to facilitate the planning, development of required infrastructure, documentation, and overall organization of the interconnected networks and equipment forming the smart grid. NIST's approach involves the division of the smart grid into seven domains, each including sub-domains, which collectively cover the spectrum of smart grid actors and applications [11]. These seven domains comprise bulk generation, transmission, distribution, operations, service provider, markets, and customers, much like the depiction provided in Figure 13.2. Moreover, the framework classifies entities into actors, encompassing components like smart meters, solar energy generators, systems including control systems, programs, and stakeholders. These entities are responsible for making decisions and facilitating the exchange of critical information required to execute various applications. Applications, in this context, represent the specific tasks carried out by one or more actors within a designated domain such as home automation, solar energy generation, energy storage, and energy management. It is worth noting that actors within the same domain share common objectives. Actors within a specific domain engage in interactions with actors from other domains, and specific domains may incorporate components from other domains. For example, a distribution utility would encompass actors from the Operations domain (such as a distribution management system (DMS)), and from the Customer domain (such as electric meters).

Demand response (DR) is an essential component of the smart grid, facilitating efficient management of electricity demand and grid stability. The Federal Energy Regulatory Commission (FERC), U.S. characterizes DR as alterations in electric consumption patterns by demand-side resources (deviating from their usual power consumption) in response to fluctuations in electricity prices over time. It can also be prompted by incentive payments structured to encourage reduced electricity consumption during periods of elevated wholesale market prices or when system reliability is at risk. One of the objectives of the U.S. Department of Energy (DOE) is to pursue the advancement of grid modernization technologies and methodologies for the effective implementation of DR strategies.

The transition toward a sustainable energy system is also evident in the European energy landscape. While the development of sustainable energy systems progresses at a relatively slower pace in some developing countries within the EU, it is gradually evolving compared to the established class of nuclear energy systems. A pivotal aspect of this transition lies in the adoption of smart energy systems within households, encompassing elements like smart meters, control mechanisms, smart appliances (SA), and home network integration. Policies are being designed within the framework of the European Strategic Energy Technology Plan (SET Plan) to provide support for smart energy solutions. It primarily focuses on the types of technologies being implemented across EU member states to monitor and optimize smart energy utilization. It is important to note that these policies also aim to provide a comprehensive assessment of the actual deployment of smart energy solutions [12]. Numerous investigations have delved into the ability of various types of loads, spanning industrial, residential, and commercial sectors, to synchronize their power consumption with RE generation, aiming to optimize the utilization of energy-derived RES. This practice is commonly referred to as DSM or DR and is already underway, albeit primarily in pilot or demonstration phases, across multiple European regions. In the context of the European Union (EU)'s transmission system, Transmission System Operators (TSOs) are collectively represented by ENTSO-E, which stands for the European Network of Transmission System Operators for Electricity. ENTSO-E is responsible for formulating roadmaps and network development plans, including the most recent "Research, Development & Innovation Roadmap 2020–2030" and the "Ten-Year Network Development Plan 2020" (TYNDP 2020). These documents outline a range of projects, encompassing transmission, battery, power-to-gas storage, and DR, all presented within a multi-scenario framework that illustrates the evolution of the EU's transmission network.

Sponsored by the Canadian Electricity Association, a comprehensive study assessed the merits and demerits associated with the adoption of various smart grid technologies, products, and services. This evaluation includes a spectrum of technologies currently being implemented in diverse regions, including distribution automation (DA), DMS, automated metering infrastructure (AMI), Smart Homes (SH), and SA. These technologies also find application in developing countries like India, China, and select suburban areas of the United States, reflecting the Canadian Electricity Associations diverse policy approach. Additionally, the study scrutinizes a range of current technologies underpinning smart grid deployments and offers insights into forthcoming trends. These include the integration of RE technologies with combined heat and power (CHP) systems, energy management and control of electric vehicle (EV) charging stations, voltage-frequency control of voltage source inverters, smart generation scheduling for wind-thermal-pumped storage systems, and optimized power system restoration.

The Nigerian electricity grid is actively pursuing the integration of smart grid technologies alongside the utilization of RES. This initiative stems from the challenging state of the power sector. The current state of power sector in Nigeria can be characterized by inefficient power plants, subpar transmission and distribution infrastructure, and an outdated metering system employed for electricity consumers. Given these circumstances, the convergence of smart grid solutions with RE generation is a promising prospect [13]. This seems to have drawn inspiration from the cutting-edge technologies and processes that are currently under development and deployment in EU member states. Additionally, the United States has had a significant influence on emerging economies like Nigeria, fostering access to a more efficient electricity system.

Given that cost is a significant factor in building a smart grid infrastructure, developing nations of the world must allocate a certain portion of their annual GDP

Table 13.2   Initial costs for smart grid (SG) deployment in emerging economies

Country	GDP (in M $)	Initial cost of SG deployment (in M $)	Percentage of GDP
Afghanistan	19,331	3,137	16.227
Cambodia	18,050	1,166	6.46
Congo Democratic Republic	35,238	11,858	33.65
Ethiopia	61,540	6,989	11.36
Haiti	8,765	2,269	25.89
Kenya	63,398	4,261	6.72
Liberia	2,053	808	39.36
Madagascar	9,739	3,072	31.54
Mali	12,747	2,537	19.9
Mozambique	14,807	3,254	21.98
Namibia	11,492	414	3.6
Niger	7,714	1,346	17.45
Papua New Guinea	16,929	357	2.11
Somalia	5,925	1,541	26.01
Swaziland	4,188	99	2.36
Uganda	27,529	2,270	8.24
Yemen Republic	37,734	3,354	8.89
Zimbabwe	14,419	1,824	12.65

toward its development. An estimate of the initial cost of smart grid deployment in various developing nations, along with their GDP figures, has been provided in Table 13.2 [14]. This table offers insights into the percentage of GDP required for smart grid development in each respective nation.

13.2.2   Global smart grid deployment initiatives

Numerous nations worldwide are actively engaged in the development of smart microgrids, pilot projects, and related initiatives. This collaborative effort has resulted in the sharing of valuable insights and experiences, ultimately contributing to more effective planning and execution of smart grid projects. Several countries, including Australia, India, Canada, Great Britain, the United States, South Korea, Ireland, and Japan, have strategically incorporated smart grid technology into their government agendas as a means to enhance energy security and reduce carbon emissions. Some of the projects related to smart grid development are listed out in the Global Smart Grid Federation Report of 2012 [15]. They are as follows:

- "In the United States, several noteworthy initiatives exemplify the integration of Smart Grid technologies. The Pacific Northwest Smart Grid Demonstration Project which operates across five states (Montana, Washington, Idaho, Oregon, and Wyoming) and involve 22 utilities, serves as a prominent pilot project showcasing seamless coordination of Smart Grid assets. In Houston, the Smart Grid is characterized by the full implementation of integrated metering systems, customer web portals, and automatic outage notifications.

Furthermore, the Smart Texas initiative demonstrates a large-scale deployment of smart meters and distribution automation.

- Canada has implemented several significant Smart Grid initiatives. These include the Transmission Dynamic Line Rating, designed to optimize the transfer capability of transmission lines, the Wide-Area Control System, which has enhanced voltage stability, and the Ontario Smart Metering Initiative, resulting in a peak shaving of 5 to 8%.

- Low Carbon London, a 4-year innovation project, has successfully integrated various low-carbon technologies, including photovoltaic installations, smart meters, electric vehicles, charging stations, and heat pumps, into the distribution network. In the town of Reading, Berkshire, 'Slough Heat & Power' has achieved a significant milestone by integrating the world's first cryogenic (low-temperature liquid) energy storage solution, known as the Cryogenic Energy Storage Pilot, into the national grid.

- In Ireland, the Commission on Energy Regulation (CER) successfully conducted a Smart Meter Trial involving approximately 9,000 homes and businesses. Additionally, the ecar Ireland project has initiated a pilot program for electric vehicle charging infrastructure, where drivers pay the electricity supplier directly rather than the charging station.

- A significant large-scale deployment of Advanced Metering Infrastructure (AMI) in Spain serves as a crucial component in the European R&D project IGREENGrid (IntegratinG Renewables in the EuropEaN Electricity Grid). This initiative involves the deployment of over 2,00,000 smart meters in the Madrid area. To prepare the regulations that promote distribution system innovation for Smart Grids in Italy, valuable field data was required. However, since these Smart Grid projects are still ongoing and field data is not yet available, numerical simulations have been developed based on a reference network in Italy for this purpose. Portugal has also deployed management and control schemes in both a pilot test site and a laboratory setting.

- Indonesia embraced the implementation of smart meters, a move met with initial skepticism primarily arising from limited consumer awareness.

- The Intelligent Networks Communities project in Australia is currently conducting tests in several areas. These tests cover network fault identification, isolation, and self-recovery, alongside power quality surveillance and distribution automation. This facilitates the development of a distribution management system at a commercial-scale.

- In Japan, Kit Carson Electric Cooperative Inc. has successfully established a network of smart meters at a cost approximately 30% lower than current market rates, thanks to the implementation of Fujitsu's wireless technology solution. Meanwhile, the Aomori Hachinohe Microgrid Demonstration Project conducted tests to evaluate the performance of a demand-supply control system in managing the impact of renewable energy on a commercial power grid. Several other noteworthy projects and pilots in Japan include the Total Energy Solutions Test Bed Project, which utilizes rooftop photovoltaic systems and lithium-ion batteries for energy generation and storage; the Distribution

Stabilizing Solution, which effectively regulates voltage for photovoltaic systems and offers rapid voltage control under intermittent conditions; the Yokohama Smart City Project, featuring a substantial introduction of electric vehicles and energy management using EVs; and the Miyako Island Mega-Solar Demonstration Research Facility, serving as a testbed to study the impact of photovoltaic generation facilities and rechargeable batteries on power system stability.

- In South Korea, the renowned Jeju Smart Grid Demonstration Complex stands out for its integration of various elements, such as wind and solar power, electric vehicles, Advanced Metering Infrastructure (AMI), energy storage, distributed automation, network monitoring, and telemetry. Additionally, the Smart Transportation initiative has implemented an electric vehicle infrastructure that relies on wireless communication. The Renewable Energy Source Operations serves as a microgrid demonstration, seamlessly incorporating large-scale wind generation and advanced battery technologies. The Consumer-Participating Smart Place concept introduces real-time electricity pricing, harnesses the potential of renewable energy sources, integrates smart appliances, and offers storage solutions, all conveniently deployed within households.

- Other Smart Grid initiatives include, but are not restricted to, the practical implementation of Smart Grids in the urban area of Milan; Smart Grid initiatives in China; mathematical modeling for optimizing the operation of the University of Genoa Smart Polygeneration Microgrid; large-scale advanced metering infrastructures, on-field applications of Smart Grids in the Italian framework; Smart Grid opportunities and applications in Turkey; laying down of criteria for Smart Grid deployment in Brazil employing the Delphi method; experimental and data collection methods for a large-scale Smart Grid deployment, the Italian Smart Grid pilot projects, focusing on the selection and assessment of the testbeds for the regulation of smart electricity distribution, and strategies for stimulating the deployment of Smart Grids through effective regulatory instruments."

The concept of the smart grid has transformed from a mere vision into a tangible goal that is gradually becoming a reality. Advancements in technology have empowered devices and systems to promote the development of a more intelligent grid. This progress is significantly bolstered by well-defined energy policies that underpin smart grid initiatives on a national scale. Interestingly, the implementation of smart grid practices across various regions does not suggest competition, but rather reflects a cohesive and interconnected global community with shared ambitions and invaluable lessons.

13.3   Country-specific smart grid scenarios

Countries around the world are making strides in upgrading their energy systems to become smarter and more efficient. The smart grid scenarios in different countries are described below, with a particular emphasis on developing nations where the development of smart grids began at a later stage. These stories highlight the global commitment to innovative energy solutions that benefit both the environment and society.

13.3.1   United States

In the past decade, energy policies of the U.S. government have been shaped by several key factors, which include energy supply security, environmental impacts of energy production, and the pursuit of energy independence to reduce reliance on energy imports. This strategy was further driven by the looming threat of an impending energy crisis, marked by a growing gap between energy consumption and domestic generation. This could have jeopardized economic growth, and the standard of living of U.S. citizens. In 2010, the United States depended on domestically generated energy for about three-fourths of its energy needs, a significant shift from being a net energy importer since the mid-1950s. It currently holds the second position of the world's largest crude oil importer, after China as of 2022.

As part of its broader agenda to curtail energy-related CO_2 emissions, the government embarked on investments to expand domestic energy resources, including renewables, while also embracing modernization initiatives targeting energy conservation and infrastructure enhancement. In recent years, residential electricity costs in the United States have surged. The rising costs tied to the upkeep and modernization of the aging energy grid, the phased decommissioning of obsolete coal-fired power facilities, and their replacement with modern natural gas and RE plants have led to this situation. It is important to highlight that there has been a sustained drive toward the production of RE, frequently governed by state-level legislation, with 29 of the 50 U.S. states formulating pertinent policies [16]. This led to the inception of a "renewable portfolio standard" (RPS), which mandates a specific proportion of RE capacity to be achieved by a designated year.

Traditionally, the U.S. electricity sector was characterized by vertically integrated monopolies, established through state statutes and governed by the "regulatory compact." The regulatory compact is an agreement where government regulators grant exclusive service territories to utilities. However, sector restructuring has been undertaken in many states. Presently, the wholesale electricity market functions within a competitive framework that provides unrestricted transmission access. Furthermore, ten states have embraced electricity retail deregulation, with an additional three states contemplating similar reforms. In the majority of states, retail electricity tariffs are subject to regulation which are established by Public Utility Commissions. Ownership of electric utilities is divided among the private sector, rural electric cooperatives, and municipal governments. The power marketers engage in buying and selling electricity, typically without direct ownership or operation of generation, transmission, or distribution assets.

In 2010, there were 663 electric utilities in the United States that collectively deployed 20,334,525 smart metering infrastructure installations, primarily targeting residential customers, accounting for approximately 90% of the total installations. The national average smart meter penetration rate in 2011 stood at approximately 14%, with seven states surpassing 25% penetration. The majority of the expenses linked to these implementations are retrieved through retail tariffs borne by consumers. The reactions to these implementations have been mixed. Utilities that adeptly conveyed the advantages of smart metering, such as Oklahoma Gas & Electric, San Diego Gas & Electric, and CenterPoint Energy in Texas, encountered a favorable response from consumers. Nonetheless, in certain regions, substantial consumer opposition has arisen, primarily stemming from health and privacy apprehensions, in addition to discontent regarding the perceived increase in electricity expenses attributed to the utilization of smart meters.

The United States demonstrates a notable degree of political and consumer sensitivity with regard to smart meters, leading to various government and industry efforts to resolve these issues. The Smart Grid Consumer Collaborative, a coalition of private sector entities, utilities, and advocacy groups, focuses on providing information and educational resources to consumers about smart grid technologies. In California, legislation has been enacted to protect the privacy of consumers' energy consumption data, offering customers the option to opt-out of smart meter installations, subject to a fee.

In contrast, the smart grid has a lower public profile compared to smart meters. The U.S. DOE established the Office of Electric Transmission and Distribution in 2003, later known as the Office of Electricity Delivery and Energy Reliability, to lead a nationwide endeavor aimed at modernizing and expanding the electricity grid. They introduced the ambitious vision of "Grid 2030" in January 2004, envisioning intercontinental power transfers, real-time data exchange with consumers, minimal economic losses from power outages, and competitive markets within the electricity industry [17].

The support for the smart grid became official with the federal Energy Independence and Security Act, 2007. The DOE administers a research and development program dedicated to smart grid technologies. State utility commissions and industry advocates are evaluating the outcomes of the DOE-funded investment program to determine best practices and assess cost-benefit implications for both consumers and utilities. The federal government allocated approximately $4.3 billion under the American Recovery and Reinvestment Act, 2009 for smart grid initiatives. These initiatives include aspects like EV manufacturing and advanced metering infrastructure implementations. Looking ahead, smart grid initiatives are anticipated to continue attracting federal attention, especially concerning issues of critical infrastructure and cybersecurity. The NIST is overseeing the development of interoperability and cybersecurity standards through a collaborative process that involves policymakers, utilities, and industry stakeholders.

13.3.2   European Union

In Europe, most electrical grids are interconnected, requiring collaborative efforts to harmonize policies and regulations across various jurisdictions in the electricity sector. Some wholesale electricity markets extend across multiple countries, such as the NordPool Spot exchange, encompassing Norway, Denmark, Finland, Sweden, and Estonia. Each EU member state maintains its regulatory authority for its respective electricity sector. To promote cross-border regulatory cooperation in gas and electricity transmission among member states, the Council of European

Energy Regulators and the Agency for the Cooperation of Energy Regulators (ACER) were established. While ACER lacks regulatory jurisdiction at national or supranational levels, it possesses the capacity to intervene if national regulators fail to collaborate effectively. On the operational front, the European Network of Transmission System Operators for Electricity (entsoe) was established to standardize grid access. This ensures effective network planning, and facilitates investments aimed at preventing power failures [12].

Since the 1990s, there has been a notable shift toward liberalization, leading to increased instances of competition in the generation and retail segments, which are being separated from transmission and distribution functions. However, many markets still witness the dominance of suppliers with nearly monopolistic control. The 2009 Third Legislative Package mandates EU member states to liberalize their electricity markets by 2011 if they haven't already done so. In 2003, EU directives introduced uniform regulations for internal electricity markets. Presently, electricity network codes are under development at the EU level to ensure fair allocation of costs and responsibilities among market participants.

The EU has instituted legislation concerning the implementation of smart grids and smart metering. Particularly, the Electricity Directive 2009/752/EC obliges EU member states to adopt smart metering systems by 2020, contingent on economic assessments demonstrating their viability. However, the specifics of smart meter rollout and cost recovery vary across EU members due to differing characteristics of their electricity sectors. Additional legislative acts relevant to smart metering include the 2009 Third Legislative Package, Directive 2005/89/EC (Security of Supply), Directive 2006/32/EC (Energy End-use Efficiency and Energy Services), and Directive 2004/22/EC (Measuring Instruments). The Electricity Directive (EU) 2019/944 outlines that the smart meters should be equipped with right technologies. Furthermore, it is essential for national regulatory authorities to monitor and ensure optimal returns on the investment made in this regard. The installed smart metering systems should be designed to serve the entire energy ecosystem, delivering benefits and satisfaction to both consumers and businesses. As of 2011, approximately 10% of EU households had smart meters installed, with an expected increase to 100 million by 2016. The EU reached 150 million smart meters by the end of 2020, achieving a 49% penetration rate across the continent.

In January 2009, the European Commission launched the Task Force on Smart Grids, aiming to provide guidance on policy, regulation, and the initial steps for implementing smart grids in accordance with the 2009 Third Legislative Package. Substantial investments have been channeled into testing the integration of smart grid technologies into the energy landscape across EU member states. The EU itself has allocated approximately €300 million for smart grid projects over the past decade. In 2010, the European Commission established the European Electricity Grid Initiative (EEGI), a nine-year research and development program with a budget of €2 billion, following the recommendations of transmission and distribution network operators [18].

The E.DSO for Smart Grids, an association representing distribution system operators from 17 EU member states, encompassing 70% of electricity supply points within the EU, is committed to advancing grid modernization to align with the EU's objectives concerning energy efficiency, greenhouse gas reduction, and RE. It functions as an intermediary between public authorities and its member entities concerning matters related to smart grid development. E.DSO has undertaken numerous initiatives to promote the advancement of smart grids.

13.3.3   India

In August 2013, the Ministry of Power (MoP), in collaboration with the India Smart Grid Forum (ISGF) and India Smart Grid Task Force (ISGTF), unveiled the "Smart Grid Vision and Roadmap for India." This visionary document aims to transform the Indian power sector into a secure, adaptable, sustainable, and digitally enabled ecosystem, ensuring reliable and high-quality energy access for all while actively involving stakeholders. This vision aligns with the MoP's overarching policy objective, which centers on providing universal access, availability, and affordability of power. To realize this ambitious vision, stakeholders were urged to formulate state-specific or utility-specific policies and programs [19]. These initiatives were supposed to be in line with the broader policies and targets established by the MoP.

During stakeholder consultations in 2013 for the Smart Grid Roadmap, a unanimous demand arose for the establishment of a NSGM. Subsequently, the MoP, with input from ISGF and ISGTF, devised the framework for NSGM, which was officially endorsed by the Government of India in March 2015.

To support the NSGM's endeavors, the Smart Grid Knowledge Center (SGKC), funded by the MoP and developed by POWERGRID, will serve as a resource center. Its primary role is to provide technical support to the Mission, encompassing technical manpower development, capacity building, outreach efforts, and recommendations for curriculum changes in technical education. The MoP has already allocated ₹9.8 crore for this purpose. The SGKC will undertake programs and activities in line with the guidance from the National Project Management Unit (NPMU), with the Head of SGKC reporting to the Director of NPMU.

The MoP delegated the responsibility to the Central Electricity Authority (CEA) to formulate functional requirements and technical specifications for domestically manufactured smart meters. In June 2013, the CEA introduced the initial edition of Smart Meter Specifications. However, during the implementation of 14 smart grid pilot projects, various distribution companies in different states issued distinct specifications, raising concerns that were subsequently brought to MoP's attention by the ISGF. MoP then approached the Bureau of Indian Standards (BIS) to develop a national standard for smart meters. BIS, in turn, assigned this task to the Technical Committee within the Electro Technical Division (ETD-13) to establish the standards for smart meters.

In August 2015, BIS published the updated Smart Meter Standard, denoted as "IS 16444: AC Static Direct Connected Watthour Smart Meter—Class 1 and 2 Specification", encompassing single-phase and three-phase energy meters, as well as single-phase and three-phase energy meters with Net Metering capabilities. Furthermore, another standard, IS 15959: Data Exchange for Electricity Meter Reading, Tariff, and Load Control—Companion Specification, was revised and released as IS 15959: Part 2—Smart Meter in March 2016. In 2016, the MoP announced the government's commitment to accelerate the deployment of smart metering, targeting customers with a monthly consumption of 200 kWh or more by December 2019. This objective is reiterated within the UDAY program and the Tariff Policy articulated by MoP. India has currently deployed 66.18 lakh smart meters, with the Central government setting a target to install 25 crore smart meters by the end of 2025 [20].

The smart grid relies on Machine to Machine (M2M) communication technologies, enabling seamless interaction between wired/wireless systems and devices without human intervention. M2M communication involves devices, like sensors or meters, capturing various events, such as motion, video, temperature, or meter readings, and transmitting this data through wireless, wired, or hybrid networks to applications, typically software programs, which translate the captured data into meaningful information. M2M is a subset of the broader Internet of Things (IoT) concept. In the Indian context, the 865–867 MHz band stands out as the most suitable frequency for outdoor applications. For indoor applications, the 2.4–2.4835 GHz band is preferred. Looking ahead, low-power radio frequency (RF) is expected to emerge as the most efficient communication technology, offering connectivity to a multitude of devices. This preference is driven by factors such as lower operating costs, reduced power consumption, minimal interference, improved signal-to-noise ratio, and enhanced penetration. However, the 865-867 MHz band, delicensed in 2005, may prove insufficient to meet the demands of IoT/M2M/Smart City applications, which involve billions of connected devices. Another delicensed sub-GHz band, 433–434 MHz, which became available in 2012, may primarily serve indoor applications due to current regulations limiting the maximum power to 10 mW. Sub-GHz frequency bands are most effectively utilized for outdoor applications within the context of neighborhood area network, field area network, and local area network (NAN/FAN/LAN) setups.

India stands as the world's fifth-largest electricity generator, boasting a total installed capacity of 331 GW as of October 31, 2017. Within this capacity, RES account for 60 GW, constituting approximately 18% of the total installed capacity. Furthermore, assessments indicate India's vast untapped solar potential of 10,000 GW and wind potential of 2,000 GW, as reported by the National Institution for Transforming India (NITI) Aayog in 2015. Recognizing the substantial benefits of RE, both the Government of India and various State Governments have undertaken a series of initiatives to foster the expansion of renewables within the nation. In 2015, the Government of India set an ambitious target to generate 175 GW from RES by 2022, which includes 100 GW from solar, 60 GW from wind, 10 GW from biomass, and 5 GW from small hydro, as outlined by the Ministry of New and Renewable Energy (MNRE) in 2015. The anticipated power requirement in 2022 is projected to reach 434 GW, with RE contributing 175 GW, equivalent to 18.9% of the total power consumption, according to the MNRE in 2015.

To realize the set targets of 175 GW from renewables, India needs a paradigm shift in its planning and governmental practices. Presently, the government has implemented several support mechanisms to facilitate the growth of RE, including Feed-In Tariff, Accelerated Depreciation, Generation Based Incentives (GBI) for Wind, Viability Gap Funding (VGF) for Solar, Net Metering, and the Renewable Energy Certificate Mechanism. These initiatives collectively play a pivotal role in India's RE development and in achieving its ambitious targets. For instance, Accelerated Depreciation has been a cornerstone for the success of wind energy projects in India, with around 70% of the utilities utilizing this incentive. The GBI for wind energy producers offer financial incentives; and VGF for solar power has led to electricity tariff reductions. Net metering policies have been adopted across all states and union territories, allowing for the seamless transfer of excess energy to the grid. Renewable Energy Certificate Mechanism regulations stipulate minimum energy purchase percentages from renewable sources, driving the adoption of solar and non-solar RE resources across the country. These initiatives collectively reflect India's commitment to fostering the growth of RE and realizing its ambitious RE targets.

Several instances of DSM implementation in Indian utilities showcase the substantial benefits reaped in various sectors. In the public sector, Maharashtra State Electricity Distribution Company Limited has undertaken energy efficiency and demand-side management initiatives in areas like HVAC systems and agricultural DSM equipment. It achieved a remarkable savings of up to 23%. The Chamundeshwari Electricity Supply Corporation Limited in Karnataka replaced 1337 pump sets with efficient ones, resulting in an annual energy saving of 56.68 lakh units (37%) with an investment of ₹5.03 crore, a project completed in March 2015. In the commercial sector, the Aranya Bhawan, an office building of the Rajasthan Forest Department in Jaipur, witnessed significant energy savings following initiatives like insulation, window replacement, installation of a high-efficiency water-cooled chiller, and the implementation of a 45 kW rooftop solar PV system. This resulted in an annual electricity savings of 2,40,000 kWh with a payback period of three years.

Within the industrial sector, ITC Limited effectively utilized energy-efficient equipment and innovative low-cost strategies at its ITC Manufacturing plant in Bangalore, yielding ₹548 lakhs in savings over a ten-year period with a total investment of ₹145 lakhs. These strategies included replacing an Air-Cooled Chiller with a Recycled Water Cooler Chiller, implementing waste heat recovery from process exhaust for steam generation, and more. In the residential sector, Godrej and Boyce conducted a Resource Energy Efficiency Retrofit of the Godrej Bhavan building, which consisted of HVAC system replacement, installation of energy-metering systems, energy audits, and the adoption of energy-efficient tube lights, incurring a total cost of ₹54 lakhs. These examples illustrate the diverse opportunities and successes in DSM across various sectors within India.

13.3.4   Australia

Australia's interest in smart grid technology dates back to 2009. The government expressed a willingness to invest approximately $100 million in this endeavor. Their primary objectives were to raise consumer awareness about energy consumption and establish a distributed demand and generation management system. To actualize these goals, five locations in New South Wales were chosen for smart grid implementation, with Energy Australia leading the initiative in collaboration with IBM, GE Energy, and Grid Net. The project aimed to construct a WiMAX-based (Worldwide Interoperability for Microwave Access) smart grid featuring automatic substations, EV integration, and the ability to support 50,000 smart meter connections [21].

In addition to this, another project was launched to test network fault detection, isolation and restoration, power quality monitoring, and the automatic distribution of electric power through a DMS. The Australian government provided incentives to encourage and invest in smart grid projects, with demand management, energy security, and energy efficiency ranking as top priorities.

Australia had set a goal to integrate 20% RE into its energy mix by 2020. Being a federal parliamentary democracy with states and territories, energy policies are under state jurisdiction, coordinated at the national level by the Council of Australian Governments (COAG). Smart meters were introduced following energy shortages in 2006 and 2007, despite the associated costs. New South Wales and the State of Victoria proceeded with smart meter deployment. Australia's commitment to the smart grid is evident through SmartGridAustralia, a non-profit organization at the forefront of modernizing the electrical system and supporting government initiatives such as the Smart Grid Smart City (SGSC) program. Australia is actively working to enhance incentives for smart grid investments and develop strategies to address demand-side regulation and time-of-use tariffs. Demand management, energy security, and energy efficiency remain high-priority areas of focus.

The electricity sector in Australia was primarily characterized by state-owned vertically integrated monopolies, except for the State of Victoria, which policy-makers aimed to deregulate in the future. Market reforms introduced in 2003 have led to free-market competition in electricity generation and retailing within the National Electricity Market (NEM) territories. However, attracting investments in new power plants has been challenging due to Australia's low electricity prices and debates surrounding the introduction of additional costs like carbon taxes.

Government or government-owned entities predominantly provide transmission and distribution services in the context of the NEM in Australia. The NEM's networks are predominantly state-owned and operated. Distribution networks are under the ownership and operation of either governmental bodies (e.g., Queensland, Western Australia, Northern Territory), private sector entities (Victoria), or a combination of both (e.g., New South Wales and the Australian Capital Territory). These organizations hold monopoly positions in their designated regions, and vertical integration of distribution and retailing is a prevalent practice within these areas. Retail prices are regulated by the Australian Energy Regulator (AER), and there are ongoing plans to introduce further market reforms to enhance competition within the electricity supply sector. In Western Australia and the Northern Territory, the government maintains ownership of the electricity distribution sector. Vertically integrated businesses cover transmission, distribution, and retail functions.

The implementation of smart meters in Australia has sparked significant political controversy. Following energy shortages in 2006–2007, the COAG committed to a national smart meter rollout, provided that the benefits outweighed the costs. Subsequent cost-benefit analyses indicated potential benefits ranging from approximately \$299 million to \$3.3 billion in net present value over a 20-year period, depending on various scenarios.

The State of Victoria initiated a mandatory smart metering infrastructure rollout, accompanied by new time-of-use pricing. All associated costs, including the smart meter expenses, were transferred to consumers, leading to a negative public response, particularly in light of rising electricity prices. As a result of this backlash, there was a temporary suspension of the rollout. However, it has since resumed despite continued consumer opposition. New South Wales has also embarked on a smart meter trial deployment. By October 2020, a total of 1.04 million smart meters were deployed within the NEM, excluding Victoria, resulting in a penetration rate of 17.4%. The Australian Energy Market Commission has proposed a set of guidelines aimed at achieving complete smart meter adoption nationwide by the year 2030, i.e., 100% penetration rate.

Under the Intelligent Network Communities project, Essential Energy, the distributor, is presently conducting trials on network fault detection, isolation, restoration, power quality monitoring, and DA issues. These trials are facilitated through the utilization of a commercial DMS. The initiative includes load control, substation monitoring, and four-quadrant interactive inverters for Volt/Var controls, making it a comprehensive smart grid endeavor. Furthermore, the project encourages customer engagement in energy management trials. Essential Energy is also assessing advanced metering infrastructure, specific customer products, education, distributed generation, and ESS as integral components of the project. Currently in the implementation phase, it is being executed on three 11 kV feeders, serving approximately 4,000 consumers, with an estimated cost of \$15 million.

The SGSC project, scheduled for completion in 2013, served as a practical testing ground for demand-side response solutions and novel supply technologies within an authentic customer setting. Its primary goal was to collect data concerning the advantages and expenses associated with various smart grid technologies in the Australian context. Customers were empowered to monitor their energy consumption, calculate energy expenditures, and assess greenhouse gas emissions, facilitated by a household energy management system enabling wireless appliance control. On the grid side, advanced monitoring and measuring devices contributed to enhanced network reliability, efficiency, and the integration of distributed energy generation, storage, and EVs. The project featured smart sensors, advanced IT systems, smart meters, and a communications network. Comprehensive performance assessments were conducted to gauge operational, environmental, and societal/consumer benefits, with an estimated project cost of \$243 million.

13.3.5   China

The smart grid plays a pivotal role in China's pursuit of its new energy objectives and the transition toward a low-carbon economy. Over recent years, China has prioritized the research and development of grid technologies. Denmark emerged as a global leader in smart grid R&D. Preliminary discussions between researchers from both nations have indicated a mutual interest in fostering further collaboration and knowledge exchange.

China's significance in the global energy landscape cannot be ignored, having surpassed the United States to become the world's largest power producer and consumer since 2010. To sustain its rapid economic growth, China has witnessed an annual increase in net power generating capacity of 13.22% from 2005 to 2010, with a projected annual growth rate of 8.5% until 2015. Notably, the share of "green" power sources, encompassing hydro, nuclear, wind, and solar, has shown substantial growth each year, while the share of thermal power, predominantly coal-based, has progressively decreased. By 2011, non-fossil power constituted 27.5% of the energy mix, with thermal power accounting for 72.5%. During this period, hydropower capacity reached 230 GW (21.9% of the total), nuclear power installed capacity reached 11.91GW (a 10% increase), and wind power capacity reached 47 GW (a notable 43.7% increase).

In recognition of the critical role smart grids play in enhancing energy efficiency and grid management, China has formulated and invested significantly in a nationwide smart grid development plan. This investment was expected to grow substantially over the next 5–10 years, with the total value of China's smart grid market forecasted to surge from $22.3 billion USD in 2011 to $61.4 billion USD in 2015, reflecting an impressive annual growth rate of 29.1% over the course of five years. China's commitment to smart grid development is integral to its broader energy and environmental sustainability efforts.

Several key government agencies in China play integral roles in shaping the landscape of smart grid development. The National Development and Reform Commission (NDRC), the foremost ministerial agency, holds a pivotal role in coordinating smart grid development within the broader context of national development plans. It exercises control over electricity pricing, sets benchmarks for feed-in tariffs and sales tariffs, and maintains authority over the review and approval of smart grid construction projects.

The National Energy Agency (NEA), a vice-ministerial department under NDRC, leads the formulation and implementation of national energy policies and development plans, which includes smart grid initiatives. The NEA is also actively involved in the approval process for smart grid projects. The State Electricity Regulatory Commission (SERC) oversees the daily operations of power generation and utility companies. An institutional reform in March 2013 had suggested merging SERC into the NEA to enhance energy supervision and management.

The China Electricity Council, a non-profit organization comprising all of China's power enterprises, operates under SERC's supervision and serves as a think tank for power policies. It wields substantial influence when shaping national smart grid plans. The Ministry of Science and Technology is responsible for cutting-edge technology research and international cooperation. Smart grid technologies held a prominent position in its 12th Five-Year Plan on National Scientific and Technological Development [22].

In the vast Chinese grid market, two major state-owned companies dominate. The State Grid Corporation of China (SGCC), the world's largest power utility company, exerts a significant influence over all aspects of China's smart grid development. SGCC takes the lead in formulating national plans, conducting technology R&D, investing in infrastructure, and piloting and testing new technologies. It also plays a vital role in establishing industrial standards. SGCC has a presence in 26 provincial power utility companies, five research institutes, and 20 other institutes. With total assets of 2077.5 billion RMB (approximately $330 billion USD) in 2012, it generated over $260 billion USD in revenue in 2011. SGCC has expanded its international operations by investing in various countries, including the Philippines, Brazil, Portugal, India, and Russia.

The China Southern Power Grid Company Limited (CSG) covers five provinces in South China and connects resource-rich western provinces with high consumption regions. CSG relies on a higher mix of hydropower (37%) and a lower mix of thermal power (56%) compared to the national average. It plays a crucial role in the nation's project to transmit power from resource-rich western China to the highly industrialized eastern region. CSG is actively involved in smart grid technology development and standards formulation, following SGCC's lead. CSG is expected to invest more than RMB 100 billion (US$15.85 billion) in smart grid development in the coming years.

By the end of 2011, SGCC had established 238 smart grid pilot projects, including technical and economic evaluations for 102 of them. These projects encompassed various aspects of smart grid implementation, from generation and transmission to distribution, consumption, and dispatching. SGCC had constructed 65 smart substations ranging from 110 KV to 750 KV, set up 243 EV charging stations, and deployed over 51 million smart meters in households [23]. These demonstration projects span the country and include high-profile initiatives like the smart grid demo at the Shanghai World Expo and the Sino-Singapore Tianjin Eco-city smart grid demo as part of integrated smart city solutions. Additionally, 28 smart communities serving 251 thousand households have been established in Beijing and Shanghai. In China, the total number of smart meters installed exceeded 450 million by 2021.

13.3.6 *African Nations*

The application of select advancements in power systems through smart grids presents an opportunity for sub-Saharan African countries to bypass conventional power systems and adopt more efficient solutions. This has the potential to expedite national and regional electrification efforts, enhance service quality, and reduce costs and environmental impact. A crucial aspect of this concept is the optimization of grid systems, the integration of high levels of RE, and the enhancement of electricity supply reliability and efficiency. In the context of sub-Saharan Africa, it is imperative to establish socially equitable power systems that ensure universal access to modern energy services without marginalizing the underprivileged.

According to the reference scenario provided by the International Energy Agency, it is anticipated that Africa's total electricity consumption will experience a twofold increase from 505 to 1012 TWh between the years 2007 and 2030. For sub-Saharan Africa, a primary focus should be on educating consumers about efficient electricity usage as they transition to smart grids, particularly among those who previously lacked access to electricity. The dissemination of training tools and resources related to cutting-edge power systems is vital. Special attention should be given to training off-grid communities, enabling them to manage and sustain mini-grid systems effectively.

The short-term objective is to primarily involve components integrated with information and communication technologies (ICT), which are integral to many smart grid systems. Sub-Saharan Africa has already witnessed successful instances of leapfrogging toward more efficient ICT solutions. The mobile phone revolution in the mid-1990s, which provided access to modern communication without the need for extensive conventional telephone networks, offers valuable lessons. Sub-Saharan Africa experienced remarkable growth in mobile phone subscriptions, with some countries, such as Kenya and Cameroon, achieving annual growth rates of up to 300%. This rapid adoption of mobile phones was partly due to the inadequacy of traditional telecommunication systems in meeting consumer demand, akin to the current inadequacies of electricity networks in the region. Moreover, the convenience and accessibility of pre-paid mobile subscriptions, used by 90% of subscribers in sub-Saharan Africa, cater to individuals with lower or irregular incomes. Smart and Equitable Grids can leverage ICT infrastructure to implement similar payment systems.

Apart from technological factors, market models that characterized the mobile phone revolution, such as shared phone usage, may serve as a precedent for smart grids. While some success factors may not directly translate to smart grids, aspects like lower initial investments and the rapid deployment of re-deployable assets contributed to the mobile phone sector's success, making assets less reliant on institutional frameworks and investor protection.

Currently, the implementation of smart grid technologies in West Africa remains limited, although discussions on this topic, particularly in relation to RE deployment, are gaining traction. A few pilot projects have been initiated. Challenges persist in the region's transmission network within individual country borders and cross-country transmission, as identified in the 2011 West African Power Pool Master Plan. These challenges encompass issues like fuel scarcity due to financial constraints, inadequate equipment maintenance, frequency control problems on interconnection lines, and the risk of slow oscillation when significant generation units are offline. Furthermore, the shortage of short-term power reserves in some zones hampers the functionality of interconnection lines, resulting in a significant portion of the installed capacity being currently unavailable.

Smart grid controls have made inroads into the West African distribution network. Some countries have started installing transformers and meters equipped with data acquisition systems, enabling remote access through Global System for Mobile Communications. These systems can detect power theft, reducing commercial losses in the distribution grid.

While DSM holds promise in theory, effective DSM in most West African countries, particularly in rural areas, is challenging. Energy consumption in such regions is typically basic, mainly encompassing lighting, fans, radios, and occasionally, TVs and refrigerators. Load is naturally concentrated in the evening when consumers are at home, leaving minimal opportunities for extensive load shedding. The possibility of achieving DSM through the established and successful telecommunications network, widely adopted across sub-Saharan Africa, is often suggested as a way forward.

Several projects have explored real-time energy consumption data access and prepaid meter top-up via mobile phones. For example, a pilot project introduced grid connections with an upper capacity limit for households. When users exceed this capacity, the system trips, and they receive a text message alert instructing them to disconnect a device for reconnection. This improved the distribution grid's reliability as the other consumers were not affected. Such models hold promise for future smart grid management through consumer involvement.

The concept of mini-grids is central to many electrification studies in West Africa. A significant portion of the population resides in remote and rural areas, making a 'bottom-up' electrification approach attractive. This approach entails progressing from household electrification systems to community and regional systems, with set standards allowing seamless connection with larger grids for enhanced reliability and power availability [24].

Moreover, smart grid technology, particularly AMI, can enhance the financial viability of mini-grids. Many existing mini-grids in West Africa struggle to remain operational due to consumer tariffs that do not cover operation and maintenance (O&M) costs. International donor funding often covers the construction of rural mini-grids, but few donors commit to ongoing payments for system upkeep. AMI can monitor O&M costs, ensuring more secure revenue collection for sustainable, long-term operation. Prepaid metering systems are recognized as a promising solution to address financial viability challenges.

13.4   Future trends and challenges

Extensive research efforts are currently underway in the field of smart grids, signifying an active pursuit of innovation and advancement. Furthermore, numerous unexplored avenues for future research within various aspects of smart grid technology remain open for exploration. These areas encompass a wide spectrum, including but not limited to forecasting techniques, optimization of power flow, communication protocols, integration of microgrids, the development of effective demand and energy management systems, the establishment of interoperability standards, considerations of scalability, economic viability, robust data encryption, and the automation of processes within the domains of power generation, transmission, and distribution.

13.4.1   Challenges in smart grid deployment

Some of the key challenges that we encounter in smart grid development have been addressed in this section. Addressing these challenges leads to a more efficient, reliable, and sustainable smart grid, ensuring energy security and environmental benefits.

The DERs are smaller power generation units and are the core component of a microgrid. DERs can be classified into two types: dispatchable and non-dispatchable. The operator does not have full control over the output of non-dispatchable DERs since they are of intermittent nature. Wind energy and solar energy are examples of these types of DERs. DERs can also be classified into AC sources and DC sources each posing their own challenges in terms of operation.

Cost considerations present a formidable challenge, particularly in the case of developing regions, when it comes to advancing and operationalizing smart grid technology. Substantial financial resources are involved, spanning the domains of transmission and distribution infrastructure, advanced metering systems, and associated technologies. The development and execution of a comprehensive financial feasibility assessment become indispensable prerequisites before embarking on a smart grid deployment process. The process of assessing financial viability must include an evaluation of the nation's capacity to bear the development expenses associated with the smart grid infrastructure. Typically, this calculation is conducted on a per consumer basis, taking into account the populace that will benefit from the smart grid services.

Effective communication is the lifeblood of a smart grid, enabling seamless data exchange among various components. It empowers real-time monitoring, control, and coordination of electricity generation, distribution, and consumption. Communication networks, both wired and wireless, link smart meters, sensors, and utility operation centers, facilitating timely responses to fluctuations and outages. Data analytics and smart algorithms harness this flow of information, optimizing grid performance, enhancing reliability, and enabling demand-response mechanisms. Robust communication within the smart grid fosters efficiency, sustainability, and adaptability to the ever-evolving energy landscape. The necessary communication protocols for controlling and coordinating different elements within a smart grid include the home area network (HAN), field area network (FAN), and wide area network (WAN), as visually represented in Figure 13.4. The HAN is responsible for managing SH, often involving multiple IoT-connected electrical devices and EVs. The FAN supports the smart meters grid and Net-zero energy buildings. Meanwhile, the WAN serves as the communication conduit for the transmission and distribution networks and smart microgrids. Figure 13.4 details the coverage range and data speeds associated with each network [25].

Smart grid security is supported by a comprehensive set of fundamental objectives that are critical for the robust operation and safeguarding of the system. These objectives should ensure the system operates safely without posing risks to technological services, public convenience, individuals, or the environment. Security measures aim to protect the system from unintended or unauthorized access, disruptions, tampering, or destruction. The reliability of the smart grid is paramount, which requires consistent performance under specified conditions. Resilience ensures that the system can endure and continue functioning as close to normal as possible even during significant disruptions. Privacy safeguards data access, allowing it only for authorized parties. Accuracy is essential for precise energy consumption calculations and efficient information distribution [26].

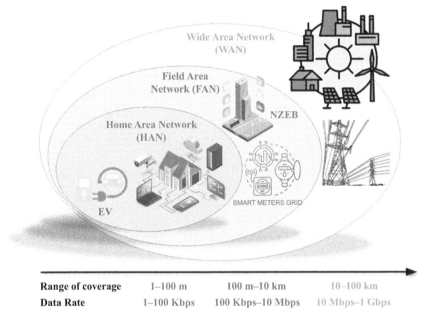

Range of coverage	1–100 m	100 m–10 km	10–100 km
Data Rate	1–100 Kbps	100 Kbps–10 Mbps	10 Mbps–1 Gbps

Figure 13.4   Comparing past and projected CO_2 emissions worldwide

Resource availability ensures both providers and consumers have access to critical data and control messages. Integrity safeguards data against unauthorized alterations. Together, these objectives form the bedrock of a resilient and reliable smart grid network, upholding the protection of critical infrastructure, data, and overall system performance. However, addressing these objectives presents a formidable challenge.

13.4.2   Future directions for smart grids

To ensure the reliability of a grid, it is imperative that all its components seamlessly collaborate throughout the entire energy delivery process, from generation to consumption. The grid comprises numerous intricate elements that require effective coordination and communication. These components are interconnected and function harmoniously through the use of specialized computer software. Consequently, the planning and implementation of grid operations have to be executed with a primary focus on achieving interoperability, enabling the grid's various components to work cohesively and efficiently as an integrated system.

Smart grids incorporate advanced technologies, including communication systems, smart meters, and control mechanisms. This technological evolution has unlocked the potential for EVs to serve not only as energy consumers but also as flexible energy sources that can store energy. Smart meters assume a pivotal role in mitigating the challenges posed by EV integration into the grid. Equipped with bidirectional communication capabilities and real-time data monitoring, these smart meters facilitate the implementation of intelligent scheduling strategies to optimize available power resources within the grid. The vehicle-to-grid (V2G) technology enables precise forecasting of power system dynamics, with particular attention to the vital aspect of EV charging [27]. Extensive research has been dedicated to the charging and discharging processes within this domain. While there have been isolated instances of grid vulnerabilities associated with V2G usage, instances of weak grid have to be addressed by harnessing the potential of RES such as solar and wind. A profound comprehension of the dynamic behavior of electric grid assumes a crucial role in predicting its reliability and efficiency when engaged in V2G operations.

The smart grid relies extensively on the data it receives, serving as both its sensory apparatus and backbone. Data is the linchpin for ensuring the reliability and efficiency of smart grid operations. The entire process involves comprehensive collection of data across power generation, transmission, transformation, and utilization stages. In essence, all decisions made within the grid hinge on this data. Moreover, data assumes a pivotal role in empowering the autonomous capabilities of the smart grid. Various challenges encountered in the realm of big data in smart grid technology include issues ranging from data storage and visualization to security. Researchers will have to direct their efforts toward amalgamating data into actionable information and valuable applications. A substantial volume of data is gathered from diverse sources, including sensors, wireless transmission, and communication networks. This data, spanning from the generation phase to utilization, has to be harnessed by various algorithms for predictive analysis and pattern recognition in power consumption. Ultimately, this will foster the development of an efficient smart energy management system. Energy-related big data not only includes data acquired from meters but also encompasses extensive datasets related to weather and environmental conditions.

While various algorithms and models have been formulated to analyze big data, substantial challenges persist in the field. These include issues related to IT infrastructure, data collection and governance, data processing and analysis, data integration and sharing, and addressing security and privacy concerns. These challenges remain central focal points for researchers in the ongoing development of smart grid technology.

Countries embracing the transition toward smarter grids will have to invariably devise their strategic deployment programs, each uniquely tailored to address pertinent challenges and hurdles. The objective of these programs should be to eliminate the barriers that stand in the way of realizing the Sustainability Targets set for 2030 and 2050. These targets should support the imperatives of sustainable grid development. Deployment priorities have to be defined that support the multifaceted nature of smart grid development. They should be in line with the distinct strategies that each nation employs along with their unique goals and circumstances. A typical set of deployment priorities consists of the following:

- **1st priority:** Focuses on the optimization of grid operation and utilization, seeking to enhance the efficiency and effectiveness of grid operations.

- **2nd priority:** Centers around the optimization of grid infrastructure, emphasizing the need to modernize and upgrade infrastructure to accommodate evolving grid demands.
- **3rd priority:** Helps to gear toward integrating large-scale intermittent generation, with a spotlight on assimilating RES with intermittent generation patterns into the grid.
- **4th priority:** Encompasses ICT, highlighting the significance of advanced ICT solutions in enhancing grid communication and control.
- **5th priority:** Pertains to active distribution networks, emphasizing the importance of flexible and responsive distribution networks.
- **6th priority:** Focuses on new marketplaces, users, and energy efficiency, emphasizing the creation of dynamic marketplaces, addressing diverse consumer needs, and promoting energy efficiency measures.

Regarding research endeavors in the field of smart grid, a cohesive approach should be followed which focus on adopting a shared perspective of the issue, with a concentrated emphasis on interoperability and the active promotion of technological standards. This strategy facilitates smooth transition from conceptualizing solutions and optimization techniques to their practical validation through field trials and deployment. Such a concerted effort accelerates the comprehensive process of actualizing the smart grid infrastructure.

13.5 Conclusion

This chapter presents an evaluation of the current state of smart grid development within the United States, Australia, India, China, the EU, and other countries. It highlights the driving forces behind smart grid implementation, particularly in the context of integrating RES, as well as ongoing innovations and enhancements in the conventional electric grid infrastructure. The smart grid concept, once a visionary idea, is gradually materializing as technology advances. Devices and systems evolve to support the creation of a more intelligent grid. Concrete energy policies have played a pivotal role in propelling smart grid initiatives forward, transcending regional boundaries and fostering a collective sense of shared goals and insights among nations.

In the realm of research activities, a concerted effort toward a unified problem-solving approach, with a strong focus on interoperability and the establishment of technology standards, is imperative. This strategic direction enabled the fast pace development of solutions and optimization techniques, followed by rigorous field testing and deployment. Despite the progress made, the potential for further refinement and expansion of this concept remains significant, marking the early stages of the modern grid era. While the full implementation of the smart grid concept remains challenging, recent research endeavors, including advancements in smart meters, demand-side management systems, self-healing technologies, and big data analytics, provide promising signals of progress within the smart grid technology.

References

[1] M. Kamran, "Introduction to smart grids," in R. E. Pomery and A. A. Khan (eds), *Fundamentals of Smart Grid Systems*, New York: Academic Press, 2023, pp. 1–22, https://doi.org/10.1016/B978-0-323-99560-3.00008-9.

[2] M. Hentea, "Smart power grid," in M. Hentea (ed), *Building an Effective Security Program for Distributed Energy Resources and Systems*, New York: Wiley, 2021, pp. 289–324, doi:10.1002/9781119070740.ch8.

[3] X. Fang, S. Misra, G. Xue, and D. Yang, "Smart grid—the new and improved power grid: A survey," *IEEE Communication Survey Tutorial*, vol. 14, no. 4, pp. 944–980, 2012, 10.1109/SURV.2011.101911.00087.

[4] R. N. Anderson, A. Boulanger, W. B. Powell, and W. Scott, "Adaptive stochastic control for the smart grid," *Proceedings of the IEEE*, vol. 99, no. 6, pp. 1098–1115, 2011, 10.1109/JPROC.2011.2109671.

[5] K. Moslehi and R. Kumar, "A reliability perspective of the smart grid," *IEEE Transactions on Smart Grid*, vol. 1, no. 1, pp. 57–64, 2010, 10.1109/TSG.2010.2046346.

[6] A. Colmenar-Santos, M.-Á. Pérez, D. Borge-Diez, and C. Pérez-Molina, "Reliability and management of isolated smart-grid with dual mode in remote places: Application in the scope of great energetic needs," *International Journal of Electrical Power & Energy Systems*, vol. 73, pp. 805–818, 2015, https://doi.org/10.1016/j.ijepes.2015.06.007.

[7] ETP SmartGrids, "SmartGrids strategic deployment document for Europe's electricity networks of the future," *European Technology Platform Smart-Grids*, 2010.

[8] A. Tarasov, "A methodological approach to the decarbonization of an isolated energy system," *2020 International Multi-Conference on Industrial Engineering and Modern Technologies (FarEastCon)*, Vladivostok, Russia, 2020, pp. 1–4, doi:10.1109/FarEastCon50210.2020.9271572.

[9] J. Conti, P. Holtberg, J. Diefenderfer, A. LaRose, J. T. Turnure, and L. Westfall, "International Energy Outlook 2016 with projections to 2040," *Independent Statistics & Analysis, U.S. Energy Information Administration*, DOE/EIA-0484, 2016.

[10] IRENA, *Global energy transformation: A roadmap to 2050* (2019 edition), International Renewable Energy Agency, Abu Dhabi, 2019.

[11] D. Heirman, "US smart grid interoperability panel (SGIP 2.0) and its testing and certification committee," *2017 IEEE International Symposium on Electromagnetic Compatibility & Signal/Power Integrity (EMCSI)*, Washington, DC, USA, 2017, pp. 1–25, doi:10.1109/ISEMC.2017.8078034.

[12] J. Crispim, J. Braz, R. Castro, and J. Esteves, "Smart grids in the EU with smart regulation: Experiences from the UK, Italy and Portugal," *Utilities Policy*, vol. 31, pp. 85–93, 2014, https://doi.org/10.1016/j.jup.2014.09.006.

[13] E. C. Nnaji, D. Adgidzi, M. O. Dioha, D. R. E. Ewim, and Z. Huan, "Modelling and management of smart microgrid for rural electrification in sub-Saharan Africa: The case of Nigeria," *The Electricity Journal*, vol. 32, no. 10, p. 106672, 2019, doi:10.1016/j.tej.2019.106672.

[14] O. Majeed Butt, M. Zulqarnain, and T. Majeed Butt, "Recent advancement in smart grid technology: Future prospects in the electrical power network," *Ain Shams Engineering Journal*, vol. 12, no. 1, pp. 687–695, 2021, https://doi.org/10.1016/j.asej.2020.05.004.

[15] S. Y. L. Lee, "The Global Smart Grid Federation Report." *Global Smart Grid Federation*, 2012.

[16] M. Cocks and N. Johnson, "Smart city technologies in the USA: Smart grid and transportation initiatives in Columbus, Ohio," in G. Nisbet and D. McClean (eds),*Smart Cities for Technological and Social Innovation*, New York: Academic Press, 2021, pp. 217–245, https://doi.org/10.1016/B978-0-12-818886-6.00012-5.

[17] Y. Zheng, J. Stanton, A. Ramnarine-Rieks, and J. Dedrick, "Proceeding with caution: Drivers and obstacles to electric utility adoption of smart grids in the United States," *Energy Research & Social Science*, vol. 93, p. 102839, 2022, https://doi.org/10.1016/j.erss.2022.102839.

[18] R. Leal-Arcas, F. Lasniewska, and F. Proedrou, "Smart grids in the European Union," in E. Payne and T. A. Lewis (eds), *Electricity Decentralization in the European Union (2nd Edition): Towards Zero Carbon and Energy Transition,* Amsterdam: Elsevier, 2023, pp. 1–109, https://doi.org/10.1016/B978-0-443-15920-6.00022-0.

[19] P. Acharjee, "Strategy and implementation of smart grids in India," *Energy Strategy Review*, vol. 1, no. 3, pp. 193–204, 2013, https://doi.org/10.1016/j.esr.2012.05.003.

[20] Archana, R. Shankar, and S. Singh, "Development of smart grid for the power sector in India," *Cleaner Energy Systems*, vol. 2, p. 100011, 2022, doi:10.1016/j.cles.2022.100011.

[21] A. M. A. Haidar, K. Muttaqi, and D. Sutanto, "Smart grid and its future perspectives in Australia," *Renewable and Sustainable Energy Reviews*, vol. 51, pp. 1375–1389, 2015, https://doi.org/10.1016/j.rser.2015.07.040.

[22] D. N.-Y. Mah, "Conceptualising government-market dynamics in socio-technical energy transitions: A comparative case study of smart grid developments in China and Japan," *Geoforum*, vol. 108, pp. 148–168, 2020, https://doi.org/10.1016/j.geoforum.2019.07.025.

[23] Y. Wang, H. Qiu, Y. Tu, Q. Liu, Y. Ding, and W. Wang, "A review of smart metering for future Chinese grids," *Energy Procedia*, vol. 152, pp. 1194–1199, 2018, https://doi.org/10.1016/j.egypro.2018.09.158.

[24] J. O. Dada, "Towards understanding the benefits and challenges of smart/micro-grid for electricity supply system in Nigeria," *Renewable and Sustainable Energy Reviews*, vol. 38, pp. 1003–1014, 2014, https://doi.org/10.1016/j.rser.2014.07.077.

[25] M. Erol-Kantarci and H. T. Mouftah, "Energy-efficient information and communication infrastructures in the smart grid: A survey on interactions and open issues," *IEEE Communications Surveys & Tutorials*, vol. 17, no. 1, pp. 179—197, 2015, doi:10.1109/COMST.2014.2341600.

[26] P. Gope and B. Sikdar, "A privacy-aware reconfigurable authenticated key exchange scheme for secure communication in smart grids," *IEEE Transactions on Smart Grid*, vol. 12, no. 6, pp. 5335—5348, 2021, doi:10.1109/TSG.2021.3106105.

[27] V. Monteiro, J. G. Pinto and J. L. Afonso, "Operation modes for the electric vehicle in smart grids and smart homes: Present and proposed modes," *IEEE Transactions on Vehicular Technology*, vol. 65, no. 3, pp. 1007—1020, 2016, doi:10.1109/TVT.2015.2481005.

Index

Printed in the USA
CPSIA information can be obtained
at www.ICGtesting.com
JSHW062101151124
73639JS00002B/5

9 781839 538049